# Recent Advances in Agronomy

# Recent Advances in Agronomy

Editor: Harry Abbott

STATES
ACADEMIC PRESS
www.statesacademicpress.com

**States Academic Press,**
109 South 5th Street,
Brooklyn, NY 11249, USA

Visit us on the World Wide Web at:
www.statesacademicpress.com



ISBN: 978-1-63989-460-4 (Hardback)



**Cataloging-in-Publication Data**

Recent advances in agronomy / edited by Harry Abbott.
p. cm.
Includes bibliographical references and index.
ISBN 978-1-63989-460-4
1. Agronomy. 2. Agriculture. 3. Crops. 4. Soil management. I. Abbott, Harry.
SB91 .R43 2022
630--dc23

# Table of Contents

# Preface

The science and technology of producing and using plants for food, fiber, fuel and land reclamation is called agronomy. It integrates the principles of soil science, plant genetics, plant physiology and meteorology. Plant breeding is an important tool of agronomy. The selective breeding of plants has resulted in the production of crops with enhanced yields and nutritional value. This has been observed in major crops such as wheat, corn and soybeans. There has also been a significant use of biotechnology to extend and expedite the development of crop varieties with desired traits. Modern agronomy also focuses on soil analysis to determine the soil that is most conducive to plant growth. These analyses aid in balancing soil nutrients for optimum results. The ever growing need of advanced agricultural technology is the reason that has fueled the research in the field of agronomy in recent times. This book is a compilation of chapters that discuss the most vital concepts and emerging trends in agronomy. Coherent flow of topics, student-friendly language and extensive use of examples make it an invaluable source of knowledge.

This book has been the outcome of endless efforts put in by authors and researchers on various issues and topics within the field. The book is a comprehensive collection of significant researches that are addressed in a variety of chapters. It will surely enhance the knowledge of the field among readers across the globe.

It gives us an immense pleasure to thank our researchers and authors for their efforts to submit their piece of writing before the deadlines. Finally in the end, I would like to thank my family and colleagues who have been a great source of inspiration and support.

**Editor**

# Preface

1

# Influence of some plant oils *Eugenia caryophyllata*, *Elettaria cardamomum* and *Citrus sinensis* on the termite species *Coptotermes heimi*

**Ayesha Aihetasham*, Aysha Anayat, Muhammad Xaaceph**
Department of Zoology, University of the Punjab, Lahore, Pakistan

*Corresponding author email: misswaqar@yahoo.com

**Abstract**
Termites are one of the most troublesome pest of plants, trees, wooden infrastructure and agricultural crops. The aim of the present study is to evaluate the population of some plant oils using *Eugenia caryophyllata, Elettaria cardamomum* and *Citrus sinensis* against *Coptotermes heimi* (Wasmann) at Punjab University, New Campus Lahore, Pakistan. All plants were effectively toxic against *C. heimi* as 100 % mortality was observed at 62.5, 125 and 250 µl/ml doses within three to four days of exposure. All plant oils extracts proved to be highly repellent to the termites even in the low concentrations. These plant oils can be used for effective termite control being less expensive and biodegradable, hence environmentally suitable.

**Keywords**: Termites, Plant extracts, Repellency

## Introduction

Termites are the most abundant pests of plants, trees and wooden structure. They severely damage agricultural crops and urban infrastructure. Although termites feed on dead wood and cellulose rich sources, they only become problematic when they attack on trees, crops and logs (Pearce, 1997; Manzoor et al., 2011; Balachander et al., 2013). They cause serious damage to different crops causing to unhealthy growth of crops (Sheikh, 2009), even almost 2800 species of the termites were found with approximately 185 possible pest's species (Lewis, 1997; Balachander et al., 2013). More than one hundred species belong to the group of the subterranean termites and just 28 species are of economic importance (Lewis, 1997). In Pakistan, there are almost 50 species which cause damage (Akhtar, 1974) but eleven species have been found destructive and need serious attention (Akhtar, 1983). As termites have become a major threat to human economy, economical and ecofriendly alternative biological control strategies such as the use of seeds extracts, essential oils, wood extracts, fruits, leaves and barks oil (Adams et al., 1998; Sakasegawa et al., 2003; Cheng et al., 2004; Jembere et al., 2005; Park and Shin, 2005; Verma et al., 2009; Ding and Hu, 2010; Rasib and Aihetasham, 2016). Their mode of action is upon causing interference in gut flora in order to repel or kill the termites (Adams et al., 1998).

Many Plant's oil extracts showed their activity even as an anti-termite activities or has a repellent effect. They were such as *Pinus roxburghii*, *Eucalyptus camaldulensis* and *Morus alba* (Rasib and Aihetasham, 2016), lemongrass, *Eucalyptus citrodora*, cedar wood, vetivergrass and clove bud (Zhu et al., 2001), *Taiwania cryptomerioides* (Chang et al., 2001) purple hop brush; *Dodonaea viscose* (Anonymous, 2001). *Cymbopogon winterianus*, *Ocimum basilicum*, *Rosmarinus officinalis* and *Cinammomum campho* (Sbeghen et al., 2002) and *Coleus ambionicus* (Singh et al., 2004) Plants oils and powder have been used extensively in the control of insect pests because they are believed to contain myriads of natural chemical that could be insecticides (Adeyemo et al., 2015).

The present research work aimed to present the antitermitic and antifeedant activities of clove (*Eugenia caryophyllata*), Elachi (*Elettaria cardamomum*) and Citrus (*Citrus sinensis*) plants essential oils against a subterranean termites *Coptotermes heimi* (Wasmann).

## Material and Methods

### Termite collection

*Coptotermes heimi* (workers and soldiers) were collected from *Populus euramericana* (populous) trees of University of the Punjab, Lahore, Pakistan. Termites were separated from the tree barks and debris in the laboratory and were kept at ±27°C only healthy individuals were taken for experiments. The apparatus used in the experiment was autoclave. Moist filter paper were placed in Petri plates for termite's survival.

### Extracts of plants oil

The essential oil samples were extracted from the collected plant material like peel of oranges (*C. sinensis*), fruits of Clove (*E. caryophyllata*) and seeds of Elachi (*E. cardamomum*) by hydro distillation for 5 hours using the process of Guenther (1949). After that these extracts were kept in glass vials and held air tight at 4°C.

### Experimental soil preparation

Soil was collected from lawn of Department of Zoology, University of the Punjab, Lahore. It was free of contamination by insecticides as there has been no known application of any chemical. The soil was used to maintain termites like natural environment. Soil was sieved through a 10- mesh oven dried at 72°C for 24 h. Petri plates were washed and sterilized at 180°C for 2h for accuracy of experiment and for avoidance of contaminants growth.

### Mortality test

For mortality test Petri dishes, filter paper and soil were sterilized. Twenty grams of sterilized soil were taken in each Petri plate and sprayed with distilled water. In each Petri plate 300 termite workers and 5 soldiers were released in arena with treated filter paper with respective dilutions. Then Petri plates were covered with black cloth. Dead termites were removed and counted after 24 hour interval to avoid fungal growth. Each experiment was replicated three times.

### Repellency test

Repellency test was performed following the method of Aihetasham et al., 2018. Filter papers cut according to the size of Petri plate and were divided into two equal halves in such a way that there was space between two halves. One half treated with oils dilutions (250, 125 and 62.5 µl/ml) and other half was served as a control using distilled water. Thirty workers were released in between treated and untreated filter paper. Numbers of termites were counted in treated and untreated zone after 60 minutes time interval.

### Data analysis

Statistical analysis and General Linear Model (GLM) Tukey's comparison analysis were conducted using software Minitab 16.

## Results and Discussion

Termiticidal activity of different plants oils was tested against the workers of *C. heimi*. Many plants extracts have been used against termites to replace used insecticides (Ahmed et al., 2011; Addisu et al., 2014; Rasib et al., 2017). Oils of both *E. caryophyllata* and *C. sinensis* on 250 and 125 ul/ml doze killed all termites (100% mortality) in two days while the same results was achieved at 62.5 µl/ml three days post-treatment. Oils of orange peel was used in USA to overcome termite species (*Coptotermes formosanus* Shiraki) in past (Raina et al., 2007). Many different compounds such as d-limonene [(R)-4-isopropenyl-1-methylcyclohexene] which was extracted from citrus peel, caused 98% mortality (Hink and Fee, 1986) of termite population (Zhu et al., 2001; Osipitan and Oseyemi, 2012). General Linear Model was used to estimate the difference between plants, days and concentrations. All parameters showed significant difference i.e plants (d.f= 2, 74; p= 0.000); days (d.f= 2, 74; p= 0.000) and concentration (d.f= 2, 74; p= 0.003). In control no mortality was recorded during experiment (Table 1). Seven species of insects had been reported toxic to orange peel (Raina et al., 2007). The Limonene oil was used to exterminate house flies (*Musca domestica*) and red imported ants (*Solenopsis invicta*). According to Hollingsworth (2005) citrus extract was used to control mealy bug.

The d- limonene vapors were used to exterminate cat flea all life *Ctenocephalides felis* (Hink and Fee, 1986), German cockroach, (*Blattella germanica*) and rice weevils, (*Sitophilus oryzae*) (Karr and Coats, 1988). The *Eugenia caryophyllata* has been used as fever reduction and mosquito repellent (Medline Plus, 2014). The clove oil has mainly composition of Eugenol (Koul et al., 2008), that attached to benzene ring by hydroxyl and ether linkage. Within the compound the aldehyde group shows high level of termite's resistant activity (Chang and Cheng, 2002). The clove oil was excellent repellent to termites, fungus and nematodes (Gupta et al., 2011). The oil of *E. cardamomum* mainly consist of 1,8-cineol (29.7%) and α-terpineol acetate (26.1%) (Korikontimath et al., 1999; Olivero-Verbel et al., 2010) but basic cardamom aroma was produced by mixing of these two major component (Lawrence, 1979). According to Abbasipour et al. (2011), the *E. cardamomum* was use to eradicate the *Callosobruchus maculatus* (bruchid beetle), *Tribolium castaneum* (red flour beetle) and *Ephestia kuehniella* (flour moth). This oil was used to fumigant toxicity against *S. granarious* (Mahmoudvand et al., 2011). *E. cardamomum* also showed different responses on egg stages and active stages of stored product insects (Huang et al., 2000). In the repellency test, the *E. cardamomum* was repellent (28.667) to termites than the other two *E. caryophyllata* and *C. sinensis* (28 and 27) at the dilution of 250 ug/ml. While at 125 ug/ml *E. cardamomum* (27.33) is more repellent than other two plants oils (26). *E. cardamomum* show equal termites repellent at even at low concentrations (d.f: 2, 6; F: 1.04; P: 0.410). In *E. caryophyllata* (d.f: 2, 6; F: 6.68; P: 0.030) and *C. sinensis* (d.f: 2, 6; F: 11.38; P: 009) repellency fades with decreases in concentration (Table. 2). The oil of *E. cardamomum* was most the effective oil extract with respect to other extracts and repellency remain constant even at low concentration. Many plants extracts was used to prevent the pest population and reported the extracts under laboratory conditions (Talukder et al., 2006; Aihetasham et al., 2016). This study shows that *E. cardamomum* was most effective once and after that *E. caryophyllata* was effective and in the end *C. sinensis* was least effective but these all plant oils can be used to suppress the termite population (*Coptotermes heimi*).

**Table 1. Average mortality of *C. heimi* (soldiers) caused by different concentrations of different plant oils**

| Plant Essential oils | Dilutions | No. of days | | | |
|---|---|---|---|---|---|
| | μl/ml | 1$^{A}$ | 2 $^{B}$ | 3 $^{C}$ | 4 |
| *C.sinensis* $^{A}$ | 250 $^{A}$ | 59.33±5.54 | 213.33±10.36 | 300.00±0.00 | |
| | 125 $^{A}$ | 117.33±21.32 | 204.00±2.64 | 300.00±0.00 | |
| | 62.5 $^{B}$ | 99.33±11.83 | 224.00±19.52 | 300.00±0.00 | |
| *E. caryophyllata* $^{B}$ | 250 $^{A}$ | 76.66±11.56 | 166.00±18.58 | 300.00±0.0 | |
| | 125 $^{A}$ | 78.00±12.16 | 172.00±11.37 | 300.00±0.00 | |
| | 62.5 $^{B}$ | 53.33±13.96 | 130.66±16.01 | 242.00±16.23 | 300.00±0.00 |
| *E. cardamomum* $^{C}$ | 250 $^{A}$ | 87.12±1.91 | 96.01±1.27 | 300.00±0.00 | |
| | 125 $^{A}$ | 68.23±2.27 | 99.00±2.21 | 300.00±0.00 | |
| | 62.5 $^{B}$ | 59.23±2.13 | 98.01±2.31 | 187.67±1.23 | 300.00±0.00 |
| Control | | 0.00±0.00 | 0.33±0.33 | 1.66±0.33 | 2.66±0.33 |

**Table 2. Repellency test of termite from different plants oil extracts at different doses. n=30**

| Dozes | Plants | | | | | |
|---|---|---|---|---|---|---|
| | *E. caryophyllata* | | *E. cardamomum* | | *C. sinensis* | |
| | T | U | T | U | T | U |
| **250μl/ml** | 27.33 $^{A}$(0.667) | 2.667 | 28.667 $^{A}$(0.882) | 1.33 | 26.00 $^{A}$(0.577) | 4.00 |
| **125μl/ml** | 25.0 $^{AB}$(1.15) | 5.00 | 26.667 $^{A}$(0.882) | 3.33 | 24 $^{AB}$(0.577) | 6.00 |
| **62.5μl/ml** | 23$^{B}$(0.557) | 7.00 | 27.33 $^{A}$(1.20) | 2.67 | 21.33$^{B}$(0.882) | 8.667 |

Abbreviations: T: treated and U: untreated and (---): standard error

## References

Abbasipour H, Mahmoudvand M, Rastegar F and Hosseinpour MH, 2011. Fumigant toxicity and oviposition deterrency of the essential oil from cardamom, *Elettaria cardamomum*, against three stored product insects. J. Insect Sci. 11-165

Adams RP, McDaniel CA and Carter FI, 1998. Termiticidal activities in the heartwood, bark/sapwood and leaves of *Juniperus* species from the United States. Biochem. System. Ecol. 16: 453-456.

Addisu S, Mohamed D and Waktole S, 2014. Efficacy of Botanical extracts against termites, *Macrotermes* spp., (Isoptera: Termitdae) under laboratory conditions. Int. J. Agric. Res. 9(2): 60-73.

Adeyemo AC, Ogungbite OC and Agbola OP, 2015. Biocidal efficacy of two medicinal plant oil extracts used as termicides against Subterranean termites *Macrotermes subhyalinus* L. Zool. Ecol. 25(3): 257-261.

Ahmed S, Zafar MI, Hussain A, Riaz MA and Shahid M, 2011. Evaluation of plant extracts on mortality and tunneling activities of subterranean Termites in Pakistan. Pesticid. Mod. World. 3: 39-54.

Aihetasham A, Bilal N and Xaaceph M, 2016. Efficacy of Seed Extracts (*Mentha arvensis, Ocimum bacilicum, Psyllium ovata, Cichorium intybus*) against *Coptotermes heimi* (Wasmann). Biologia (Lahore). 62(1): 103-109.

Aihetasham A, Aziz H and Rasib KZ, 2018. Efficacy of slow acting toxicants on *Heterotermes indicola* (Wasmann) (Isoptera: Rhinotermitidae). Pak. J. Zool. 50(1): 373-375.

Akhtar MS, 1974. Zoogeography of termites of Pakistan. Pak. J. Zool. 6: 85-106.

Akhtar MS, 1983. Wood destroying termites (Isoptera: Rhinotermitidae) of Pakistan. Mater. Organ. 188:277–291.

Anonymous, 2001. Termite control without chemicals. HDRA-The organic organisation, pp: 1-14.

Balachander M, Ramadevi Ok and Sasidharan TO, 2013. Dissemination of *Metarhizium anisopliae* infection among the population of the *Odontotermus obesus* by augmenting the fungal conidia with attractant. J. Asia-Pacific Entomol. 16:199-208.

Chang ST, Cheng SS and Wang SY, 2001. Antitermitic activity of essential oils and components from Taiwania (*Taiwania cryptomerioides*). J. Chem. Ecol. 27(4): 1267-1274.

Chang ST and Cheng SS, 2002. Antitermitic/ activity of leaf essential oils and components from *Cinnamomum osmophleum*. J. Agric. Food Chem. 50:1389-1392.

Cheng MJ, Wu CC, Tsai IL and Chen IS, 2004. Chemical and antiplatelet constituents from the stem of *Zanthoxylum beecheyanum*. J. Chinese Chem. Society. 51:1065-1072.

Ding Wand Hu XP, 2010. Antitermitic effect of the *Lantana camara* plant on subterranean termites (Isoptera: Rhinotermitidae). Insect Sci. 7(5): 427-433.

Guenther E, 1949. The essential oils, Vol. 2, Van Nostrand, Princeton, 112.

Gupta A, Sharma S and Naik SN, 2011. Biopesticidal value of selected essential oils pathogenic fungi, termites and nematodes. Int. Biodeterior. Biodegrad. 10: 1-5.

Hink WF and Fee BJ, 1986. Toxicity of D-limonene, the major component of citrus peel oil, to all life stages of the cat flea, *Ctenocephalides felis* (Siphonaptera: Pulicidae). J. Med. Entomol. 23: 400–404.

Hollingsworth RG, 2005. Limonene, a citrus extract, for control of mealybugs and scale insects. J. Econ. Entomol. 98: 772–779.

Huang Y, Lam SL and Ho SH, 2000. Bioactivities of essential oil from *Elletaria cardamomum* (L.) Maton. To *Sitophilus zeamais* Motschulsky and *Tribolium castaneum* (Herbst). J. Stored Prod. Res. 36:107–117.

Jembere B, Getahun D, Negash M and Sevoum E, 2005. Toxicity of Birbira (*Milletia ferruginea*) seed crude extracts to some insect pests as compared to other botanical and synthetic insecticides. 11th NAPRECA (Natural Products and Drug Delivery) Symposium Book of Proceeding, Astanarivo, Madagaskar, pp. 88-96.

Karr LL and Coats JR, 1988. Insecticidal properties of d-limonene. J. Pest. Sci. 13: 287–289.

Korikontimath VS, Mulge R and Zachariah JT, 1999. Variations in essential oil constituents in high yielding selections of cardamom. J. Plant. Crops. 27: 230–232.

Koul O, Walia S and Dhaliwal GS, 2008. Essential oils as green pesticides: Potential and constraints. Biopesticides Int. 4: 63–84.

Lawrence BM, 1979. Major tropical spices cardamom (*Eletteria cardamomum*). Essential Oils. Allured Publishing Corporation. p. 104.

Lewis VR, 1997. Alternative Control strategies for Termites. J. Agric. Entomol.14:291-307.

Mahmoudvand M, Abbasipour H, Rastegar F, Hosseinpour MH and Basij M, 2011. Efficacy of some plants as a post-harvest protectant against some major stored pests. Arch. Phytopathol. Plant Protection. 45(7): 806-811.

Manzoor F, Chaudhary M, and Sheikh N, 2011. Diversity and proportion of termite species in garden trees and wheat crop in District Bhakkar, Pakistan. Pak. J. Zool. 43: 537-541.

MedlinePlus, U.S. National Library of Medicine and National Institutes of Health. 2014. Retrieved August 18, 2014.

Olivero-Verbel J, González-Cervera T, Güette-Fernandez J, Jaramillo-Colorado B and Stashenko E, 2010. Chemical composition and antioxidant activity of essential oils isolated from Colombian plants. Rev. Bras. Farmacogn. 20: 568–574.

Osipitan AA and Oseyemi AE, 2001. Evaluation of the Bio-insecticides potential of some trophical plant extracts against termites (Termitidae: Isoptera) in Ogun state, Nigeria. J. Entomol. 9: 257-265.

Park IK and Shin SC, 2005. Fumigant activity of plant essential oils and components from garlic (*Allium sativum*) and clove bud (*Eugenia caryophyllata*) oils against the Japanese termite (*Reticulitermes speratus* Kolbe). J. Agric. Food Chem. 53: 4388-7392.

Pearce MJ, 1997. Termites biology and pest management. CAB International. pp: 172.

Raina A, Bland J, Doolittle M, Lax A, Boopathy R, and Folkins M, 2007. Effect of Orange Oil Extract on the Formosan Subterranean Termite (Isoptera: Rhinotermitidae). J. Econ. Entomol. 100(3): 880-885.

Rasib KZ and Aihetasham A, 2016. Constituents and termiticide potential of some wood extracts against *Coptotermes heimi* (Wasmann) (Isoptera: Rhinotermitidae). Türk. Entomol. Derg. 40(2): 165-174.

Rasib K, Hidayat W and Aihetasham A, 2017. Feeding Preferences and Control of a Pakistani Termite *Odontotermes obesus* (Rambur) (Isoptera, Rhinotermitidae). Annu. Res. Rev. Biol. 18(5): 1-13.

Sakasegawa M, Hori K and Yatagi M, 2003. Composition and anti-termite activities of essential oils and Melaleuca species. J. Wood Sci. 49:181-187.

Sbeghen AC, Dalfovov V, Serafini LA and De-Barros NM, 2002. Repellency and toxicity of basil, citronella, ho-sho and rosemary oils for the control of termite, *Cryptotermes brevis* (Isoptera: Kalotermitidae). Sociobiol. 40(3): 585-594.

Sheikh N, Irfan S and Naz N, 2009. Effect of the *Nerium oleander* to the subterranean termites; *Coptotermes heimi* (Wasmann). Pak. J. Sci. 61:29-31.

Singh G, Singh OP, Prasad YR, De-Lampasona MP and Catalan C, 2004. Chemical and insecticidal investigations on leaf oil of *Coleus amboinicus* Lour. J. Flav. Frag. 17(6): 440-442.

Talukder FA, Islam MS, Hossain MS, Rahman MA and Alam MN, 2006. Toxicity effects of botanicals and synthetic insecticides on *Tribolium castaneum* (Herbst) and *Rhyzopertha dominica* (F.). Bangladesh J. Environ. Sci. 10(2): 365-371.

Verma M, Sharma S and Prasad R, 2009. Biological alternatives for termite control: A review. Int. Biodeterior. Biodegrad. 63(8): 959-972.

Zhu BC, Henderson G, Chen F, Fei H and Laine RA, 2001. Evaluation of vetiver oil and seven insect-active essential oils against the Formosan subterranean termite. J. Chem. Ecol. 27(8):1617-1625.

# 2

# Blood cholinesterase level and its association with neurobehavioral performance due to insecticide exposure among male cocoa farmers in Pahang and Perak, Malaysia

**Norsyazwani Mohammad, Emilia Zainal Abidin*, Nor 'Aqilah Amani Zainal Mubarik, Vivien How, Sarva Mangala Praveena, Zailina Hashim**
Department of Environmental and Occupational Health, Faculty of Medicine and Health Sciences, Universiti Putra Malaysia, 43400 UPM Serdang, Selangor, Malaysia.

## Abstract

**Background:** Chlorpyrifos (group II) is an example of organophosphate (OP) insecticides used in cocoa farm that contains moderately hazardous substances that may be dangerous, toxic and can cause severe effects to human health. It is a concern considering cocoa plantations uses large amount of pesticides. There is a lack of study focusing on cocoa farmers' exposure to organophosphate and data is needed to allow preventive measures to be planned and implemented in order to protect the health and safety of the workers in this specific agricultural sector.

**Objective:** This study aims to determine blood cholinesterase level due to exposure to organophosphate pesticides and identify its link with neurobehavioral performance of cocoa farmers in cocoa plantations across Perak and Pahang, Malaysia.

**Method:** This is a cross-sectional study among participants that includes cocoa farmers as the exposed group and fishermen as the comparison group. Structured questionnaire were administered to obtain background information and occupational usage of insecticides. Blood samples were collected via finger prick technique and seven tests of World Health Organization (WHO) Neurobehavioral Core Test Battery (NCTB) were used to measure the neurobehavioral performance of the respondents. Data collected were entered into statistical software and were analysed according to objectives.

**Results:** A total of 178 respondents were recruited in this study. This study had reversed results where the blood cholinesterase level of the comparison group was below than the normal range compared to the exposed group. The standard scores of Pursuit Aiming and Trail Making tests showed that the exposed group had significantly poorer performance than the comparison group. There was no significant correlation between the blood cholinesterase with neurobehavioral performance scores for the exposed group. Income, years of handling were significant predictor factors for Pursuit Aiming Test whereas age, use of gloves and education for Trail Making Test.

**Conclusion:** The findings of this study showed that neurobehavioral assessment conducted have provided insight into the early effects of OP pesticide exposures when results detected impairment of cocoa farmers' motor steadiness and visual motor coordination. There is a need to encourage commitment of employers to protect workers through conducting medical surveillance, reviewing existing training modules, assist in the amendment of the existing guideline for use and provide adequate and suitable personal protective equipment for the use of the workers in order to fulfil the objectives of the Occupational Safety and Health Act.

*Corresponding author email: za_emilia@upm.edu.my

**Keywords:** Organophosphate, blood cholinesterase level, neurobehavioral performance, cocoa farmers

## Introduction

Malaysia is one of the countries in Asia that produces cocoa for the manufacturing of cocoa-based products such as cocoa butter, cocoa paste, cocoa powder and chocolate. The total size of cocoa plantation in Malaysia is approximately 29,951 hectare in 2004 (Malaysia Cocoa Board, 2004) where majority of cocoa is cultivated in Sabah and Sarawak. However, in Peninsular Malaysia, cocoa farms are more concentrated in Pahang, Perak and Kelantan. Organophosphate (OP) insecticides that has been widely used in local cocoa farms contains hazardous substances that may be dangerous, toxic and can cause severe effects such as acute affect (eye irritation, skin irritation) and chronic effect (birth defect, neurodevelopment problems) to the human health. It has been established by the Insecticide Resistance Action Committee (IRAC) that OP insecticide had been classified as Groups 1A Cholinesterase Inhibitor which will increase cholinergic effects of neurotransmitter, acetylcholine in body and depolarize neural transmission, resulting in damage to neurological functioning (Weiss et al. 2004). Evidence have shown associations between OP exposure and neurobehavioral impairment in adult populations of farmers and farmworkers (Baldi et al. 2011; Kamel et al. 2003).

The Malaysian Cocoa Board have reported that farms use approximately 10-15 litres of pesticides for one hectare of plantation in a three months' cycle of cocoa planting in the Peninsular Malaysia (Malaysia Cocoa Board, 2016). However, occupational exposures and published data regarding pesticide exposure among cocoa farmers is lacking in literature. Considering that pesticides including OP pesticides are used in plantation fields, the biggest concern would be the exposure experienced by cocoa farm workers if there were no proper emphasis for occupational health and safety. The legislation requirement as specified in the Occupational Safety and Health Act 514 1994 (Use and Standard of Exposure of Chemicals Hazardous to Health) Regulation 2000 in Malaysia have made it mandatory that workers who are exposed to OP to undergo routine medical surveillance of cholinesterase activity in red cells and plasma. Although trainings such as Standard of Practice (SOP) for spraying pesticides and personal protective equipment (PPE) use have been reported to be provided, the effectiveness of these control methods have not been reported in literature. Furthermore, there is a need to assess the health effects among cocoa farmers from the usage of those pesticides to enhance the importance of the trainings provided. Currently, in Malaysia there are limited studies which has been done to identify the effects of OP on blood cholinesterase level and neurobehavioral performance specifically among cocoa farmers unlike other agricultural sectors as reported in literature (Lokhman et al., 2015; How et al., 2014). As such, this study aims to determine the blood cholinesterase level and its association with neurobehavioral performance of cocoa farmers in cocoa plantations across Perak and Pahang, Malaysia due to exposure to OP pesticides. This study will hopefully provide basic information on pesticide exposures among this sub group of agricultural workers.

## Materials and Method

This study employs a cross-sectional design to determine blood cholinesterase levels and neurobehavioral performance and its link with occupational OP exposures among cocoa farmers in Sungai Sumun, Perak and Jengka, Pahang. The locations were selected because these are the only two plantations and research centre for cocoa in Peninsular Malaysia. Blood cholinesterase levels in this study act as the biomarker of exposure to OP. Before the study was conducted, institutional ethical approval was sought. This study had received approval from the Universiti Putra Malaysia Ethics Committee Board prior to data collection. The Malaysian Cocoa Board granted the approval for the study to be conducted.

The sampling frame were cocoa farmers who have been working in Cocoa Plantation and Research Centre at Pahang and Perak and those who handles pesticides in the workplace. A list of 96 cocoa farmers in Peninsular Malaysia was obtained from the Malaysian Cocoa Board. Purposive sampling was used to select samples as only certain cocoa farmers with inclusive and exclusive criteria that could provide desired information related to OP use were invited to take part in this study. The inclusion and exclusion criteria used in this study is as presented in Table 1 and Table 2 respectively. This study uses several types of methodology, namely questionnaire, cholinesterase test kit and seven tests of neurobehavioral performance instruments.

**Table 1: Inclusion and Exclusion Criteria of the Study – Exposed Group**

| Inclusion criteria | Exclusion criteria |
|---|---|
| • Male adults<br>• 20 – 60 years old<br>• Have at least two years of farming experience.<br>• Particularly involved in handling of OP pesticides.<br>• No diagnosis of chronic disease such as cancer, kidney failure, liver failure, chemotherapy.<br>• Not mentally and physically affected such as not mentally retarded, able to see, read, write, and speak and response well. | • Work as part time pesticides handler.<br>• Work outside other than in cocoa farms.<br>• Have history of head injury.<br>• Has alcohol intake. |

**Table 2: Inclusion and Exclusion criteria of the study – Comparison Group**

| Inclusion criteria | Exclusion criteria |
|---|---|
| • Male adults.<br>• 20 – 60 years old.<br>• No direct contact or history with handling of pesticides specifically OP in farm.<br>• No diagnosis of chronic disease such as cancer, kidney failure, liver failure, chemotherapy.<br>• Able to see, read, write, and speak and response well.<br>• Malaysian | • Not able to read, write, speak and response well.<br>• Reported alcohol intake. |

**Study instruments**

***Questionnaire***

The questionnaire used in this study was adapted from several studies (Holtman, 2013; Al-Haddad and Al Sayyad, 2013; Lorenz, 2011; Nordin et al. 2002; Yassin et al. 2002; World Health Organization, 1986) and was modified according to the setting of this study. The interview-based questionnaire was provided in Malay. There were two different sets of questionnaires; namely for the exposed and the comparison group. The questionnaire for exposed group (cocoa farmers) consisted of eight parts namely Part A: Demographic Information, Part B: Background information, Part C: Background Information on Working History, Part D: Information on Use of Pesticides at Workplace, Part E: Information on Use of Personal Protective Equipment (PPE), Part F: Work Practices when Handling Pesticides, Part G: Other Information and Part H: Health Information. On the other hand, the questionnaire for the comparison group only consisted of Part A, B, C and H.

Pre-test of questionnaires were conducted prior to data collection among representatives of cocoa farmers, to observe whether the questionnaire can be understood and to avoid any unfamiliar terms. The questionnaire were also tested for its reliability.

***Cholinesterase Test Kit***

In this study, a competent person conducted all blood collection from all subjects. About 10 μl of blood from each respondent was obtained using finger prick technique. Blood collection was only taken once for baseline and only to observe chronic exposure. Then the blood samples were analysed using LOVIBOND AF267 Cholinesterase Test Kit. All containers containing biological specimen were labelled with only identification code to ensure confidentiality.
Specimen collected were stored in refrigerator or freezer and were allowed to stand for at least 20 to 30 minutes to reach room temperature before analysis. The kits and specimens were stored in room temperature (not exceeding 30°C). The expiration date of the test kit and chemicals were ensured before use.

The substrate solutions were freshly prepared daily and any samples or solutions that had air bubbles were eliminated. Blood cholinesterase levels were categorised according to the specific range given by the manufacturer. The three categories were normal (100-75%), overexposure (74.9-50%) and serious overexposure (49.9-25%).

***Neurobehavioral Core Test Battery (NCTB)***

The neurobehavioral Core Test Battery (NCTB) assessment was used to determine the health effects measured in this study. The full set of test took about 30-45 minutes to complete and its results were compared between the two groups. NCTB includes the following subtests: Benton Visual Retention, Digit Symbol, Digit Span, Pursuit Aiming, Trail Making, Santa Ana Manual Dexterity and Simple Reaction Time. Each subtest was administered to all subjects by only one specific researcher. All testing procedures were understood and practiced by the researchers to avoid any bias. Furthermore, testing was conducted in an area where it was free from distracting noise, had adequate lighting and comfortable room temperature. All instructions given to respondents were loud, concise and clear. In terms of scoring, the standard score below 50 indicates that a respondent had poor neurobehavioral performance whereas standard score above 50 indicates good neurobehavioral performance level (WHO, 1986).

**Statistical Analysis**

All analysis was performed using the IBM SPSS (Statistical Package for Social Science), version 22. The significance level was set at $p<0.05$. The $\chi^2$ and Fisher's Exact tests were performed to compare socio demographic characteristics, blood distribution and neurobehavioral scores while Mann-Whitney U test was performed to identify significant differences for blood cholinesterase concentration. Pearson and Spearmen correlations were used to evaluate associations between organophosphate exposures with neurobehavioral performance level. To identify significant predictors of the selected neurobehavioral performance, multiple linear regression was performed.

## Result

**Background Information of the Respondents**

Data collection was conducted between January to May 2017. The number of respondents who participated in this study was 178 people. For the exposed group, a total of 90 from 96 cocoa farmers were recruited while for the comparison group, 88 out of 94 fishermen agreed to participate in this study. The response rate for this study was 93%.

All respondents were males and were of Malay ethnicity. Three-quarter of the exposed group were of younger age group (less than 40 years of age), more than 90% of them had secondary level of education, half of them had income of more than RM3000 and above, almost 80% of them had an occupational tenure of less than 15 years. There were higher percentage of the comparison group who were living within 5 km of agricultural area when compared to exposed group. For both groups of respondents, more than 70% of them smokes and about one-third of them had normal BMI. Table 3 presents the distribution of socio-demographic characteristics of the respondents.

**Blood Cholinesterase Level**

The average (mean and standard deviation, SD) for blood cholinesterase concentration (in percent) for the exposed group was 86.7 (13.5) % and for comparison group was 79 (17.7) %. Lower level of cholinesterase concentration was observed for the exposed group (cocoa farmers). Table 4 presents the blood cholinesterase levels of the two groups of respondents in this study.

**Distribution of Blood Cholinesterase Concentration**

Blood cholinesterase concentration of the respondents were categorised into three groups. It was noted that this study had reversed results whereby there were higher percentage of the comparison group in the overexposure and serious overexposure category (21.6%) when compared to exposed group (8.9%). The distribution of blood cholinesterase is as presented in Table 5.

**Neurobehavioral Performance**

For the neurobehavioral performance, only Pursuit Aiming ($\chi^2$: 39.66; $p < 0.05$) and Trail Making test ($\chi^2$:7.676; $p < 0.05$) showed a significant difference between the exposed and comparison group. Higher percentage of respondents in the exposed group (98.9% for Pursuit Aiming, 51.1% for Trail Making) performed poorer in both tests compared to the comparison group (61.4% for Pursuit Aiming, 30.7% for Trail Making) that scored below the score of 50. Table 6 presents the cross-tabulation of the distribution of neurobehavioral test scores between the two groups.

**The Relationship between Blood Cholinesterase Level and Neurobehavioral Performance**

Table 7 shows the correlation between blood cholinesterase levels with NCTB test scores. There was no significant correlation between blood cholinesterase levels with all neurobehavioral performance scores for the exposed group.

**Factors Influencing Neurobehavioral Impairment**

Table 8 showed the selected variables included in the regression analysis for Pursuit Aiming test. Simple linear regression was carried out prior to selecting variables to be included in further analysis. Variables that were included had p-value of less than 0.25 in the Simple Linear Regression or are determined to be clinically important. From the overall selected variables, income and years of handling organophosphate pesticides were the significant predictor factors ($p<0.05$) for Pursuit Aiming test of the exposed group. In this study, the variables explained 14.1% of the outcome.

Table 9 presents the regression analysis for Trail Making test. Variables such as age, education level and use of gloves were the significant predictor factors ($p<0.05$) for Trail Making test. In this test, 25.7% of the variables explained the outcome of the Trail Making test. An increase of age were inversely linked with the standard score of the Trail Making test while and an increase of education level and the use of gloves were linked were linked to increased score of the Trail Making test.

**Table 3: Cross-tabulation of socio-demographic characteristics of the exposed (farmers) and comparison (fishermen) groups**

| Variables | Exposed N (%) (n=90) | Comparison N (%) (n=88) | $\chi^2$ | *p-value* |
|---|---|---|---|---|
| **Age (years)** | | | | |
| 20 – 29 | 27 (30) | 5 (5.7) | | |
| 30 – 39 | 34 (37.8) | 12 (13.6) | | |
| 40 – 49 | 15 (16.7) | 29 (33) | 44.1[a] | <0.05* |
| 50 – 60 | 14 (15.6) | 42 (47.7) | | |
| **BMI** | | | | |
| Underweight | 5 (5.6) | 4 (4.5) | | |
| Normal | 39 (43.3) | 30 (34.1) | | |
| Overweight | 29 (32.2) | 36 (40.9) | 2.09[b] | 0.55 |
| Obesity | 17 (18.9) | 18 (20.5) | | |
| **Education Level** | | | | |
| Primary school | 5 (5.6) | 23 (26.1) | | |
| Secondary school | 78 (86.7) | 60 (68.2) | 14.2[a] | <0.05* |
| STPM/Diploma | 7 (7.8) | 5 (5.7) | | |
| **Income (RM)** | | | | |
| <1500 (USD381.60) | 47 (52.2) | 68 (77.3) | | |
| 1501(USD381.85) – 3000 (USD763.20) | 38 (42.2) | 18 (20.5) | 12.2[b] | <0.05* |
| >3000 (USD763.20) | 5 (5.6) | 2 (2.3) | | |
| **Working Years (years)** | | | | |
| <15 | 70 (77.8) | 42 (47.7) | | |
| 16 – 30 | 18 (20) | 30 (34.1) | | |
| >30 | 2 (2.2) | 16 (18.2) | 20.9[a] | <0.05* |
| **Distance from home to any agricultural area** | | | | |
| Less than 5 kilometres | 48 (53.3) | 80 (90.9) | | |
| 5-10 kilometres | 31 (34.4) | 6 (6.8) | 31.1[a] | <0.05* |
| More than 10 kilometres | 11 (12.2) | 2 (2.3) | | |

| | | | | |
|---|---|---|---|---|
| **Smoking Status** | | | | |
| Smoker | 64 (71.1) | 68 (77.2) | 0.88[a] | 0.35 |
| Non-smoker | 26 (28.9) | 20 (22.7) | | |
| **Number of Cigarettes (per day) (n=64)** | | | | |
| 1 – 20 | 58 (90.6) | 61 (89.7) | 2.39[b] | 0.28 |
| 21 – 40 | 4 (6.3) | 7 (10.3) | | |
| 41 - 60 | 2 (3.1) | 0 (0) | | |
| **Years of smoking (years) (n=64)** | | | | |
| <15 | 29 (45.3) | 6 (8.8) | 22.6[a] | <0.05* |
| 16 – 35 | 29 (45.3) | 53 (77.9) | | |
| >35 | 6 (9.4) | 9 (13.2) | | |

N=178 **p*-value significant at 0.05 level

[a] Chi-square tests [b] Fisher's Exact Test with Freeman-Halton extension

**Table 4: Blood Cholinesterase Concentrations of the Respondents**

| Variable | Median (IQR) | | Mean (SD) | | z | *p-value* |
|---|---|---|---|---|---|---|
| | Exposed | Comparison | Exposed | Comparison | | |
| **Blood Cholinesterase Level (%)** | 87.5 (25) | 75 (25) | 86.7 (13.5) | 79 (17.7) | -2.889 | 0.004* |

N=178

**p*-value significant at 0.05 level

Blood cholinesterase level nearing value of 100 indicates normal blood cholinesterase level.

IQR= Interquartile range SD = Standard Deviation

Mann-Whitney U Test

**Table 5: Distributions and Comparison of Blood Cholinesterase Concentration**

| Group | Frequency (%) | | | $\chi 2$ | *p-value* |
|---|---|---|---|---|---|
| | **Normal** | **Overexposure** | **Serious Overexposure** | | |
| **Exposed** | 82 (91.1) | 7 (7.8) | 1 (1.1) | 5.64 | 0.04* |
| **Comparison** | 69 (78.4) | 17 (19.3) | 2 (2.3) | | |
| **Both Groups** | 151 (84.8) | 24 (13.5) | 3 (1.7) | | |

N=178

**p*-value significant at 0.05 level

Fisher's Exact Test with Freeman-Halton extension

**Table 6: Comparison of Category of Neurobehavioral Performance Level between Two Groups**

| | | | | |
|---|---|---|---|---|
| **Benton Visual Retention Test** | | | | |
| | **Below 50 (%)** | **Above 50 (%)** | $\chi^2$ | *p-value* |
| **Group** | | | 1.804 | 0.179 |
| Exposed | 33 (36.7) | 57 (63.3) | | |
| Comparison | 41 (46.6) | 47 (53.4) | | |
| **Digit Span Test** | | | | |
| | **Below 50 (%)** | **Above 50 (%)** | $\chi^2$ | *p-value* |
| **Group** | | | 1.900 | 0.168 |
| Exposed | 46 (51.1) | 44 (48.9) | | |
| Comparison | 54 (61.4) | 34 (38.6) | | |

| **Digit Symbol Test** | | | | |
|---|---|---|---|---|
| | **Below 50 (%)** | **Above 50 (%)** | $\chi^2$ | *p-value* |
| **Group** | | | 0.026 | 0.871 |
| Exposed | 48 (53.3) | 42 (46.7) | | |
| Comparison | 48 (54.5) | 40 (45.5) | | |
| **Time Reaction Test** | | | | |
| | **Below 50 (%)** | **Above 50 (%)** | $\chi^2$ | *p-value* |
| **Group** | | | 2.706 | 0.100 |
| Exposed | 52 (57.8) | 38 (42.2) | | |
| Comparison | 40 (45.5) | 48 (54.5) | | |
| **Pursuit Aiming Test** | | | | |
| | **Below 50 (%)** | **Above 50 (%)** | $\chi^2$ | *p-value* |
| **Group** | | | 39.66 | <0.05* |
| Exposed | 89 (98.9) | 1 (1.1) | | |
| Comparison | 54 (61.4) | 34 (38.6) | | |
| **Trail Making Test** | | | | |
| | **Below 50 (%)** | **Above 50 (%)** | $\chi^2$ | *p-value* |
| **Group** | | | 7.676 | <0.05* |
| Exposed | 46 (51.1) | 44 (48.9) | | |
| Comparison | 27 (30.7) | 61 (69.3) | | |
| **Santa Ana Dominant Test** | | | | |
| | **Below 50 (%)** | **Above 50 (%)** | $\chi^2$ | *p-value* |
| **Group** | | | 1.092 | 0.296 |
| Exposed | 40 (44.4) | 50 (55.6) | | |
| Comparison | 46 (52.3) | 42 (47.7) | | |
| **Santa Ana Non-Dominant Test** | | | | |
| | **Below 50 (%)** | **Above 50 (%)** | $\chi^2$ | *p-value* |
| **Group** | | | 0.028 | 0.868 |
| Exposed | 49 (54.4) | 41 (45.6) | | |
| Comparison | 49 (55.7) | 39 (44.3) | | |

N=178

*p-value significant at 0.05 level

Greater neurobehavioral performance denoted by score above 50, poor neurobehavioral performance denoted by score below 50.

Chi-Square Test

**Table 7: The Relationship between Blood Cholinesterase Level and Neurobehavioral Performance**

| **Variables** | **Blood Cholinesterase Level (%)** | | | |
|---|---|---|---|---|
| | **Exposed** | | **Comparison** | |
| | *r* | *p-value* | *r* | *p-value* |
| **Benton Visual Retention Test[b]** | 0.111 | 0.297 | -0.124 | 0.251 |
| **Digit Span Test[b]** | -0.011 | 0.916 | -0.251 | 0.018* |
| **Digit Symbol Test[a]** | 0.160 | 0.131 | -0.201 | 0.060 |
| **Time Reaction Test[b]** | 0.065 | 0.541 | -0.105 | 0.330 |
| **Pursuit Aiming Test[b]** | -0.200 | 0.059 | 0.018 | 0.866 |
| **Trail Making Test[b]** | -0.006 | 0.954 | -0.194 | 0.071 |
| **Santa Ana Dominant Test[a]** | 0.125 | 0.240 | -0.142 | 0.187 |
| **Santa Ana Non-Dominant Test[a]** | 0.175 | 0.098 | -0.282 | 0.008** |

N=178
*$p$-value significant at 0.05 level
**$p$-value significant at 0.01 level
[a] Pearson Correlation
[b] Spearman Correlation

**Table 8: Regression analysis for selected variables for Pursuit Aiming Test**

| Variables | SLR[a] | | MLR[b] | | | | |
|---|---|---|---|---|---|---|---|
| | B (95% CI) | *p-value* | Adjusted B | t-statistics | *p-value* | F | *p-value* |
| **(Constant)** | | | 50.032 | 34.224 | <0.05* | | |
| **Blood Cholinesterase Level** | -0.005 (-0.028, 0.017) | 0.64 | 0.001 | 0.072 | 0.94 | | |
| **BMI** | | | 0.275 | 1.572 | 0.12 | | |
| Underweight | -0.176 (-1.507, 1.154) | 0.79 | | | | | |
| Normal | -0.249 (-0.862, 0.364) | 0.42 | | | | | |
| Overweight | -0.093 (-0.746, 0.559) | 0.78 | | | | | |
| Obesity | 0.592 (-0.177, 1.361) | 0.13 | | | | | |
| **Income (RM)** | | | 0.967 | 3.203 | <0.05* | | |
| <1500 (USD381.60) | -0.126 (-0.736, 0.484) | 0.68 | | | | | |
| 1501 (USD381.85) – 3000 (USD763.20) | -0.471 (-1.080, 0.139) | 0.13 | | | | | |
| >3000 (USD763.20) | 2.788 (1.595, 3.981) | <0.05* | | | | | |
| **Smoking Status** | | | -0.080 | -0.159 | 0.87 | | |
| No | 0.577 (-0.085, 1.239) | 0.09 | | | | | |
| Yes | -0.577 (-1.239, 0.085) | 0.09 | | | | | |
| **Years of Smoking (years)** | -0.020 (-0.045, 0.005) | 0.12 | -0.014 | -0.663 | 0.51 | 2.351 | 0.017* |
| **Years of Living in Current Home (years)** | -0.017 (-0.040, 0.007) | 0.16 | -0.011 | -0.990 | 0.33 | | |
| **Years of Handling Pesticides (years)** | -0.042 (-0.088, 0.004) | 0.07 | -0.086 | -2.912 | <0.05* | | |
| **Use of PPE** | | | -0.384 | -0.592 | 0.56 | | |
| No | -0.176 (-1.507, 1.154) | 0.79 | | | | | |
| Yes | 0.176 (-1.154, 1.507) | 0.79 | | | | | |
| **Taking bath after handling pesticides** | | | -0.184 | -0.317 | 0.75 | | |
| No | -0.320 (-1.390, 0.749) | 0.55 | | | | | |
| Yes | 0.320 (-0.749, 1.390) | 0.55 | | | | | |
| **Time of change to new clothes after handling pesticides** | | | -0.242 | -0.889 | 0.38 | | |
| Immediately | 0.500 (-0.103, 1.103) | 0.10 | | | | | |
| Before lunch time | -0.415 (-1.019, 0.189) | 0.18 | | | | | |
| At the end of work | -0.388 (-1.717, 0.941) | 0.56 | | | | | |

N=90 *$p$-value significant at 0.05 level [a] Simple Linear Regression [b] Multiple Linear Regression
Income and Years of Handling Pesticides were significantly linked with Pursuit Aiming Test score of exposed group

**Table 9: Regression analysis for selected variables for Trail Making Test**

| Variables | SLR[a] | | MLR[b] | | | | |
|---|---|---|---|---|---|---|---|
| | B (95% CI) | *p-value* | Adjusted B | t-statistics | *p-value* | F | *p-value* |
| **(Constant)** | | | 52.555 | 7.881 | <0.05* | 2.739 | 0.006* |
| **Blood Cholinesterase Level** | -0.015 (-0.141, 0.110) | 0.81 | 0.007 | 0.112 | 0.91 | | |
| **Age** | -0.233 (-0.389, -0.077) | <0.05* | -0.205 | -2.134 | <0.05* | | |
| **Education Level** | | | 4.675 | 2.133 | <0.05* | | |
| Primary School | -2.000 (-9.341, 5.341) | 0.59 | | | | | |
| Secondary School | -2.628 (-7.552, 2.295) | 0.29 | | | | | |
| STPM/Diploma | 5.697 (-0.475, 11.869) | 0.07 | | | | | |
| **Income (RM)** | | | -0.368 | -0.224 | 0.82 | | |
| <1500 (USD381.60) | 2.771 (-0.549, 6.091) | 0.10 | | | | | |
| 1501(USD381.85)–3000 (USD763.20) | -2.677 (-6.040, 0.685) | 0.12 | | | | | |
| >3000 (USD763.20) | -0.729 (-8.081, 6.622) | 0.84 | | | | | |
| **Smoking Status** | | | 2.528 | 1.349 | 0.18 | | |
| No | -2.440 (-1.318, 0.191) | 0.19 | | | | | |
| Yes | 2.440 (1.318, 0.191) | 0.19 | | | | | |
| **Tenure (years)** | -0.219 (-0.451, 0.014) | 0.07 | 0.228 | 0.896 | 0.37 | | |
| **Years of Handling Pesticides (years)** | -0.260 (-0.515, -0.006) | <0.05* | -0.223 | -0.832 | 0.41 | | |
| **Use of Gloves** | | | 3.768 | 2.334 | <0.05* | | |
| No | -3.889 (-7.227, -0.551) | <0.05* | | | | | |
| Yes | 3.889 (0.551, 7.227) | <0.05* | | | | | |
| **Location of washing hands after handling pesticides** | | | -0.391 | -0.767 | 0.45 | | |
| Near the farm area with bottled water | 0.069 (-4.141, 4.280) | 0.97 | | | | | |
| Toilet located at pesticide storage | 3.757 (-0.081, 7.594) | 0.06 | | | | | |
| Others | -3.822 (-7.092, -0.552) | <0.05* | | | | | |
| **Location of taking bath after handling pesticides** | | | -2.170 | -1.911 | 0.06 | | |
| Toilet located at pesticide storeroom | 3.580 (-1.983, 9.143) | 0.20 | | | | | |
| Toilet at home | 2.533 (-0.999, 6.066) | 0.16 | | | | | |
| Others | -5.894 (-10.604, -1.364) | <0.05* | | | | | |
| Not applicable | -1.799 (-7.705, 4.107) | 0.55 | | | | | |

N=90 **p*-value significant at 0.05 level [a] Simple Linear Regression [b] Multiple Linear Regression

Age, Education Level and Use of Gloves were significantly linked with Trail Making Test score of exposed group

## Discussion

### Blood Cholinesterase Level

This study was able to assess the cholinesterase levels and measure neurobehavioral performance tests of 178 respondents from the population of cocoa farmers and fishermen in Perak and Pahang. In general, this study showed contradictory results by which the comparison group had lower blood cholinesterase level than the exposed group. It was found that all the comparison respondents in Perak area had blood cholinesterase level in normal range but results from the Pahang area was contrary to the hypothesis of this study, where the blood cholinesterase level of those who were not occupationally exposed to organophosphate was below than the normal range for most of them.

To explain this contrary results, several factors were considered. For the inhibition of blood cholinesterase, there are other chemicals that could cause similar neurotoxicity mechanism like organophosphate such as polycyclic aromatic hydrocarbons, detergents, heavy metals and contaminant mixtures (Bressler and Goldstein, 1991). There could be possibility of the presence of other chemicals that were not expected such as heavy metals (Abdullah, 2016). Animal study by Akinrinade et al. (2015) reported that excessive aluminium intake may induce the progression of cell deaths which leads to inhibition of AChE. Besides, other study by Phyu and Tangpong (2014) on sensitivity of AChE to environmental pollutants reported that other heavy metals such as lead may cause inhibition due to free radicals released as studied on painters and rat (Khan et al., 2009; Tsakiris, 2000). An alternative explanation also includes that fact that more than 90% of the comparison group were living within 5 km of intensive agricultural area. There may be the possibility that there are environmental cross-contamination that may have affected the blood cholinesterase level of the comparison groups living in Kuantan, Pahang to be depressed.

Neurobehavioral performance testing using NCTB is a method for early detection of nervous system impairment due to any effect of neurotoxic chemical exposure on human being. Ismail et al. (2012) and Ross et al. (2010) had conducted a meta-analysis studies that focused on OP exposure among adult agricultural workers and pesticide applicators. These two studies reported deficits in motor skills and slower reaction times. They found consistent findings across many studies on impaired neurobehavioral function concerning cognitive functions following low-level exposures to OP.

Based on the results of the present study, there were significant link between occupational exposures to OP with the neurobehavioral performance of the respondents in the exposed group. Pursuit Aiming and Trail Making tests were the neurobehavioral batteries that were found to exhibit the effects. The standard scores for both tests showed that the exposed group were linked to significantly poorer performance when compared to the comparison group. The poorer performance indicates poor or impaired motor steadiness (Pursuit Aiming Test) and visual and motor coordination (Trail Making Test).

The results of this study is aligned with a previous study by London et al. (1997) that reported an association of OP exposure with NCTB Pursuit Aiming test among fruit farm workers. The study by London et al. (1997) were conducted on 163 male sprayers and 84 male non-spraying workers of fruit farms in South Africa. Like the present study, Farahat et al. (2003) also revealed that the Trail Making Test were one of the sub-tests of NCTB where the exposed group exhibited significantly lower performance than the control group after adjusting the confounders of age and education level. Both studies were cross sectional studies which determine the long term of OP exposure with the neurobehavioral outcomes. The study by Farahat et al. (2003) were performed among 52 male cotton crop workers in Turkey who were being compared with 50 unexposed male controls that were similar in age, socioeconomic class, and years of education.

In the present study, it was found that the blood cholinesterase level of the comparison group was lower than the exposed group. Despite that, even though it was may be due to environmental cross contamination, 91.1% of the cocoa farmers had blood cholinesterase level above the normal range. As for the association of blood cholinesterase level, there was no any significant correlation between the blood cholinesterase levels with any of the neurobehavioral performance test scores. This study had aligned findings with study by Farahat et al. (2003) where acetylcholinesterase was not significantly correlated with either neurobehavioral performance or neurological abnormalities.

Although there was no significant correlation between blood cholinesterase levels with any of the neurobehavioral performance test scores in this study, but through monitoring and assessment of blood cholinesterase level, the cocoa farmers were still at risk of having depressed blood cholinesterase level as some of them have blood cholinesterase level in the range of overexposure and serious over exposure. Occupational exposure does occur and during the questionnaire administration session with the cocoa farmers, many of them said that they felt burning sensation on their face when they sprayed the pesticides as the wind carried the sprayed pesticide through their direction. The cocoa farmers also have differing frequency of pesticide spraying activities for each phases of cocoa harvesting in which it ranged from one, two or four times in a month. A cross-sectional study such as the present study may not be able to full assess the extent of the exposures to pesticides at just one point of time.

It is also important to underline that there was no reported baseline reading of blood cholinesterase level among these cocoa farmers to enable comparison to be made. Baseline measurement of blood cholinesterase level among those handling organophosphate pesticides should be conducted prior to employment (Pesticide Management Education Programme, 1993). Thus, this study may provide data that is useful for the relevant administration to enable safety precautions or other implementations regarding the protection of farmer's health takes place.

Further analysis was performed on two of the NCTB subtests which were Pursuit Aiming and Trail Making tests. The most important factors affecting the scores of Pursuit Aiming Test were income and years of handling OP pesticides. Results showed that years of handling pesticides had an inverse relationship with Pursuit Aiming test score. These findings were aligned with the previous studies by Roldan-Tapia et al. (2005) and Kamel et al. (2003) as they reported decrement of neurobehavioral performance was significantly related with increasing years of working in agriculture field.

Income was one of the significant predictor factors as it has significant positive relationship with Pursuit Aiming test. In other words, the higher the income of the cocoa farmers, the higher the standard score of their Pursuit Aiming test. Previous study reported total income were one of the confounders that were associated with the outcome of neurobehavioral performance (Starks et al., 2012). Similar to the study by Marston (2013) which revealed that those who had low income may had problem with their neurobehavioral development as chronic stress due to low income can contribute to significant problems towards executive functioning, working memory, social comprehension and emotion regulation.

In addition, it was found that three variables which were the use of gloves, age and education level were the significant predictors of the Trail Making test. An increase of age was inversely linked with the standard score of the Trail Making test while an increase of education level and the use of gloves were linked to increased score of the Trail Making test. Impairment of neurobehavioral functions may elevate if gloves are not used or unsuitable type of gloves are used instead, as OP pesticides in formulation of solvents may enhance greater uptake by the skin and is difficult to be removed from the body (International Program on Chemical Safety (INCHEM), 2017). For age, this study on cocoa farmers were able to support previous findings which reported that when age increased, the standard score of Trail Making test will be significantly decreased. Younger group of respondents may perform well in this test, as human neurological functioning will deteriorate as age increases (Rohlman et al., 2007b). Based on the results obtained, an increase of the education level will also significantly increase the standard score of Trail Making test. This was consistent with the findings from Anger (2003), which presented significantly poorer scores among newly literate adult Nicaraguan males compared with non-neurotoxicant exposed Hispanic groups in the United States.

## Conclusion

The findings of this study have shown that neurobehavioral assessment conducted have provided an insight into the early effects of OP pesticide exposures when results detected impairment of cocoa farmers' motor steadiness and visual motor coordination. The blood cholineterase level in this study have provided baseline health data for the relevant administration of these farmers and can be used by the relevant authority to encourage commitment of employers to protect workers through conducting medical surveillance, reviewing existing training modules, assist in the amendment of the existing guideline for use and provide adequate and suitable personal protective equipment for the use of the workers in order to fulfil the objectives of the Occupational Safety and Health Act.

## Acknowledgement

Our deepest gratitude to Malaysia Cocoa Board for granting the permission to conduct the research and all the staffs who were directly or indirectly involved in this study. This study was made possible due to the financial support provided by Universiti Putra Malaysia under the *Inisiatif Putra Siswazah* grant scheme (IPS No. 9466400).

## References

Abdullah NH, Mohamed N, Sulaiman LH, Zakaria TA and Rahim DA, 2016. Potential Health Impacts of Bauxite Mining in Kuantan. Malays. J. Med. Sci. 23(3): 1-8.

Akinrinade ID, Memudu AE and Ogundele OM, 2015. Fluoride and aluminium disturb neuronal morphology, transport function, cholinesterase, lysosomal and cell cycle activities. Pathophysiology. 22(1): 105-115.

Al-Haddad SA and Al-Sayyad AS, 2013. Pesticide handlers' knowledge, attitude and practice. Bahrain Med. Bull. 35(1): 10-13.

Anger WK, 2003. Neurobehavioral tests and systems to assess neurotoxic exposures in the workplace and community. Occup. Environ. Med. 60(1): 531-538.

Balali-Mood M and Saber HR, 2012. Recent advances in the treatment of organophosphorus poisonings. Iranian J. Med. Sci. 37(2): 74-91.

Bressler JP and Goldstein GW, 1991. Mechanisms of lead neurotoxicity. Biochem. Pharmacol. 41(4): 479–484.

Farahat TM, Abdelrasoul GM, Amr MM, Shebl MM, Farahat FM and Anger WK, 2003. Neurobehavioural effects among workers occupationally exposed to organophosphorous pesticides. Occup. Environ. Med. 60(4): 279-286.

Holtman "unpublished results", 2014. Neurobehavioural effects of pesticide exposure among emerging farmers in the Western Cape. Ph.D. dissertation, Univesity of Cape Town, South Africa.

How V, Hashim Z, Ismail P, Md Said S, Omar D and Mohd Tamrin SB, 2014. Biological monitoring of genotoxicity to organophosphate pesticide exposure among rice farmers: Exposure-effect continuum study. JOHE. 2 (1-2): 27-36.

International Program on Chemical Safety (INCHEM). Organophosphorus pesticides. http://www.inchem.org/documents/pims/chemical/pimg001.htm.

Ismail AA, Bodner TE and Rohlman DS, 2012. Neurobehavioral performance among agricultural workers and pesticide applicators: a meta-analytic study. Occup. Environ. Med. 69: 457-464.

Kamel F, Rowland AS, Park LP, Anger WK, Baird DD, Gladen BC, Moreno T, Stallone L and Sandler DP, 2003. Neurobehavioral performance and work experience in Florida farmworkers. Environ. Health Perspect. 111(14): 1765-1772.

Khan MI, Mahdi AA, Islam N. Rastogi SK and Negi MPS, 2009. Assessment of erythrocyte acetylcholine esterase activities in painters. Indian J. Occup. Environ. Med. 13(1): 23-27.

Lionetto MG, Caricato R, Calisi A, Giordano ME and Schettino T, 2013. Acetylcholinesterase as a biomarker in environmental and occupational medicine: new insights and future perspectives. BioMed Research International. 2013(2013): 1-8.

Lokhman NN, Hashim Z, How V and Chokeli R, 2015. Blood cholinesterase concentration and neurobehavioral performance of primary schoolchildren at Tanjung Karang, Malaysia. Aust. J. of Basic Appl. Sci. 9 (22): 63-70.

London L, Myers JE, Nell V, Taylor T and Thompson ML, 1997. An investigation into neurologic and neurobehavioral effects of long-term agrichemical use among deciduous fruit farm workers in the Western Cape, South Africa. Environ. Res. 73(1-2): 132-145.

Lorenz A "unpublished results", 2011. Assessment of Pesticide Knowledge, Attitudes and Practices among Pregnant Women in Northern Thailand. M.S. thesis, College of William and Mary, Virginia, United States.

LOVIBOND Tintometer. (n.d.). AF267 CHOLINESTERASE TEST KIT. http://www.lovibondcolour.com/instrument/af267- cholinesterase-test-kit.

Marston DC, 2013. Neurobehavioral effects of poverty. http://www.apa.org/pi/ses/resources/indicator/2013/01/povertybehaviors.aspx

Nordin R, Araki S, Sato H, Yokoyama K, Wan Muda WAM and Kyi DW, 2002. Effect on safety bahaviours with pesticides use on occurrence of acute symptoms in male and female tobacco-growing Malaysian farmers. Industrial Health. 40(2): 182-190.

Occupational Safety and Health (Use and Standard of Chemicals Hazardous to Health) Regulation 2000, 2000. http://forfreedownloadonly.com/cat8/usechh-regulations-2000-pdf.php

Phyu MP and Tangpong J, 2014. Sensitivity of acetylcholinesterase to environmental pollutants. J. Health Res. 28(4): 277-283.

Rohlman DS, Anger WK and Lein PJ, 2011a. Correlating neurobehavioral performance with biomarkers of organophosphorous pesticide exposure. Neurotoxicology. 32(2): 268-276.

Rohlman DS, Lasarev M, Anger WK, Scherer J, Stupfel J and McCauley L, 2007b. Neurobehavioral performance of adult and adolescent agricultural workers. Neurotoxicology. 28(2): 374-380.

Roldan-Tapia L, Parron T, Sanchez-Santed F, 2005. Neurobehavioral effects of long-term exposure to organophosphate pesticides. Neurotoxicol. Teratol. 27(2): 259-266.

Ross SJM, Brewin CR, Curran HV, Furlong CE, Abraham-Smith KM and Harrison V, 2010. Neuropsychological and psychiatric functioning in sheep farmers exposed to low levels of organophosphate pesticides. Neurotoxicol. Teratol. 32(4): 452-459.

Starks SE, Gerr F, Kamel F, Lynch CF, Jones MP, Alavanja MC and Hoppin JA, 2012. Neurobehavioral function and organophosphate insecticide use among pesticide applicators in the agricultural health study. Neurotoxicol. Teratol. 34(1): 168-176.

Tsakiris S, Angelogianni P, Schulpis KH and Stavridis JC, 2000. Protective effect of L-phenylalanine on rat brain acetylcholinesterase inhibition induced by free radicals. Clin. Biochem. 33(2): 103-106.

World Health Organization (WHO, 1986. Operational guide for the WHO Neurobehavioral Core Test Battery (NCTB). World Health Organization Office of Occupational Health, Geneva, Switzerland.

Yassin MM, Abu Mourad TA and Safi JM, 2002. Knowledge, attitude, practice and toxicity symptoms associated with pesticide use among farm workers in the Gaza strips. Occup. Environ. Med. 59(6): 387-393.

# 3

# Sensitivity of *Macrophomina phaseolina* (Tassi) Goid isolates of maize (*Zea mays* L) to different temperature and pH levels

**Waqas Ashraf[1], ShahbazTalib Sahi[2], Amer Habib[2], Atta Ur Rehman Khan[3], Muhammad Ahmad Zeshan*[4], Anum Intisar[4], Absar Ahmad[5]**

[1]Department of Plant Pathology, University College of Agriculture and Environmental Sciences, The Islamia University of Bahawalpur, Bahawalpur, Pakistan

[2]Department of Plant Pathology, University of Agriculture, Faisalabad, Pakistan

[3]Department of Plant Pathology, PMAS-Arid Agriculture University Rawalpindi, Pakistan

[4]Department of Plant Pathology, University College of Agriculture, University of Sargodha, Sargodha, Pakistan

[5]University of Veterinary and Animal Sciences, Lahore, Pakistan

*Corresponding author email: ahmd_1566@yahoo.com

## Abstract

*Macrophomina phaseolina* is the devastating fungus of many crops. For this study, infected maize samples were collected from four districts (Kasur, Okara, Sahiwal and Pakpatan). Twenty-four isolates of *M. phaseolina* were isolated from infected maize samples and their growth was evaluated at 20, 25, 30, 35 and 40°C as well as at 5.0, 5.5, 6.0, 6.5 and 7.0 pH. The growth of fungal isolates was significantly affected by different levels of pH and temperature. Higher mean dry mycelial weight was observed at 35°C followed by 40°C. Higher mean dry mycelial weight at pH 6.5 and 7.0, clearly indicated the preference of isolates to particular range of pH. Mean dry mycelial weight was increased with increase in pH and temperature.

**Keywords:** Global warming, Mycelial weight, *Macrophomina phaseolina*, Temperature, pH

## Introduction

Global warming and its effects on plant disease induced by varying climate have become a serious concern with tentative predictions. The increase in global temperature might affect the plants and the pathogen eventually affecting their interaction (Garrett et al., 2006). The crop yield is badly affected by biotic and abiotic factors especially diseases which depends upon crop germplasm (Bashir and Malik, 1988). The fungus *Macrophomina phaseolina* is the devastating pathogen and distributed in varying climatic conditions (Iqbal and Mukhtar, 2014). Physiological factors like temperature, moisture and heat affect the prevalence of the pathogen (Dhingra and Sinclair, 1978). *M. phaseolina* is a soil and seed-borne fungus that cause disease by forming microsclerotia/pycnidia (Pun et al., 1998). *M. phaseolina* shows very high morphological (Mayek-Perez et al., 1997), pathogenic (Su et al., 2001), physiological (Mihail and Taylor, 1995) and genetic (Babuet al., 2007) variation due to heterokaryosis (Beas-Fernandez et al., 2006). It cause charcoal rot, stem canker, root rot, stem rot and seedling blight in different crops (Sanei and Razavi, 2011). The infection caused by the pathogen is greatly influenced by environmental factors (Maholay, 1992). *M. phaseolina* becomes destructive during summer at soil temperature of 20-40°C (Yang and Navi, 2003). Its pathogenicity increases with increase in temperature (Saleh et al., 2010). Low moisture level enhances the growth rate and survival ability of the fungus (Zazarini et al., 1985). Despite extensive studies on the effect of temperature on *M. phaseolina* growth attributes no precise range of epidemiological factors has been described up till now (Csondes et al. 2012). The temperature that affects the growth of the fungus varies with crop species (Das et al., 2008). The optimum temperature for microsclerotial development also varies with region to region (Das, 1988). At lower pH the growth of *M. phaseolina* becomes slower than at higher pH (Kulkarni, 2000). These issues are responsible for planning of efficient and suitable management approaches for the management of disease. Current reports of charcoal rot disease and global warming emphasizes the effect of different factors on pathogen biology. In the above background, the experiment was carried out to find the effect of different temperature and pH ranges on the *M. phaseolina*.

## Materials and Methods

### Collection of diseased plants

A random survey of major maize growing districts in Punjab (Kasur, Pakpatan, Sahiwal and Okara) was conducted during March 2015 and different villages were chosen for sample collection. Three fields were randomly selected in each village. The selected districts, villages and fields had variations in soil properties, cropping pattern and field history. The charcoal rot infected and healthy plant counts were made from 4m × 4m area of four randomly selected spots in each field. The stem samples with microsclerotia from the symptomatic plants were collected for further studies. The samples were taken in paper bags and then in polythene bags until use (Lotfalinezhad et al., 2013).

### Isolation, purification and maintenance of the isolates

The diseased portions of plants were collected from surveyed areas, washed in tap water and dried on sterilized blotting paper. Infected portion was cut into pieces (5 mm) and surface sterilized by dipping in 1% sodium hypochlorite for 30 sec. Sterilized pieces were washed to remove excess sterilant and transferred to 1% (w/v) agar and incubated for 72 hours at 27±1 °C. Hyphal tips were cut with cork borer and transferred to Petri dishes having potato dextrose agar (PDA) medium and identified by following Barnett and Hunter (1972). The isolates were transferred to PDA slants and stored at 4±1°C for further studies (Csondes et al. 2012).

**Table 1: Isolates of *Macrophomina phaseolina* collected from maize plants from districts of Punjab**

| Sr.No. | Isolate code | District | Locations |
|---|---|---|---|
| 1 | K-1 | KASUR | Bheala |
| 2 | K-2 | | Garewala |
| 3 | K-3 | | Talwandi |
| 4 | K-4 | | Atari |
| 5 | K-5 | | Noor Pur |
| 6 | K-6 | | KhudianKhas |
| 7 | O-1 | OKARA | Aktharaabad |
| 8 | O-2 | | Ahmadabad |
| 9 | O-3 | | BasirPur |
| 10 | O-4 | | HavaliLakha |
| 11 | O-5 | | Hujra Shah Muqeem |
| 12 | O-6 | | Renala Khurd |
| 13 | S-1 | SAHIWAL | Kassowal |
| 14 | S-2 | | Chak 42/12 L |
| 15 | S-3 | | Chak 21/11 L |
| 16 | S-4 | | Chak 44/12 L |
| 17 | S-5 | | AddePur |
| 18 | S-6 | | Bashera |
| 19 | P-1 | PAKPATAN | Chak 17 SP |
| 20 | P-2 | | JamanBodla |
| 21 | P-3 | | Bunga Hayat |
| 22 | P-4 | | Malka Hans |
| 23 | P-5 | | Chak 50 SP |
| 24 | P-6 | | Chak 30 SP |

### Effect of different pH on fungal mycelial weight

Isolates of *M. phaseolina* were grown on the Potato Dextrose Broth in selected pH levels of 5.0, 6.0, 6.5 and 7.0. The pH was maintained by 1N NaOH or HCl. Conical flasks having 30 ml medium at different pH levels were inoculated with 7 days old mycelial discs of the isolates. Three replicates were used and flasks incubated at 27± 1°C (Sukanya and Jayalakshmi, 2017).

### Effect of temperature on fungal mycelial weight

Growth of each isolate was tested at 20, 25, 30, 35 and 40°C. Thirty ml of potato dextrose broth was poured into 150 ml conical flasks and sterilized. Ten days old, 5 mm mycelial discs of the isolates were inoculated separately into conical flasks. Three replications were maintained and incubated at selected temperatures (Bekadda et al. 2008).

### Measurement of fungal dry mycelial weight

The broth was centrifuged at 12,000 × g for 15 min at 4±1°C after 10 days of incubation. Pre-weighed Whatman filter paper No. 1 was used for filtration of broth. The filter paper along with fungal mat was oven dried at 60±1 °C for 24 hours. The dried fungal mycelium was kept in desiccators and weighed using weighing balance (Iqbal and Mukhtar, 2014).

### Statistical Analysis

Collected data was interpreted by statistical analysis. Analysis of variance (ANOVA) technique was selected for data analysis by using MINITAB/STAT statistical analysis software (Minitab, 2010).

## Results

### Effect of different pH on fungal biomass

Variation in *M. phaseolina* biomass accumulation due to change in hydrogen ion concentration (pH) was recorded. The fungal mass of all the isolates differed with changing pH (Fig. 1). Fungal biomass and pH are directly proportional to each other. Mean dry mycelial weight increased at pH range 6.5 and 7 that more clearly indicated the preference of isolates to high pH. Highest mean (321 mg) dry mycelial weight was observed at pH 7.0 followed by 282 mg at pH 6.5. Least growth (83 mg) was observed at pH 5.0 indicating its inability to support the growth of *M. phaseolina* isolates. Out of 24 isolates, 9 were grouped in the lowest range of mycelial weight (132-170 mg), 4 were ranked in the middle range (171-207 mg) and 11 were in the range of (208-244 mg).

### Effect of temperature on fungal biomass

The growth of all fungal isolates was significantly affected by varying temperatures. However, isolates showed different response in terms of dry mycelial weight at all temperature ranges. The individual effects of temperatures on the fungal biomass of different isolates are given in (Fig.1). Among the different temperatures tested at 35°C was most favorable (331 mg) and it was closely followed by 40°C (297 mg) for the growth of *M. phaseolina*. At 20°C, 25°C and 30°C poor growth was observed indicating isolates' preference towards higher temperature for the growth. Fungal biomass increased with increasing temperatures up to 35°C but at 40°C it suddenly decreased. Among 24 isolates, 15 were in the range of (155.9-198.0mg) and 9 isolates grouped in the range (199.0-241.1mg) (Table. 2).

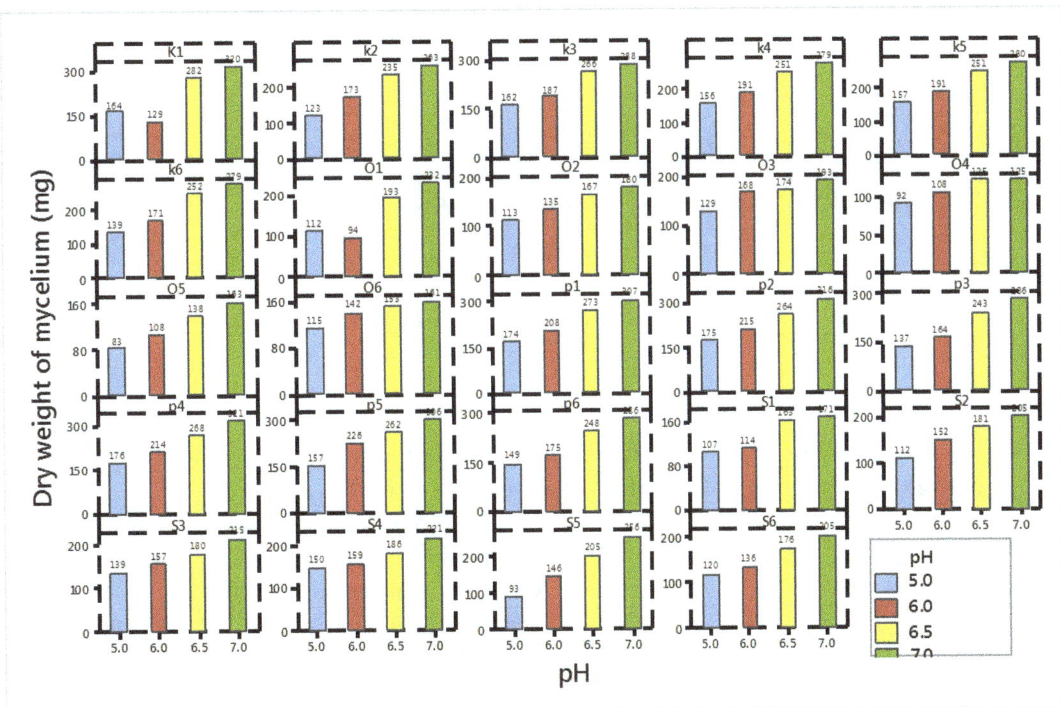


**Fig 1: Effect of pH on dry weight of *M. phaseolina* isolates**

**Table 2: Grouping of isolates of *Macrophomina phaseolina* based on dry mycelial weight at different pH levels**

| Group | Range (mg) | Number of isolates | Location of Isolates |
|---|---|---|---|
| 1 | 207.49-244.76 | 11 | Bheala, Talwandi, Atari, Noor Pur, Khudian Khas, Chak 17 SP, Jaman Bodla, Bunga Hayat, Malka Hans, Chak 50 SP, Chak 30 SP |
| 2 | 170.21-207.48 | 4 | Garewala, Chak 21/11 L, Chak 44/12 L, AddePur |
| 3 | 132.93-170.20 | 9 | Aktharaabad, Ahmadabad, Basir Pur, Havali Lakha, Hujra Shah Muqeem, Renala Khurd, Kassowal, Chak 42/12 L, Bashera |

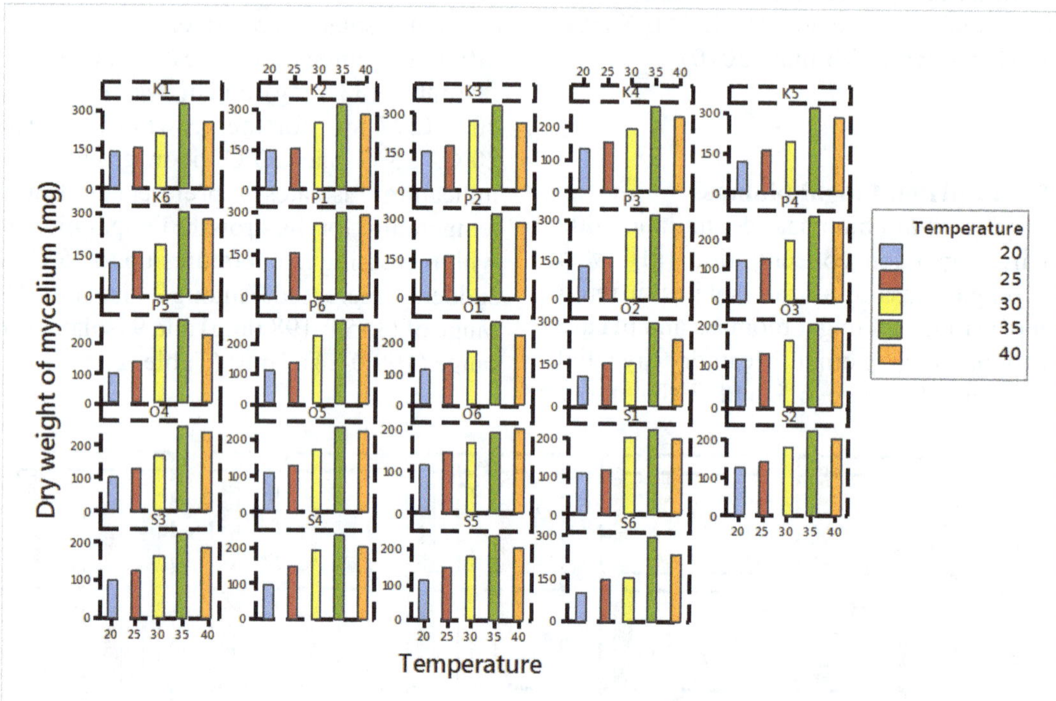


**Fig 2: Effect of temperature on dry weight of *M. phaseolina* isolates**

**Table 3: Grouping of isolates of *Macrophomina phaseolina* based on dry mycelial weight at different temperatures**

| Group | Range (mg) | Number of isolates | Location of Isolates |
|---|---|---|---|
| 1 | 199.0-241.1 | 9 | Bheala, Noor Pur, Khudian Khas, Aktharaabad, Ahmadabad, Basir Pur,Havali Lakha, Hujra Shah Muqeem, Renala Khurd |
| 2 | 155.9-198.00 | 15 | Garewala,Talwandi, Atari, Chak 17 SP, Jaman Bodla, Bunga Hayat, Malka Hans, Chak 50 SP, Chak 30 SP, Kassowal, Chak 42/12 L,Chak 21/11 L,Chak 44/12 L, AddePur,Bashera |

## Discussion

Temperature and pH ranges varied greatly for *M. phaseolina* in different experiments, which may be the effect of crop specie, growing media and growing conditions (Ratnoo and Bhatnagar, 1991). The most favorable temperature for *M. phaseolina* growth was 35°C followed by 40°C while reduced growth rate was recorded at 20 °C and 25 °C. Csondes et al. (2012) stated that infection caused by *M. phaseolina* in sunflower growing areas is greatly affected by the environmental variable such as pH and temperature. These results are supported by the findings of (Vianaet al., 2002) who recorded 35°C as the optimum temperature for *M. phaseolina* isolates obtained from common bean (*Phaseolus vulgaris* L.) fields. According to Maholay (1992) mycelia growth and microsclerotial development was best at 30°C. High temperature (25 to 35°C) enhanced *M. phaseolina* dry weight (Sharma et al., 2004). In Hungary, *M. phaseolina* isolates were subjected to different temperature ranges and best growth was recorded at 25-35 °C (Csondes et al., 2007). Manici et al. (1995) obtained fungal isolates from different ecological zones of Italy and found that 30-35°C was temperature for growth. High temperature may exert selection pressure on the pathogen which adapt to survive at higher or lower temperatures.

A survey was conducted in different climatic zones of Varanasi and described that due to increased activities of fungus at high temperature and in drought conditions high charcoal rot disease incidence was recorded (Kaur et al., 2012a). At high temperature, the concentration of hydrolytic enzymes were increased in microsclerotia of *Macrophomina* (Kaur et al., 2012b). The metabolic activities, such as transformation of substrate into products are carried out with the help of biological catalysts (enzymes) that requires particular range of pH. Jha and Dubey (2000) isolated *M. phaseolina* from okra to check the effect of pH and maximum fungal dry weight was recorded at pH 7. The mycelium of *M. phaseolina* isolated from muskmelon was developed best at pH 5.0 and pH 6.0 (Singh and Chohan, 1982). Nischwitz et al. (2004) used different irrigation types in melon fields and found pH 4.0 and 6.0 as the best for *M. phaseolina* growth. These results contradict with present study which may be due to soil conditions, crop specie and cultural practices.

## Conclusion

The present investigation suggests adaptability of the pathogen to wide range of temperature and pH that increases fitness in particular niche. It might also be implicated that predicted global warming is likely to increase the range and severity of *Macrophomina* charcoal rot disease. Edaphic factors, such as temperature and pH critically affect the survival of *M. phaseolina*as well as influence the disease incidence in various crops.

## Acknowledgements

The authors highly acknowledge the financial assistance of Higher Education Commission for the completion of current study.

## References

Barnett HL and Hunter BB, 1972. Illustrated genera of imperfect fungi. Burgress Publishing Company, Minneapolis MN, p.241.

Bashir MA and Malik BA, 1988. Diseases of major pulse crops in Pakistan. Trop. Pest Manag. 34: 309-314.

Babu BK, Srivastava AK and Arora DK, 2007. Identification and detection of *Macrophomina phaseolina* by using species-specific oligonucleotide primers and probe. Mycologia. 99: 797-803.

Beas-Fern´andez R, De Santiago-de Santiago A, Hern´andezDelgado S and Mayek-P´erez N, 2006. Characterization of Mexican and non-Mexican isolates of *Macrophomina phaseolina* based on morphological characteristics, pathogenicity on bean seeds and endoglucanase genes. J. Plant. Pathol. 88: 53-60.

Csondes I, Cseh A, Taller J and Poczai P, 2012. Genetic diversity and effect of temperature and pH on the growth of *Macrophomina phaseolina* isolates from sunflower fields in Hungary. Mol. Biol. Rep. 39: 3259-3269.

Csondes I, Sandor K and Richard G, 2007. Growth of *Macrophomina phaseolina* isolates depend on different temperature. AnaleleUniversităţii din Oradea, Fascicula: ProtecţiaMediului, 12: 31-34.

Das ND, 1988. Effect of different sources of carbon, nitrogen and temperature on the growth and sclerotial production of *Macrophomina phaseolina* (Tassi) Goid, causing root rot/charcoal rot disease of castor. Ind. J. Plant Pathol. 6: 97-98.

Das IK, Fakrudin B and Arora DK, 2008. RAPD cluster analysis and chlorate sensitivity of some Indian isolates of *Macrophomina phaseolina* from sorghum and their relationships with pathogenicity. Microbiol. Res. 163:215-224.

Dhingra OD and Sinclair JB, 1978. Biology and pathology of *Macrophomina phaseolina.* Minas Gerais, Brazil; Universidade Federal de Vicosa. p.166.

Garrett KA, Dendy SP, Frank EE, Rouse MN and Travers SE, 2006. Climate change effects on plant disease: genomes to ecosystems. Ann. Rev. Phytopathol. 44: 489-509.

Iqbal U and Mukhtar T, 2014. Morphological and pathogenic variability among *Macrophomina phaseolina* isolates associated with mungbean (*VignaradiataL.*) Wilczek from Pakistan. Sci. World J. http://dx.doi.org/10.1155/2014/950175.

Jha AK and Dubey SC, 2000. Occurrences of collar rot of okra in the plateau region of Bihar. J. Res. Birsa Agric. Univ. 12: 67-72.

Kaur S, Chauhan VB, Singh JP and Singh RB, 2012a. Status of *Macrophomina* stem canker disease of pigeonpea in eastern Uttar Pradesh. J. Food. Legumes. 25: 76-78.

Kaur S, Dhillon GS, BrarSK and Chauhan VB, 2012b. Carbohydrate degrading enzyme production by the plant pathogenic mycelia and pycnidia strains of *Macrophomina phaseolina* through koji fermentation. Indus. Crops Prod. 36: 140-148.

Kulkarni S, 2000. Biology and management of dry stalk rots of maize (*Zea mays* L.) caused by *Fusarium moniliformae* shield and *Macrophomina phaseolina* (Tassi) Goid. Ph.D. Thesis, University of Agricultural Sciences, Dharwad, pp. 162-164.

Lotfalinezhad E, Mehri Z and Sanei SJ, 2013. Temperature response of *Macrophomina phaseolina* isolates from different climatic in Iran. Annu. Rev. Res. Biol. 3(4): 724-734.

Maholay MN, 1992. *Macrophomina* seed and pod rot of butter bean (*Phaseolus lunatus* L.). Ind. J. Mycol. Pl. Pathol. 22: 220-226.

Manici LM, Caputo F and Cerato C, 1995. Temperature responses of isolates of *Macrophomina phaseolina* from different climate regions of sunflower production in Italy. Plant Dis. 79: 834-838.

Mayek Pérez, N, LópezCastañeda C and Gallegos JA, 1997. Variation on in vitro cultural characteristics of *Macrophomina phaseolina* isolates and its virulence on common bean. Agrociencia. 31: 187-195.

Mihail JD and Taylor SJ, 1995. Interpreting variability among isolates of *Macrophomina phaseolina*in pathogenicity, pycnidium production and chlorate utilization. Can. J. Bot.73: 1596-1603.

Minitab 17 Statistical Software, 2010. State College, PA: Minitab, Inc.USA.

Nischwitz C, Olsen M and Rasmussen S, 2004. Effect of irrigation type on inoculums density of *Macrophomina phaseolina*in melon fields in Arizona.J. Phytopathol. 152: 133-137.

Pun KB, Sabitha D and Valluvaparidasan V, 1998. Studies on seed-borne nature of *Macrophomina phaseolina* in okra. Plant Dis. Res.13: 249-290.

Ratnoo RS and Bhatnagar MK, 1991. Effect of temperature and pH on growth and sclerotia formation of Macrophomina phaseolina. Ind. J. Mycol. Plant Pathol. 21(3): 279-280.

Saleh AA, Ahmed HU, Todd TC, Travers SE, Zeller KA, Leslie JF and Garrett KA, 2010. Relatedness of *Macrophomina phaseolina* isolates from tallgrass prairie, maize, soybean and sorghum. Mol. Ecol. 19(1): 79-91.

Sanei SJ and Razavi SE, 2011. Charcoal rot in nursery of olive in Golestan province of Iran. Int. Res. J. Agric. Sci. Soil Sci. 1: 211-217.

Sharma VK, Gaur RB and Bisnoi HR, 2004. Cultural, morphological and physiological variability in *Macrophomina phaseolina.* J. Mycol. Plant Pathol. 34: 532-534.

Singh RS and Chohan JS, 1982. Physio-pathological studies of *Macrophomina phaseolina* causing charcoal rot in muskmelon. Ind. J. Mycol. Plant Pathol. 12: 81-82.

Su G, Suh SO, Schneider RW and Russin JS, 2001. Host specialization in the charcoal rot fungus, *Macrophomina phaseolina.* Phytopathol. 91: 120-126.

Sukanya R and Jayalakshmi SK, 2017. Response of inoculation to seed and seedling infection by *M. phaseolina* in sorghum. Adv. Plants Agric. Res. 6(1): 198.

Viana FMP, de SNL and De-Souza NL, 2002. Effect of temperature and water tension of the substrate on germination of microsclerotia or *M. Phaseolina*. Phytopathol. 23:236-239.

# 4

# Physiological Studies on Colletotrichum Gloeosporioides Associated with Wither Tip Disease of Citrus and its Chemical Control

**Salman Ghuffar*[1], Muhammad Zeshan Ahmed[1], Muhammad Farooq Aslam[2], Luqman Amrao[1], Sajjad Hyder[2]**

[1]Department of Plant Pathology, University of Agriculture Faisalabad, Pakistan

[2]Department of Plant Pathology, PMAS-Arid Agriculture University Rawalpindi, Pakistan

*Corresponding author email: mominsalman3610@gmail.com

## Abstract

Citrus is the second largest fruit produced in the world and Pakistan is among the 12 large producers of the citrus fruit. It is grown in tropical and subtropical climate all over the world. Besides its high economical & nutritional values citrus is attacked by different pathogen such as fungi, bacteria, viruses and nematodes. Among all the pathogens *Colletotrichum gloeosporiodes* causing citrus wither tip disease is one of the major constrain in citrus production. Therefore current study was conducted to investigate the different physiological characters on the mycelial growth of *C. gloeosporiodes* and its chemical control. Among different fungal nutrient media Potato Dextrose Agar (PDA) gave maximum mycelial growth (7.9 cm) followed by Citrus leaf extract Agar (CLEA) (4.7 cm) and corn meal Agar (CMA) (3.3 cm).Temperature of 30°C favored maximum colony growth (8 cm) followed by 25°C (7.4 cm), 35°C (4.2 cm), 20°C (3.5 cm) and 15°C (2.3cm). A pH level of 6 favored maximum colony growth (7.9 cm) followed by 5 (7.5 cm), 4.5 (5.3cm), 4 (4.8 cm), 7(3.3cm) and 3(2.4 cm).Among the application of different fungicides such as Topsin-M, Copper oxychloride and Aliette at different concentration (300, 600 and 900 ppm). Topsin-M gave maximum result to inhibit the Maximum mycelial growth inhibition of *C. gloeosporioides* (1.1 cm) was produced by Topsin-M 9 days after incubation at 30 °C as compared to control (7.6 cm).

**Keywords:** Physiological studies, *Colletotrichum gloeosporioides,* chemical control

## Introduction

Citrus belongs to *Rutaceae* family, grown in tropical and subtropical domains of the world. China, India and Himalayan region are the major citrus fruit producing regions (NIHORT, 2003). According to (FAO, 2015) Pakistan stands 30th position in citrus production among citrus growing countries and annual fruit production is 2007 thousand tones. As compare to all provinces in Punjab above 95% of citrus fruits are produced and 70 percent production is Kinnow (Niaz*etal*, 2004). Numerous factors are responsible for pre and post-harvest losses in citrus production of which disease play very vital role. Citrus fruits production is drastically reduced by wither tip disease in tropical and subtropical domains (Rohlf,2000).*Colletotrichum gloeosporioides* remains a universal pathogen that cause diseases on many fruits such as mango, papaya, and apple and especially wither tip disease caused on citrus due to this pathogen (Simmonds,1965).Physiological studies on *C. gloeosporioides* was carried out by many researchers as (Sudhakar, 2000) reported that maximum mycelial growth of *C. gloeosporioides* was observed in five culture medias (Potato dextrose agar, Richard's agar, Brown's agar, Sabouraud's agar, and Oat meal agar). *In vitro* studies revealed that maximum mycelia growth of *C. gloeosporioides* reached after 09 days incubation on PDA, at 25-35 °C (Annonymous, 2001). According to the (Narendra Kumar, 2006) maximum fungal mycelial growth was recorded at pH 6.5 and 6.0 while minimum growth was noticed at pH 4.0.*In vitro* condition the optimum pH for the growth of pathogenic fungai ranges between 5.5 and 7.0.Inhibitory effect of various chemicals viz., benomyl, prochloraz, chlorothalonil, captan and thiram was studied for the control of *C. gloeosporioides* causing wither tip disease of citrus (Eikelenboom, 1964). The present study aimed to investigate the pathogen under different media, temperature, and pH levels requirements for mycelial growth and its chemical control.

## Materials and Methods

**Collection of diseased sample from Citrus orchard**
Samples of the infected plants parts viz. (twigs and leaves) were collected from citrus orchard located in University of Agriculture Faisalabad. The samples were brought in mycology lab wrapping in the new polythene bags and stored in refrigerator at 4 C° for further use.

Physiological (Nutrient media, pH and temperature) studies of the *C. gloeosporiodes* associated with wither tip disease in citrus

**Effect of selective Nutrient Medias on *C. gloeosporiodes***
The cultural characters of *C. gloeosporiodes* were examined on PDA, Corn Meal Agar (CMA) and Host leaf extract Agar which was prepared according to the prescribed recipies. After that 5 cm discs from *C. gloeosporioides* cultures were cut by using a sterilized transfer needle and a single disc was placed at the centre of glass cavity blocks. Each set of experiment was replicated thrice and the plates were incubated at 30±1°C for a week. The cultural characters such as colony diameter, colour, type of margin, aerial growth were recorded.

**Effect of pH levels on physiology of *C. gloeosporioides***
PH effect on the growth of *C. gloeosporioides* was studied on PDA media. The pH of the medium was adjusted to various levels namely 3, 4, 4.5, 5, 6 and 7 and pH level in each treatment was maintained by electronic pH meter. Medium with known pH level was poured into conical flasks and sterilized. Flasks containing sterilized culture medium were poured into petriplates allow them to solidify and inoculated with fresh culture of *C. gloeosporioides* and incubated for 9 days .Each treatments was replicated thrice. Optimum pH was determined by measuring fungal mycelial growth (cm).

**Effect of various temperature levels on *C. gloeosporioides***
Medium was added to 100 ml flasks and sterilized $121^0$C and 15psi pressure for 20 minutes. After sterilization media poured into the petriplates was inoculated with fresh culture of *C. gloeosporioides* and kept at temperature levels viz., 15, 20, 25, 30 and 35°C for fungal colony development and three replications were maintained for each treatment. Most suitable temperature for *C .gloeosporioides* was determined by measuring mycelial growth (cm).

***In vitro* testing of selected fungicides against *C. gloeosporodes***
Efficacy of selected fungicides (Copper oxychloride, Topsin-M and Aliette) was tested by Poisoned food technique (Hawamdeh and Ahmad, 2001) against *C. gloeosporioides* at various concentrations of 300, 600 and 900 ppm respectively with three replications for each treatment. The actively growing periphery of *C.gloeosporioides* was shifted aseptically to the centre of each petri dish having poisoned medium and incubated at 30 C°. The fungal mycelial growth was noted 3, 6 and 9 days after incubation and percentage mycelial growth inhibition (*I*) was determined by using the following formula:

$$\text{Inhibition of the mycelial radial growth (\%)} = \frac{C - T}{C} \times 10$$

(Perrucci*et al.*, 2003)

C: colony diameter of Control, T: colony diameter of treatment.

Data was analyzed by ANOVA test and differences within treatments were determined by LSD test at P= 0.05 (steel *et al.*, 1997).

## Results and discussion

### Physiological study of *C. gloeosporioides*

**Nutrient media effect on mycelial growth of *C. gloeosporiodes***

The nutrient media plays influential role on the growth of *C. gloeosporiodes* as illustrated in Table. 1. According to result revealed that pathogen grew well in a different culture media but the highest colony diameter (7.9) was observed after 9 days on PDA medium followed by host leaf extract medium (4.7 cm) and the minimum colony diameter (3.3 cm) was observed on corn meal agar.

**Effect of pH levels on radial growth of *C. gloeosporiodes***

*C. gloeosporioides* growth was observed on various pH levels on various time intervals as shown in table 2. After 9 days of incubation result indicated that at pH level 6.0 the colony growth (7.9 cm) was maximum followed by 5.0, 4.5, 4.0 and 7. The growth was observed minimum at pH 3. Fungal growth increases at 6.0 by increasing pH up to a certain level after this mycelial growth was decreased at 7.0 pH. According to result at pH levels 5.0 and 6.0 were best among all for radial growth of *C. gloeosporioides* but the pH levels 3.0, 4.0 and 7.0 were not favorable for colony growth.

**Effect of various temperature levels on *C. gloeosporiodes***

Radial growth of *C. gloeosporioides* has shown on various temperature levels along with duration in Table 3. Result revealed that maximum radial growth(8.0 cm) was observed at 30°C followed by 20, 25, 35°C while minimum growth was recorded at 15°C and 35°C. The temperature ranged 25 to 30°C was significant for *C. gloeosporiodes*. Statistically significant relationship between radial colony growth and temperature levels was observed.

***In vitro* evaluation of fungicides against *C. gloeosporioides***

The results after 3, 6 and 9 days shown in table 4, 5 and 6 revealed that selected fungicides showed different efficacy levels in inhibiting the radial growth of *C. gloeosporioides* respectively. Statistical data indicated that Topsin-M was found most effective for controlling the growth of pathogen that showed colony diameter 1.6, 1.3 and 1.1 cm at 300, 600 and 900 ppm after 9 days incubation followed by Copper oxychloride and Aliette respectively.

**Table 1: Effect of Nutrient media on fungal colony growth of *C. gloeosporioides***

| Media | Duration (days) Colony growth (cm) | | | Means (cm) |
|---|---|---|---|---|
| | 3 days | 6 days | After 9 days | |
| PDA | 2.6 e | 3.9 c | 7.9 a | 4.8 a |
| CLEA | 1.5 g | 2.4 f | 4.7 b | 2.8 b |
| CMA | 0.9 h | 1.5 g | 3.3 d | 1.9 c |
| Means | 1.6 c | 2.6 b | 5.3 a | |

Various characters are likely to be followed by the level of statistical significance of 5% LSD
Alpha = 0.05LSD value for (M) = 0.0841 (D) = 0.0732 and (MxD) = 0.1301

**Table 2: Effect of pH Level on mycelial growth of *C. gloeosporioides***

| pH | Duration (days) | | | Means (cm) |
|---|---|---|---|---|
| | After 3 days Colonygrowth (cm) | After 6 days Colonygrowth (cm) | After 9 days Colonygrowth (cm) | |
| 3 | 0.7 n | 1.2 m | 2.4 j | 1.4 f |
| 4 | 1.6 l | 2.5 ij | 4.8 d | 2.9 d |
| 4.5 | 1.8 k | 2.7 h | 5.3 c | 3.2 c |
| 5 | 2.5 ij | 3.7 f | 7.5 b | 4.5 b |
| 6 | 2.6 hi | 4.0 e | 7.9 a | 4.8 a |
| 7 | 1.1 m | 1.6 l | 3.3 g | 2.0 e |
| Means | 1.7 c | 2.6 b | 5.2 a | |

Various characters are likely to be followed by the level of statistical significance of 5% LSD
Alpha = 0.05LSD value (pH) = 0.0636LSD value (Duration) = 0.0450LSD value (PxD) = 0.1102

**Table 3: Temperature effect on mycelia colony growth of *C. gloeosporioides***

| Temperature | Duration (Hours) Colony growth (cm) | | | Means (cm) |
|---|---|---|---|---|
| | 3 days | 6 days | 9 days | |
| 15°C | 0.6 m | 1.1 l | 2.3hi | 1.3e |
| 20°C | 1.2 l | 1.7 j | 3.5f | 2.1d |
| 25°C | 2.4 h | 3.7e | 7.4b | 4.5b |
| 30°C | 2.7 g | 3.9d | 8.0a | 4.8a |
| 35°C | 1.4 k | 2.2i | 4.2c | 2.6c |
| Means | 1.6c | 2.5b | 5.0a | |

Various characters are likely to be followed by the level of statistical significance of 5% LSD
Alpha = 0.05 LSD value (Temperature) = 0.0908 (Duration) = 0.0702 and (DxT) = 0.1564

**Table 4:*In vitro* evaluation ofselected fungicides 3 days after treatment against *C. gloeosporioides***

| Fungicides | Concentrations (ppm) Colony growth (cm) | | | Means (cm) |
|---|---|---|---|---|
| | 300 ppm | 600 ppm | 900 ppm | |
| Copper oxychloride | 0.80 | 0.60 | 0.40 | 0.60 c |
| Topsin-M | 0.60 e | 0.40 | 0.30 | 0.43 d |
| Aliette | 1.20 | 1.10 | 0.90 | 1.06 b |
| Control | 2.50 | 2.40 | 2.60 | 2.50 a |
| Means | 1.27 a | 1.12 b | 1.05 b | |

Various characters are likely to be followed by the level of statistical significance of 5% LSD
Alpha = 0.05 LSD value (F) = 0.0963 LSD value (Conc) = 0.0834 LSD value (FxC) = 0.1667

**Table: 5 *In vitro* evaluation of selected fungicides 6 days after treatment against *C. gloeosporioides***

| Fungicides | Concentrations (ppm) Colony growth (cm) | | | Means (cm) |
|---|---|---|---|---|
| | 300 ppm | 600 ppm | 900 ppm | |
| Topsin-M | 1.10 h | 0.90 j | 0.70 k | 0.90 d |
| Copper oxychloride | 1.50 f | 1.30 g | 1.0 i | 1.26 c |
| Aliette | 2.50 c | 2.20 d | 1.70 e | 2.13 b |
| Control | 5.2 a | 5.0 b | 5.2 a | 5.13 a |
| Means | 2.57a | 2.35 b | 2.15 c | |

Various characters are likely to be followed by the level of statistical significance of 5% LSD
Alpha = 0.05 LSD value (F) = 0.0282 LSD value (Conc) = 0.0244 LSD value (FxC) = 0.0489

**Table: 6 *In vitro* evaluation of selected fungicides 9 days after treatment against *C. gloeosporioides***

| Fungicides | Concentrations (ppm) Colony growth (cm) | | | Means (cm) |
|---|---|---|---|---|
| | 300 ppm | 600 ppm | 900 ppm | |
| Topsin-M | 1.6 fg | 1.3 h | 1.1 i | 1.33 d |
| Copper oxychloride | 2.1 e | 1.7 f | 1.5 g | 1.76 c |
| Aliette | 3.7 b | 3.3 c | 2.8 d | 3.26 b |
| Control | 7.6 a | 7.5 a | 7.6 a | 7.56 a |
| Means | 3.75 a | 3.45 b | 3.25 c | |

Various characters are likely to be followed by the level of statistical significance of 5% LSD
Alpha = 0.05 LSD value (Fungicides) = 0.0717 LSD value (Concentration) = 0.0621 LSD value (FxC) = 0.1242

## Discussion

Citrus is considered as the "common man's fruit" crops. The fruits are immensely important as they are delicious, have high food value, rich in vitamin A and C and also fats and occupy a very prominent place in the diet of human being. Citrus suffers from several diseases and wither tip is one of them cause severe loses. According to the (Deshmukh*et al., 2012*) in an experiment on evaluation of 15 solid media, maximum mycellial growth of *C. gloeosporioides* was noticed on Potato Dextrose Agar media (88.74 mm) followed by Richard's Agar (85.43 mm), Oat Meal Agar (81.51 mm), Malt Extract Agar media (80.25 mm) andCorn Meal Agar (79.40 mm) but no growth was observed on Sach's agar media (0.00 mm). Maximum mycelial growth of *C. gloeosporioides* in Potato dextrose agar was also recorded by (Pandey *et al*., 2012a). For the confirmation of maximum growth at different pH level (Sangeetha, 2003) observed that maximum radial growth of *C. gloeosporioides* at 6.0 pH. Chandra *et al* (2004) observed that the optimal colony growth of *C. gloeosporioides* ranges from 5.2 to 6.4 pH. Among all the physiological factors temperature plays a significant role for increasing the growth of pathogens. It can affect almost all stages of fungi including its growth, spore germination and reproduction. Favourable temperature for the growth of *C. gloeosporioides* is 30°C observed by (Sangeetha and Rawal, 2008). After the evaluation of five fungicides against *C. gloeosporioides* topsin-M is one the most effective at 0.15 per cent concentration for 100% fungal growth inhibition, followed by Mancozeb (78.51%) at 0.35% (Biradar, 2002).

## References:

Anonymous, 2001. Physiological Requirements of *Colletotrichum gloeosporioides* (Penz.) Penz. and Sac. A Causal Agent of wither tip disease of citrus, http// www.Phdeb.org.pkPhytopathol. 2: 95-97.

Biradar SV, 2002. Studies on *Colletotrichum gloeosporioides* (Penz.) Penz and Sacc., *Botryodiplodia theobromae* Pat. And *Fusarium semitectum* Berk. And Rav. Causing fruit rot of custard apple (*Annonasquamosa*L.). *M.Sc. (Agri.) Thesis*, Uni. Agric. Sci., Dharwad, India.

Deshmukh A J, Mehta B P, Sabalpara A N and Patil V A, 2012. In vitro effect of various nitrogen, carbon sources and pH regimes on the growth and sporulation of *Colletotrichum gloeosporioides* Penz. And Sacc causing anthracnose of Indian bean. J. Biopest., 5: 46-49.

Ding D I, Jian G U and Zhuang Q I J, 2007. Effect of Different Factors on Growth and Spore Production of a Colletotrichum sp., a Potential Biocontrol Agent for Water hyacinth. J. Aquat. Plant Manage. 45: 129-131

Eikelenboom C, 1964. A microbioassay for fungicide residues.Chemical abstract, 61:6261.

FAO, 2015. FAO Production Year Book, FAO, Rome, Italy.

Hawamdeh, A S and Ahmad S, 2001. *In vitro* control of Alternariasolani, the cause of early blight of tomato. J. Biological Sci. 949-950.approach. 2nd Ed., McGraw Hill Inter. Book Co. Tokyo, Japan.

Hegde Y R, Hegde R K and Kulkarni S, 1993. Physiological studies on *Colletotrichum gloeosporioides* (Penz) Penz and Sacc. causing anthracnose of arecanut, *Karnataka* J. Agric. Sci. 6:411- 412.

Hiremath, 1993. Physiological studies on *Colletotrichum gloeosporioides* causing anthracnose of arecanut. Kar. J. Agric. Sci. 6: 411-412.

Narendra Singh, 1992. Biological control of red rot disease of sugarcane. *Indian Phytopathol.*43: 64.

Niaz AC, Maken M N and Malik S A, 2004. Native historical background and importance of citrus fruits in Pakistan. Proceedings of the 1st Int. Conf on Citriculture.University of Agriculture, Faisalabad.48-56.

NIHORT, 2003. Nigerian Institute of Horticultural Research. Commercial Crop production Guide Series in Nigeria. 8: 10.

Pandey A, Yadava L .P, Manoharan M, Chauhan U K and Pandey B K, 2012 a. Effectiveness of cultural parameters on the growth and sporulation of *Colletotrichum gloeosporioides* causing anthracnose disease of mango (*Mangifera indica* L.). Online J. Biological Sci., 12(4) : 123-133.

Perrucci S, Mancianti F, Ciont PL, Flamini G, Morelli I and Macchioni G ,2003. *In vitro* anti fungal activity of essential oils against some isolates of Micro sporumcanis and M gypseum. Planta Medi 60:184–187.

Rohlf, 2000 numerical taxonomy and multivariate analysis system.Version2.1.Setauket, New York: Exeter Publishing Ltd.

Sangeetha, C G, 2003. Studies on anthracnose of mangocaused by *Colletotrichum gloeosporioides* (Penz.) Penz. andSacc., *PhD Thesis,* Univ. Agric. Sci.,Bangalore, pp. 77-156

Sangeetha C G and Rawal R D, 2008. Nutritional studies of *Colletotrichum gloeosporioides* (Penz.) Penz. andSacc. theincitant of mango anthracnose. World J. Agric. Sci., 4(6) : 717-720.Chandra M., Thind T.S.,Prem R. and Arora J.S. 2004.Plant Disease Research 19 (2):200-203.

Simmonds J H, 1965. A study of the species of Colletotrichum causing ripe fruit rots in Queensland. Queensland J. Agric. and Animal Sci. 22: 437-459.

Steel L.R.D, Torrie J.H and Dickey D A, 1997. Principles and Procedures of Statitics. biometrical approach.3rd Edition, McGraw Hill Book Co., New York, USA.

Sudhakar, 2000. Biology and management of Stylosanthes anthracnose caused by *Colletotrichum gloeosporioides* (Penz.) Penz. And Sacc. M.Sc (Agri.) Thesis, Uni. Agric. Sci., Dharwad,India.

# 5

# Intra-specific variations among wheat genotypes for phosphorus use efficiency

**Muhammad Abbas*, Muhammad Irfan, Javaid Ahmed Shah, Muhammad Yousuf Memon**
Soil and Environmental Sciences Division, Nuclear Institute of Agriculture (NIA), Tando Jam, Sindh, Pakistan

*Corresponding author email: alimadad51214@gmail.com

## Abstract

Phosphorus (P) deficiency in Pakistani soils severely limits the crop production, hence, the development of P-efficient cultivars seems inevitable to uphold the productivity of P-impoverished agricultural systems. Ten wheat genotypes (AA-V1, AA-V2, AA-V3, MSH-3, MSH-5, BWQ-4, EST-28/11, EST-29/9, ESW-9525 and NIA-Sunder) were evaluated for P-efficiency and responsiveness under two phosphorus levels viz., 20 and 200 µM in hydroponic culture. Completely randomized design with two way factorial arrangement was used and the treatments were replicated five times. Seven days old seedlings were shifted to solution culture and grown for four weeks after transplanting. The plants were then harvested for recording growth and phosphorus related attributes. Low P level markedly reduced shoot and root dry weight, P concentration and accumulation in wheat genotypes. However, P utilization index of shoot and root were improved by 30 and 13% at low P supply. Shoot P utilization index was positively correlated ($r > 0.43$; $n = 50$) with shoot dry weight and shoot P accumulation at both P levels. Genotypes were grouped into four classes by regressing shoot dry weight at low P level and physiological P-use efficiency. Genotypes AA-V3 and NIA-Sunder were categorized as efficient and responsive (ER) genotypes, while AA-V1, BWQ-4 and EST-28/11 were identified as efficient but non-responsive (ENR). Genotypes AA-V2 and MSH-5 formed the non-efficient but responsive (NER) group. Non-efficient and non-responsive (NENR) category included MSH-3, EST-29/9 and ESW-9525. The results of present study indicated that AA-V3 and NIA-Sunder were the most P-efficient and responsive genotypes as they have the potential to yield more under varying levels of P availability. However, the results should be confirmed under filed conditions.

**Keywords:** Genotypes, Hydroponics, P-efficiency, Plant nutrition, Wheat

## Introduction

World's food production is critically dependent on the continued supply of phosphorus (P) fertilizers. Phosphatic fertilizers are primarily derived from phosphate rock which is a finite and non-renewable resource and is expected to deplete in the near future. Almost 90 % of world's rock phosphate is being consumed to synthesize P fertilizers and animal feed additions (Cordell et al., 2009). Dwindling rock phosphate reserves, high costs and low recovery of P fertilizers make it imperative to increase phosphorus use efficiency at plant and farm level (Simpson et al., 2011). Agronomic management strategies like precision P fertilization, polymer coated P-fertilizers, and recycling of P from domestic, agricultural and industrial wastes can be helpful in improving P use at farm level (Ma et al., 2009). At plant level, development of P-efficient cultivars can greatly reduce

P fertilizer requirements in modern agriculture (Richardson et al., 2009). Improved P-efficiency can be conferred by high P uptake and/or efficient internal use of P in the plants. Strategies by which plants adapt to P-deficient soils include root structural and architectural modifications (Hammond and White, 2011), mycorrhizal associations (Smith and Read, 2008), release of organic acids and phosphatases into rhizosphere (Gahoonia and Nielsen, 2004; Wasaki et al., 2009), and efficient P use at cellular level (Rengel and Marschner, 2005).

The strategy of breeding P-efficient cultivars and their distribution to the farmers is gaining recognition as the most cost-effective and sustainable measure for improved P use and crop production. The fundamentals for evolving P-efficient cultivars include i) useable genetic diversity for P efficiency should be present, ii) characteristics underlying P efficiency should be recognized and iii) screening background should allow the manifestation of the desired characters in a reliable and repeatable way. Plant species and even genotypes within species exhibit considerable variations in their growth response to phosphorus deficiency stress (Wang et al., 2010; Irfan et al., 2017). Substantial genotypic variations in wheat for P use efficiency have been reported by many researchers (Osborne and Rengel, 2002; Ozturk et al., 2005; Mittal and Sethi, 2005; Gunes et al., 2006; Yaseen and Malhi, 2009; Korkmaz et al., 2010; Abbas et al., 2016), which can be effectively utilized in the future breeding programs. Despite the presence of extensive genetic variability for P efficiency, little progress has so far been made in developing P-efficient cultivars in wheat and other cereal crops (Rose and Wissuwa, 2013). The reason behind this slow progress is the complexity of P efficiency traits, screening of limited number of genotypes, poor understanding of screening environment and mechanisms underlying P efficiency.

Wheat is the staple food grain of Pakistani people. It accounts for 2 % of GDP of the country and 9.9 % of the value added in agriculture. The crop was cultivated on an area of 9.26 million ha during 2015-16 with production of 25.5 million tons. About 0.83 million tons of phosphorus were purchased from the market during 2015-16, half of which was consumed by the wheat crop (Pakistan economic survey, 2015-16). Improving P use efficiency in wheat crop alone can have a substantial impact on fertilizer demand in future. In this experimentation, growth response of ten wheat genotypes was studied under low and high P conditions in solution culture.

## Materials and Methods

### *Plant culture and growth conditions*

Healthy seeds of ten wheat genotypes viz., AA-V1, AA-V2, AA-V3, MSH-3, MSH-5, BWQ-4, EST-28/11, EST-29/9, ESW-9525 and NIA-Sunder were obtained from Plant Breeding and Genetics Division, Nuclear Institute of Agriculture (NIA), Tando Jam. Wheat seeds were disinfected with 2% sodium hypochlorite solution for 30 seconds and subsequently rinsed with distilled water. The seeds were germinated on plastic mesh contained in plastic cups and soaked with distilled water. After 7 days of seed germination, seedlings were shifted to polythene lined iron tubs (25-L capacity) containing modified Johnson's nutrient solution (Johnson et al., 1957). The iron tubs were placed in a rain protected net house with no control over temperature and humidity. The agro-meteorological conditions of the net house during the course of experiment are given in the Table -1. Two P levels, i.e. low (20 μM) and high (200 μM) were maintained with ammonium dihydrogen phosphate salt ($NH_4H_2PO_4$), while keeping concentration of rest of the nutrients same. The difference in nitrogen concentration between two P levels was adjusted with $NH_4NO_3$ salt. Completely randomized design with factorial arrangement of ten genotypes and two P levels was employed. There were five replicates of each treatment. The solution was continuously aerated and its pH was maintained at 5.5 ± 0.5 with hydrochloric acid and sodium hydroxide on daily basis. The experiment was conducted during Rabi 2015-16.

### *Plant harvest*

After four weeks of transplanting, the plants were harvested, rinsed with distilled water and then divided into shoot and root portions. The samples were then oven-dried at 70 °C for 72 hours and weighed. The shoot and root biomass was expressed as mg plant$^{-1}$.

### *Phosphorus assays and P efficiency indicators*

Oven-dried shoot and root samples were milled and a known quantity of the grinded material was digested in di-acid mixture of $HNO_3$ and $HClO_4$ (Miller, 1998). Phosphorus concentration in the digested material was then analyzed spectrophotometrically at 420 nm wavelength (Chapman and Pratt, 1961). Total P accumulation in tissue was calculated by multiplying

P concentration with dry weight of that particular tissue. The following formula was used to compute phosphorus stress factor (PSF):

$$\text{PSF (\%)} = \frac{\text{Shoot dry weight at high P} - \text{Shoot dry weight at low P}}{\text{Shoot dry weight at high P}} \times 100$$

Shoot/root phosphorus utilization index (PUI) was determined by the formula of Siddiqi & Glass (1981).

$$\text{Shoot/ root PUI } (g^2 \text{ DW mg}^{-1} \text{ P}) = \frac{\text{Shoot/root dry weight (g plant}-1)}{\text{Shoot/root P concentration (mg P g}-1)}$$

The formula given by Fageria (1998) was employed to compute physiological phosphorus use efficiency (PPUE).

$$\text{PPUE (mg SDW mg}^{-1} \text{ P)} = \frac{\text{Shoot dry weight (mg) at high P} - \text{Shoot dry weight (mg) at low P}}{\text{Shoot P accumulation (mg) at high P} - \text{Shoot P accumulation (mg) at low P}}$$

Where SDW stands for shoot dry weight.

***Statistical analysis of data***

The collected data were subjected to the analysis of variance according to completely randomized factorial design and the treatment means were differentiated by the Least Significant Difference (LSD) method (Gomez and Gomez, 1984). Further, the means of genotypes at each P level were separated individually using LSD at 1% probability level by employing completely randomized design. Separate statistical analysis for genotypes at each P level has been suggested by da Silva and Gabelman (1992) due to heterogeneity of error variances at two P levels.

## Results

Statistical analysis of data revealed significant interactive effects of P levels and genotypes on all growth and P related parameters (Table -2, 3, 4, 5).

***Biomass production***

Shoot and root dry weights of 10 wheat genotypes are presented in Table -2. Significant variations ($P < 0.01$) among genotypes were observed for shoot and root dry biomass at both P levels. Under low P conditions, shoot dry weight ranged from 157 mg plant$^{-1}$ (MSH-5) to 309 mg plant$^{-1}$ (AA-V3) with an average value of 220 mg plant$^{-1}$. The difference was 97% between the lowest and the highest shoot dry weight at low P level. Shoot dry weight of MSH-5 was statistically at par with that of ESW-9525 at low P. At high P level, AA-V3 again produced the highest shoot dry weight (935 mg plant$^{-1}$) while ESW-9525 produced the lowest shoot dry matter (555 mg plant$^{-1}$) and the later was statistically at par with EST-29/9, EST-28/11 and NIA-Sunder for shoot dry weight. On an average, shoot dry weight was increased 3 folds when P concentration in culture solution was increased from low to high level. Genotypes also revealed substantial differences in root dry weight at both P levels (Table -2). Genotype AA-V3 was statistically superior to all other genotypes in root dry weight production at low P supply, while MSH-5 produced the lowest root dry weight (158 mg plant$^{-1}$), statistically equivalent to that of NIA-Sunder, ESW-9525 and EST-28/11. At high P level, AA-V1 produced the highest root dry matter (517 mg plant$^{-1}$), followed by AA-V3 (367 mg plant$^{-1}$) and AA-V2 (365 mg plant$^{-1}$). Genotype NIA-Sunder had the lowest value of root dry weight (212 mg plant$^{-1}$) at high P supply. Root shoot ratio (RSR) also differed significantly ($P < 0.01$) among various genotypes at both levels of P supply (Table -2). It varied between 0.74 to 1.10 at low P and 0.32 to 0.63 at adequate P supply. Overall, two fold increase was noted in root shoot ratio when P supply in growth medium was lowered from 200 to 20 µM. At low P supply, AA-V2 exhibited higher RSR which was statistically identical to those of MSH-3, MSH-5, BWQ-4, EST-29/9 and ESW-9525. Under high P conditions, AA-V1 and AA-V2 had higher RSR. Numerically, NIA-Sunder revealed the lowest RSR at both P levels.

Phosphorus stress factor (PSF) indicates the relative decrease in SDW when P supply in rooting medium was decreased from high to low level. In this study, genotypes could not produce statistically significant variations ($P > 0.05$) in PSF or comparative tolerance to phosphorus stress (Figure 1). Phosphorus stress factor ranged between 65 to 77% for different genotypes with an average value of 68%.

***Phosphorus concentration and accumulation***

Phosphorus concentration in root and shoot tissues varied significantly ($P < 0.01$) among wheat genotypes at both phosphorus levels (Table -3). Shoot P concentration ranged between the highest value (2544 µg P g$^{-1}$) for NIA-Sunder and the lowest (1775 µg P g$^{-1}$) for AA-V2 when the plants were grown with low P level. Genotypes viz., MSH-3, MSH-5 and EST-28/11 had statistically equivalent shoot P concentration with that of NIA-Sunder at low P supply. Under adequate P conditions, genotypes showed three times higher P concentration in shoot than when they were grown at low P supply. Genotype EST-29/9 had the highest P concentration (8333 μg P $g^{-1}$) in its shoot, while AA-V3 accumulated the lowest shoot P concentration of 6426 μg P $g^{-1}$.

Root P concentration ranged between 1849 μg P $g^{-1}$ (AA-V2) to 2483 μg P $g^{-1}$ (MSH-5) at low P level and from 6053 μg P $g^{-1}$ (AA-V3) to 7625 μg P $g^{-1}$ (EST-29/9) at high P level. On average, P concentration in shoot and root tissues decreased by 71% and 69%, respectively when P supply in growth solution was lowered from 200 to 20 μM.

Significant variations ($P < 0.01$) in shoot and root P accumulation/uptake were manifested by various genotypes at each P level (Table -4). Genotype AA-V3 had the highest shoot P accumulation (586 μg P $plant^{-1}$) followed by NIA-Sunder (568 μg P $plant^{-1}$), AA-V1 (551 μg P $plant^{-1}$) and EST-28/11 (517 μg P $plant^{-1}$), all being statistically identical in shoot P accumulation at low P. Genotypes EST-29/9, AA-V2, ESW-9525 and MSH-5 were least efficient in accumulating P in their shoot tissues. At high P level, AA-V1 had the highest shoot P accumulation (6714 μg P $plant^{-1}$) and was statistically superior to all other genotypes except AA-V3. The least P accumulation in shoot was observed in ESW-9525. The P accumulation in root tissues of various genotypes varied from 371 to 499 μg $palnt^{-1}$ at low P level and from 1509 to 3523 μg $palnt^{-1}$ at high P level. On an average, P accumulation in shoot and root tissues increased by almost 11 and 5 folds, respectively when phosphorus supply in the growing medium was increased from 20 to 200 μM.

***Phosphorus efficiency indices***

Shoot and root P utilization index was significantly improved under low P supply (Table -5). Averaging across the genotypes, P utilization index of shoot and root was increased by 30 and 13%, respectively when plants were grown with low P level. Shoot P utilization index (SPUI) varied from 0.06 $g^2$ SDW $mg^{-1}$ P for MSH-5 to 0.16 $g^2$ SDW $mg^{-1}$ P for AA-V3 with an average value of 0.13 $g^2$ SDW $mg^{-1}$ P under low P condition. At high P level, SPUI varied from 0.07 $g^2$ SDW $mg^{-1}$ P (EST-29/9) to 0.15 $g^2$ SDW $mg^{-1}$ P (AA-V3). Genotype AA-V3 proved to be most efficient in shoot P utilization at both P levels. Root P utilization index (RPUI) ranged between 0.07 to 14 $g^2$ SDW $mg^{-1}$ P at low P level and from 0.03 to 0.08 $g^2$ SDW $mg^{-1}$ P at high P level. At low P level, AA-V3 was statistically superior in RPUI to the rest of genotypes and at high P level, AA-V1 appeared to be most the efficient in root P utilization. Physiological phosphorous use efficiency (PPUE) varied from 88 mg SDW $mg^{-1}$ P for EST-29/9 to 118 mg SDW $mg^{-1}$ P for MSH-5 with an average value of 102 mg SDW $mg^{-1}$ P (Figure 2). There was a variation of about 34% between the highest and the lowest PPUE.

**Table – 1: Agro-meteorological conditions during the course of experiment (15-12-2015 to 11-01-2016)**

| | Minimum temperature (°C) | Maximum temperature (°C) | Relative humidity (%) | Sunshine hours (hrs) | Wind speed (km $hr^{-1}$) | Pan evaporation (mm $day^{-1}$) |
|---|---|---|---|---|---|---|
| **Range** | 4-12 | 22-32 | 50-71 | 6.6-9.3 | 0.4-2.1 | 2.3-4.2 |
| **Mean** | 8.35 | 25.77 | 58.00 | 8.60 | 1.18 | 3.12 |

**Table – 2: Growth performance of ten wheat genotypes at low and high P level**

| Genotypes | Shoot dry weight (mg plant$^{-1}$) | | Root dry weight (mg plant$^{-1}$) | | Root shoot ratio | |
|---|---|---|---|---|---|---|
| | $P_{20\ \mu M}$ | $P_{200\ \mu M}$ | $P_{20\ \mu M}$ | $P_{200\ \mu M}$ | $P_{20\ \mu M}$ | $P_{200\ \mu M}$ |
| AA-V1 | 265 b† | 835 b | 207 bc | 517 a | 0.78 d | 0.63 a |
| AA-V2 | 206 cd | 698 c | 225 b | 365 b | 1.10 a | 0.53 ab |
| AA-V3 | 309 a | 935 a | 266 a | 367 b | 0.86 b-d | 0.40 cd |
| MSH-3 | 212 cd | 659 cd | 208 bc | 249 c-e | 0.98 a-c | 0.39 cd |
| MSH-5 | 157 e | 695 c | 158 d | 272 cd | 1.01 ab | 0.39 cd |
| BWQ-4 | 229 c | 662 cd | 224 b | 277 c | 0.98 a-c | 0.42 b-d |
| EST-28/11 | 226 c | 649 c-e | 183 cd | 256 c-e | 0.81 cd | 0.38 cd |
| EST-29/9 | 190 d | 588 de | 201 bc | 257 c-e | 1.06 a | 0.45 bc |
| ESW-9525 | 185 de | 555 e | 170 d | 225 de | 0.92 a-d | 0.40 cd |
| NIA-Sunder | 224 c | 649 c-e | 165 d | 212 e | 0.74 d | 0.32 d |
| **F values for analysis of variance** | | | | | | |
| Genotype (G) | 36.40*** | | 63.32*** | | 7.97*** | |
| P levels (P) | 3622.89*** | | 489.57*** | | 771.63*** | |
| G×P | 10.91*** | | 32.73*** | | 6.19*** | |

†Means in the same column followed by the same letter(s) are statistically not different at the 1% probability level by LSD test. NS = non-significant at $P > 0.05$, * = significant at $P < 0.05$, ** = significant at $P < 0.01$, *** = significant at $P < 0.001$

**Table – 3: Phosphorus concentration in shoot and root tissues of wheat genotypes at low and high P level**

| Genotypes | Shoot P concentration (µg P g$^{-1}$) | | Root P concentration (µg P g$^{-1}$) | |
|---|---|---|---|---|
| | $P_{20\ \mu M}$ | $P_{200\ \mu M}$ | $P_{20\ \mu M}$ | $P_{200\ \mu M}$ |
| AA-V1 | 2051 b-d† | 8055 ab | 2318 ab | 6866 a-c |
| AA-V2 | 1775 d | 6988 b-d | 1849 c | 6286 bc |
| AA-V3 | 1903 cd | 6426 d | 1908 c | 6053 c |
| MSH-3 | 2219 a-c | 7925 a-c | 1957 bc | 6171 bc |
| MSH-5 | 2499 a | 7031 a-d | 2483 a | 7160 a-c |
| BWQ-4 | 1948 b-d | 7240 a-d | 2174 a-c | 7480 a |
| EST-28/11 | 2289 ab | 7913 a-c | 2013 bc | 7047 a-c |
| EST-29/9 | 1912 cd | 8333 a | 2102 a-c | 7625 a |
| ESW-9525 | 2113 b-d | 7378 a-d | 2196 a-c | 6678 a-c |
| NIA-Sunder | 2544 a | 6665 cd | 2424 a | 7333 ab |
| **F values for analysis of variance** | | | | |
| Genotype (G) | 4.09*** | | 5.47*** | |
| P levels (P) | 2601.96*** | | 2401.30*** | |
| G×P | 4.79*** | | 2.24* | |

†Means in the same column followed by the same letter(s) are statistically not different at the 1% probability level by LSD test. NS = non-significant at $P > 0.05$, * = significant at $P < 0.05$, ** = significant at $P < 0.01$, *** = significant at $P < 0.001$

**Table – 4: Phosphorus accumulation in shoot and root tissues of wheat genotypes at low and high P level**

| Genotypes | Shoot P accumulation (µg P plant$^{-1}$) | | Root P accumulation (µg P plant$^{-1}$) | |
|---|---|---|---|---|
| | $P_{20\ \mu M}$ | $P_{200\ \mu M}$ | $P_{20\ \mu M}$ | $P_{200\ \mu M}$ |
| AA-V1 | 551 ab† | 6714 a | 469 a-c | 3523 a |
| AA-V2 | 366 f | 4870 b-d | 419 b-d | 2300 b |
| AA-V3 | 586 a | 6024 ab | 499 a | 2216 bc |
| MSH-3 | 475 b-d | 5185 bc | 409 cd | 1543 ef |
| MSH-5 | 397 d-f | 4980 cd | 391 d | 2005 b-d |
| BWQ-4 | 450 c-e | 4810 cd | 490 ab | 2075 b-d |
| EST-28/11 | 517 a-c | 5099 bc | 371 d | 1757 d-f |
| EST-29/9 | 363 f | 4916 cd | 425 b-d | 1883 c-e |
| ESW-9525 | 395 ef | 4102 d | 372 d | 1509 f |
| NIA-Sunder | 568 a | 4349 cd | 398 cd | 1576 ef |
| **F values for analysis of variance** | | | | |
| Genotype (G) | 11.12*** | | 45.82*** | |
| P levels (P) | 3666.39*** | | 3106.91*** | |
| G×P | 8.86*** | | 38.16*** | |

†Means in the same column followed by the same letter(s) are statistically not different at the 1% probability level by LSD test. NS = non-significant at $P > 0.05$, * = significant at $P < 0.05$, ** = significant at $P < 0.01$, *** = significant at $P < 0.001$

**Table -5. Phosphorus utilization index of ten wheat genotypes at low and high P level**

| Genotypes | Shoot P utilization index ($g^2$ SDW mg$^{-1}$ P) | | Root P utilization index ($g^2$ RDW mg$^{-1}$ P) | |
|---|---|---|---|---|
| | $P_{20\ \mu M}$ | $P_{200\ \mu M}$ | $P_{20\ \mu M}$ | $P_{200\ \mu M}$ |
| AA-V1 | 0.13 b† | 0.10 b | 0.09 de | 0.08 a |
| AA-V2 | 0.12 b-d | 0.10 bc | 0.12 b | 0.06 b |
| AA-V3 | 0.16 a | 0.15 a | 0.14 a | 0.06 ab |
| MSH-3 | 0.10 de | 0.08 c-e | 0.11 c | 0.04 c |
| MSH-5 | 0.06 f | 0.10 bc | 0.06 f | 0.04 c |
| BWQ-4 | 0.12 bc | 0.09 b-d | 0.10 cd | 0.04 c |
| EST-28/11 | 0.10 c-e | 0.08 c-e | 0.09 de | 0.04 c |
| EST-29/9 | 0.10 c-e | 0.07 e | 0.10 cd | 0.03 c |
| ESW-9525 | 0.09 e | 0.08 de | 0.08 ef | 0.03 c |
| NIA-Sunder | 0.09 e | 0.10 bc | 0.07 f | 0.03 c |
| **F values for analysis of variance** | | | | |
| Genotype (G) | 36.28*** | | 35.31*** | |
| P levels (P) | 21.98*** | | 814.39*** | |
| G×P | 7.08*** | | 12.74*** | |

†Means in the same column followed by the same letter(s) are statistically not different at the 1% probability level by LSD test. NS = non-significant at $P > 0.05$, * = significant at $P < 0.05$, ** = significant at $P < 0.01$, *** = significant at $P < 0.001$

**Table – 6: Correlation matrix among various growth and P-related parameters of wheat genotypes at low and high P level**

| | | RDW | RPC | RPAC | RPUI | RSR | SDW | SPC | SPAC |
|---|---|---|---|---|---|---|---|---|---|
| **RPC** | Low P | -0.49 ** | | | | | | | |
| | High P | -0.24 NS | | | | | | | |
| **RPAC** | Low P | 0.69 *** | -0.13 NS | | | | | | |
| | High P | 0.94 *** | 0.01 NS | | | | | | |
| **RPUI** | Low P | 0.91*** | -0.79*** | 0.53** | | | | | |
| | High P | 0.91*** | -0.52** | 0.80*** | | | | | |
| **RSR** | Low P | 0.25 NS | -0.13 NS | -0.01 NS | 0.23 NS | | | | |
| | High P | 0.80 *** | -0.12 NS | 0.82 *** | 0.75*** | | | | |
| **SDW** | Low P | 0.69 *** | -0.30 NS | 0.61 *** | 0.62*** | -0.51 ** | | | |
| | High P | 0.72 *** | -0.33 NS | 0.62 *** | 0.71*** | 0.27 NS | | | |
| **SPC** | Low P | -0.63 *** | 0.58 *** | -0.42 * | -0.69*** | -0.33 NS | -0.28 NS | | |
| | High P | 0.06 NS | 0.35 NS | 0.16 NS | -0.12NS | 0.33 NS | -0.28 NS | | |
| **SPAC** | Low P | 0.27 NS | 0.11 NS | 0.31 NS | 0.13NS | -0.68 *** | 0.76 *** | 0.28 NS | |
| | High P | 0.77 *** | -0.13 NS | 0.77 *** | 0.75*** | 0.50 ** | 0.75 *** | 0.18 NS | |
| **SPUI** | Low P | 0.83*** | -0.50** | 0.67*** | 0.80*** | -0.22NS | 0.88*** | -0.69*** | 0.43* |
| | High P | 0.50** | -0.42* | 0.37* | 0.58*** | 0.04NS | 0.89*** | -0.67*** | 0.49** |

RDW: Root dry weight, RPC: Root P concentration, RPAC: Root P accumulation, RSR: Root shoot ratio, SDW: Shoot dry weight, SPC: Shoot P concentration, SPAC: Shoot P accumulation, RPUI: Root P utilization index, SPUI: Shoot P utilization index. NS = non-significant at $P > 0.05$, * = significant at $P < 0.05$, ** = significant at $P < 0.01$, *** = significant at $P < 0.001$ (n = 50)

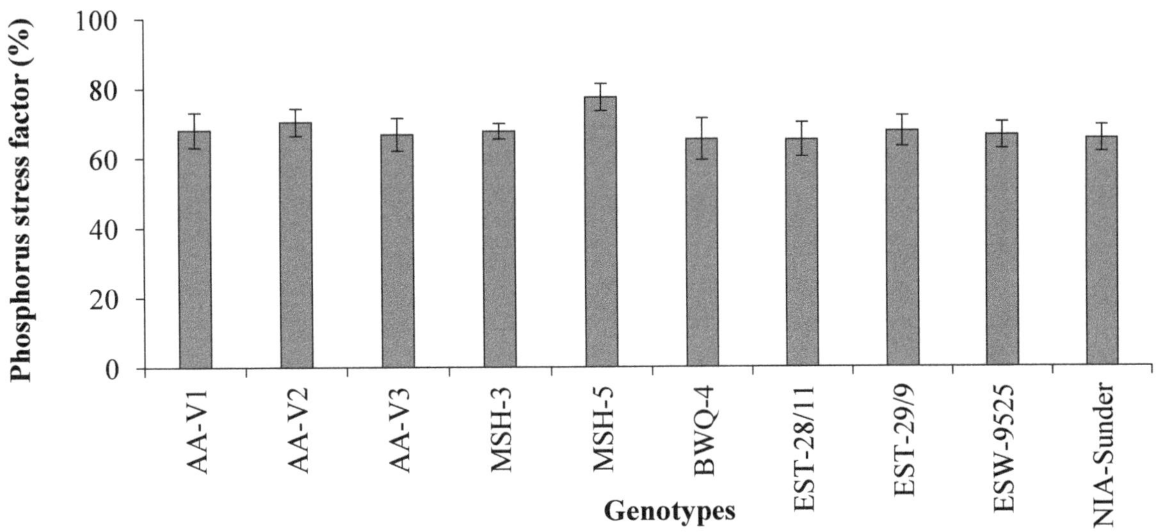


**Figure – 1: Relative reduction in shoot dry weight (phosphorus stress factor) of wheat genotypes in response to low P supply**

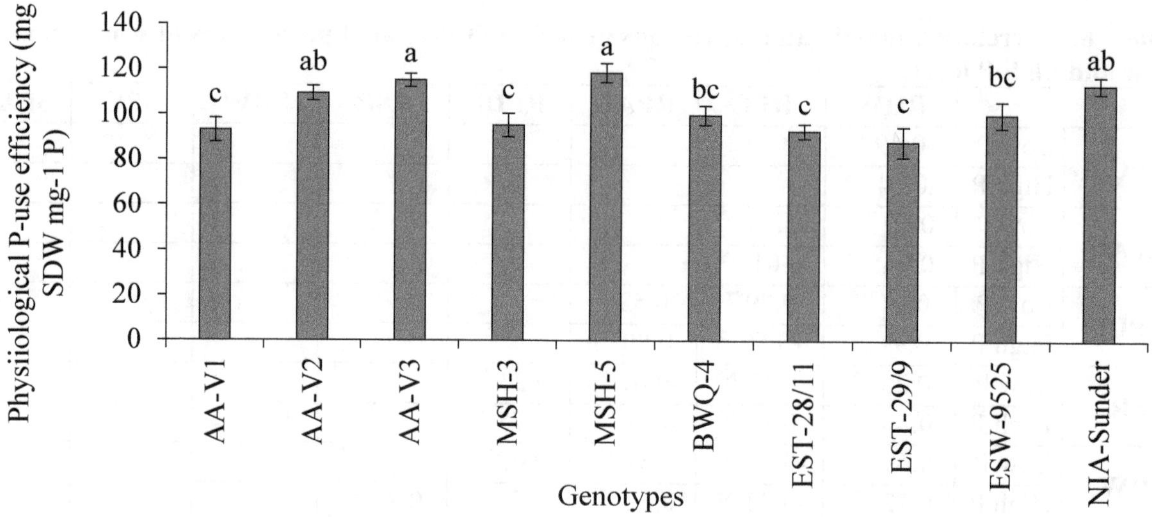


**Figure – 2: Physiological P-use efficiency of wheat genotypes**

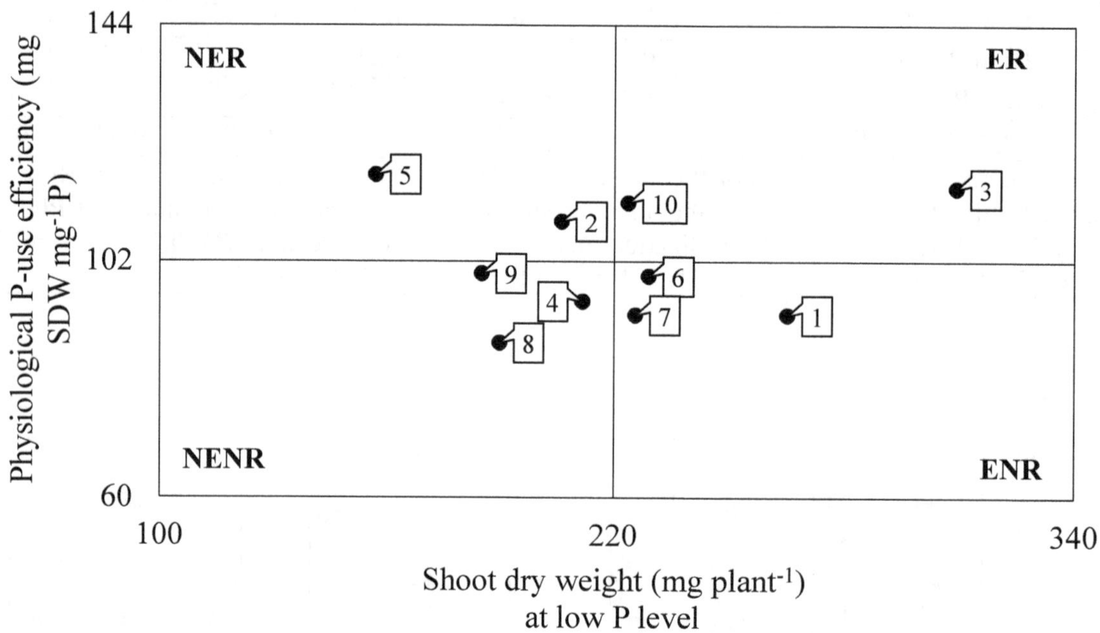


**Figure – 3: Relationship between shoot dry weight at low P level and responsiveness to P, measured as physiological phosphorus use efficiency for different wheat genotypes.**

The intersecting horizontal and vertical lines represent average shoot dry weight and physiological P-use efficiency of 10 wheat genotypes. ER: Efficient and responsive, NER: Non-efficient but responsive, ENR: Efficient but non-responsive, NENR: Non-efficient and non-responsive.1. AA-V1, 2. AA-V2, 3. AA-V3, 4. MSH-3, 5. MSH-5, 6. BWQ-4, 7. EST-28/11, 8. EST-29/9, 9. ESW-9525, 10. NIA-Sunder

## Discussion

The main thrust of this study was to explore the extent of genotypic variations present among the genotypes for P use efficiency and to identify ideal genotypes that can have better adaptability to situations of varying P availability. Incorporation of P-efficient genotypes into future breeding programs will improve the pace of breeding efforts aimed at evolving P-efficient cultivars for the P-impoverished agricultural systems of the

Pakistan. Highly significant interactive effects of P × G on SDW production clearly indicate that useful genetic variations exist among wheat genotypes for P efficiency and responsiveness. The SDW of wheat is an important parameter and is linked to grain/economic yield of this crop in most cases. Production of higher shoot biomass under P deficient conditions is considered a reliable tool for evaluating P use efficiency at early growth stages of wheat crop (Alloush, 2003). In this study, manifestation of substantial differences among wheat genotypes for SDW at two phosphorus levels was observed and such genetic variability can successfully be exploited in breeding programs to develop new P-efficient and high yielding cultivars in future. Abbas et al. (2016) employed the similar criterion to categorize wheat genotypes for P-efficiency. Dependence of shoot dry weight on various growth and P-related parameters at both P levels was determined in terms of 'r' values (Table -6). Shoot dry weight had significant and positive relationship (r = 0.76*** and r = 0.75***) with shoot P accumulation at low and high P level, which implies that genotypes with higher shoot P accumulation produced higher shoot biomass by the efficient utilization of accumulated P. However, P concentration in both shoot and root had negative impact on the production of shoot dry biomass. A negative correlation of shoot P concentration with SDW (r = -0.28) at both P levels suggests that the genotypes with lower shoot P concentration were more efficient in P utilization than those having high shoot P concentration. Abbas et al. (2016) have disclosed similar results while evaluating P utilization efficiency of various wheat cultivars. Root dry weight had positive relationship with shoot dry weight and shoot P accumulation at both P levels, indicating the significance of root development in increased production of SDW and shoot P uptake. Root shoot ratio negatively affected SDW of genotypes, although the relationship had statistical significance at low P level only. Under low P conditions, plants tend to maintain their root development at the expense of shoot growth and more biomass is partitioned towards roots under low P availability which helped the plants to establish more extensive root system and absorb more phosphorus from P-deficient medium (Marschner, 1995; Lambers et al., 2006).

Phosphorus stress factor or relative reduction in SDW in response to P stress can also be helpful in assessing the comparative tolerance of genotypes to P deficient conditions. Genotypes with relatively low PSF are considered as more P-efficient and high yielding under low P conditions, therefore, can be selected for P deficient soils. However, PSF is not a reliable parameter for evaluating P stress tolerance, since it only shows the extent of reduction in SDW and the absolute values of SDW cannot be ignored (Abbas et al., 2016). In the present study, genotypes EST-29/9 and ESW-9525 showed PSF value (67%) similar to that of AA-V3 but their inefficiency for shoot biomass production when compared with later one make them unacceptable choice for limited P conditions. Genotypes with high PSF value (e.g. MSH-5) have low utility in P limited soils, however these genotypes are considered suitable candidates for high input agricultural systems because of their higher yield potentials.

Phosphorus utilization index denotes the amount of shoot/root biomass produced per unit of P concentration in that particular tissue. Genotypes studied in this experiment significantly differed for shoot and root P utilization index at low and high P levels. Further, a strong and positive relationship of SPUI with SDW and shoot P accumulation was revealed at both P levels which inferred that higher SDW and improved P uptake translated into high P efficiency of the genotypes. Conversely, genotypes with higher SPUI revealed lower shoot P concentration than those with lower SPUI, hence, a negative relationship (r = -0.29) was observed between SPUI and shoot P concentration. Similar outcomes were revealed by Aziz et al. (2006) and Abbas et al. (2016).

Screening of crop genotypes under low P conditions and subsequent grouping based on their growth performance forms the cornerstone of any breeding program designed for improving phosphorus efficiency in any crop. Such type of grouping/categorization will lead to the identification of genotypes appropriate for growing on soils with varying P levels. In the present study, genotypes were grouped into four classes by regressing PPUE against shoot dry weight at low P level as proposed by Fageria and Baligar (1993). Efficient and responsive (ER) group of genotypes included AA-V3 and NIA-Sunder (Figure 3). These genotypes produced SDW and PPUE higher than their respective averages computed for ten wheat genotypes. Genotypes producing more than average SDW under low P conditions, but exhibiting PPUE lower than the average were classified as efficient but non-responsive (ENR).

Genotypes AA-V1, BWQ-4 and EST-28/11 fell into this category. The third group is categorized as non-efficient but responsive (NER) and includes AA-V2 and MSH-5. Members of this group had characteristics of low SDW and high P responsiveness. Genotypes exhibiting SDW and P response less than their respective averages were identified as non-efficient and non-responsive (NENR). Genotypes MSH-3, EST-29/9 and ESW-9525 fell into this group. Among the four classes of genotypes, ER category is the most important because its members can produce more under the conditions of low as well as high P availability. Such genotypes are better adapted to growing environments with varying levels of available phosphorus. The other most important category is efficient but non-responsive and its members can be grown successfully on P-impoverished soils. The NER genotypes can be occasionally incorporated into hybridization schemes for their P-response characters, while non-efficient and non-responsive genotypes have no significance from practical point of view.

## Conclusion

The results of this study indicated that considerable genetic diversity exists among wheat genotypes for P efficiency and responsiveness. Among various wheat genotypes studied in this experiment, AA-V3 and NIA-Sunder proved to be ideal ones which can be successfully grown on soils of low and high P availability. However, the results should be confirmed under filed conditions.

## References

Abbas M, Aslam M, Shah JA, Depar N and Memon MY, 2016. Relative growth response of hydroponically grown wheat genotypes to deficient and adequate phosphorus levels. Pak. J. Agric. Agri. Engg. Vet. Sci. 32 (2): 169-181.

Alloush GA, 2003. Responses of hydroponically grown chickpea to low P: pH changes, nutrient uptake rates and root morphological changes. Agron. 23: 123-133.

Aziz T, Rahmatullah, Maqsood MA, Tahir MA, Ahmad I and Cheema MA, 2006. Phosphorus utilization by six Brassica cultivars (*Brassica juncea* L.) from tri-calcium phosphate; a relatively insoluble P compound. Pak. J. Bot. 38: 1529-1538.

Chapman HD and Pratt PF, 1961. Phosphorus, p. 160-170. In: Methods of analysis for soil, plants and waters. Division of Agricultural Sciences, University of California, USA.

Cordell D, Drangert JO and White S, 2009. The story of phosphorus: global food security and food for thought. Global Environ. Change. 19: 292-305.

da Silva AE and Gabelman WA, 1992. Screening maize inbred lines for tolerance to low P stress conditions. Plant Soil. 146:181-187.

Fageria NK and Baligar VC, 1993. Screening crop genotypes for mineral stresses, pp. 142–162. In: Adaptation of plants to soil stress. INTSORMIL Publication No. 94-2, University of Nebraska, Lincoln, USA.

Fageria NK, 1998. Optimizing nutrient use efficiency in crop production. Revista Brasileira de Engenharia Agricola e Ambiental. 2: 6-16.

Gahoonia TS and Nielsen NE, 2004. Barley genotypes with long root hairs sustain high grain yields in low-P field. Plant Soil. 262: 55-62.

Gomez KA and Gomez AA, 1984. Statistical procedures for agricultural research. John Wiley and Sons, New York, USA.

Gunes A, Inal A, Alpaslan M and Cakmak I, 2006. Genotypic variation in phosphorus efficiency between wheat cultivars grown under greenhouse and field conditions. Soil. Sci. Plant Nutr. 52: 470-478.

Hammond JP and White PJ, 2011. Sugar signaling in root responses to low phosphorus availability. Plant Physiol. 156(3): 1033-1040.

Irfan M, Shah JA and Abbas M, 2017. Evaluating the performance of mungbean genotypes for grain yield, phosphorus accumulation and utilization efficiency. J. Plant Nutr. doi:10.1080/01904167.2017.1381718.

Johnson CM, Stout PR, Broyer TC and Carlton AB, 1957. Comparative chlorine requirements of different plant species. Plant Soil. 8: 327-353.

Korkmaz K, Ibrikci H, Karnez E, Buyuk G, Ryan J, Oguz H and Ulger AC, 2010. Responses of wheat genotypes to phosphorus fertilization under rain-fed conditions in the Mediterranean region of Turkey. Sci. Res. Essays. 5(16): 2304-2311.

Lambers H, Shane MW, Cramer MD, Pearse S and Veneklaas EJ, 2006. Root structure and functioning for efficient acquisition of phosphorus: Matching morphological and physiological traits. Ann. Bot. 98: 693-713.

Ma Q, Rengel Z and Rose TJ, 2009. The effectiveness of deep placement of fertilizers is determined by crop species and edaphic conditions in Mediterranean-type environments: a review. Aust. J. Soil Res. 47:19-32.

Marschner H, 1995. Mineral nutrition of higher plants. Academic Press, San Deigo, CA., USA.

Miller RO, 1998. Nitric-perchloric wet digestion in an open vessel, pp. 57-62. In: Y. P. Kalra (ed.), Handbook of reference methods for plant analysis. CRC Press, Washington, DC., USA.

Mittal RK and Sethi GS, 2005. Genetic variability in-triticale x bread wheat derivatives under normal and phosphorus stress regimes. J. Environ. Biol. 26: 105-107.

Osborne LD and Rengel Z, 2002. Genotypic differences in wheat for uptake and utilisation of P from iron phosphate. Aust. J. Agric. Res. 53: 837-844.

Ozturk L, Eker S, Torun B and Cakmak I, 2005. Variation in phosphorus efficiency among 73 bread and durum wheat genotypes grown in a phosphorus-deficient calcareous soil. Plant Soil. 269: 69-80.

Pakistan Economic Survey, 2015-16. Ministry of Finance, Government of Pakistan, Islamabad, Pakistan.

Rengel Z and Marschner P, 2005. Nutrient availability and management in the rhizosphere: exploiting genotypic differences. New Phytol. 168: 305-312.

Richardson AE, Hocking PJ, Simpson RJ and George TS, 2009. Plant mechanisms to optimize access to soil phosphorus. Crop Pasture Sci. 60:124-143.

Rose TJ, Liu L and Wissuwa M, 2013. Improving phosphorus efficiency in cereal crops: is breeding for reduced grain phosphorus concentration part of the solution? Front. Plant Sci. 4: 444. doi:10.3389/fpls.2013.00444.

Siddiqi MY and Glass AD, 1981. Utilization index: A modified approach to the estimation and comparison of nutrient utilization efficiency in plants. J. Plant Nutr. 4: 289-302.

Simpson RJ, Oberson A, Culvenor RA, Ryan MH, Veneklaas EJ, Lambers H, Lynch JP, Ryan PR, Delhaize E, Smith FA, Smith SE, Harvey PR and Richardson AE, 2011. Strategies and agronomic interventions to improve the phosphorus-use efficiency of farming systems. Plant Soil. 349: 89-120.

Smith SE and Read DJ, 2008. Mycorrhizal symbiosis, 3rd ed. Academic Press and Elsevier, London.

Wang X, Shen J and Liao H, 2010. Acquisition or utilization, which is more critical for enhancing phosphorus efficiency in modern crops? Plant Sci. 179: 302-306.

Wasaki J, Maruyama H, Tanaka M, Yamamura T, Dateki H, Shinano T, Ito S and Osaki M, 2009. Overexpression of the LaSAP2 gene for secretory acid phosphatase in white lupin improves the phosphorus uptake and growth of tobacco plants. Soil Sci. Plant Nutr. 55: 107-113.

Yaseen M and Malhi SS, 2009. Differential growth performance of 15 wheat genotypes for grain yield and phosphorus uptake on a low phosphorus soil without and with applied phosphorus fertilizer. J. Plant Nutr. 32: 1015-1043.

6

# Eco-Friendly Management of Pulse Beetle (*Callosobruchus Chinensis*) by using Different Plant Materials

**Rukhsana Kausar[1], Muhammad Amjad Bashir[1,*], Muhammad Shahid Nisar[1], Muhammad Ishaq Asif Rehmani[2], Muhammad Sadiq Hashami[3], Sagheer Atta[1]**

[1]Department of Plant Protection, Faculty of Agricultural Sciences, Ghazi University DG Khan, Punjab, Pakistan
[2]Department of Agronomy, Faculty of Agricultural Sciences, Ghazi University DG Khan, Punjab, Pakistan
[3]Department of Economics, Institute of Southern Punjab, Multan, Pakistan

*Corresponding author email: amjadhajbani@hotmail.com

## Abstract

Legumes (pulses) are ancient crops of the Sub-continent, Middle East and various parts of the Africa. These provide high quality protein and considered to be the best food for vegetarian population in India, South Asia, West Asia and Southern European countries. Chickpea is valued for its nutritive grains with high protein content of 25-29%. These grain legumes are, however, susceptible to different species of beetles of the family Bruchidae in storage commonly known as Dhora. At current, insect manage procedures frequently rely on synthetic insecticides and fumigants. Other than substance safety actions may be result as like numerous severe drawbacks. Insecticides have wide and unsystematic application causes natural imbalance confrontation to insect, insect recovery and epidemic of secondary insect creates phyto-toxicity. Insecticide residues in diet and nourish. Furthermore, non-stop use of insecticides leads to harmful result on insect pollination, biological agent like different Predators, parasitoids and too become cause of the ecological contamination. Owing to these drawbacks, internationally scientists are annoying to approve substitute methods of insect manage. The make use of close by obtainable native place resources in the management of harmful insects are a primeval technology. Now in modern world, in numerous parts of the earth a range of crop of vegetation have been tried lately by researchers with an elevated amount of achievement as granule. Protestants next to pulse beetle to decrease plague in storage space. Observance these views in brain, the in attendance learning was conducted to examine the pesticide efficiency of a number of native botanicals powders as granule protestant next to *Callosobruchus chinensis* legumes grains. Reviewed shows that different plant material which are friendly with ecosystems have significant results against pulse beetle and serve as an alternative methods for the pest control.

**Keywords:** Dhora beetle; pulse beetle; pest management; environment friendly pest control

## Introduction

Legumes engaged in recreation a fundamental part in the food of ordinary citizens of third earth state as well as Pakistan. These are also food source of protein for those who cannot take meat and because of less amount of protein from desire level will suffer from disease like thiamin, niacin, calcium, and iron (Sharma, 1984; Bhalla et al., 2008; Egesa et al., 2016; Rehman et al., 2016). Special types of legumes are developed during the cool weather in Pakistan create an entire 2, 05,000 million tunes of legumes. Along with the legumes, chickpea only in use 4.06% region of legumes farming and contributed seven thousand metric tons of yearly production (BBS, 2008). However one of the main restrictions for rising legumes manufacture is victims of germ capability and harm of legumes from insect nuisance attack in storage space. legumes in rising countries bear lofty qualitative and quantitative losses from the harass of Dohra beetle main nuisance invasion in storeroom (Ahmed et al.,2003; Aslam, 2004). They cause damage to legumes both in the field and also storeroom but swarm is further vital in stored form. The adult beetles do not effect spoil to the legumes via feeding but them assemble and oviposition on legume and pollute through excreta. The larva is only dependable for the legume spoil the larvae devastate grains by feeding indoor to some extent or entirely and build them in poor condition for human being use on 4 to 98% defeat of legume seeds might be practical owing to the invasion through the Dohra beetle in cargo space (Mookherjeeet *et al.*, 1970).

Pest seldom effect broad harm to stored food which could sum to 5-10% in the moderate and 20-30% in the temperate region (Nakakita, 1998). Foodstuff particle fatalities are extremely region particular and far above the ground heat and moisture generally help the development, of loss causing organisms. In India; storage losses to various food grain commodities, stored for 6 months after harvest, have been estimated 2.5% due to insect pests alone. Legume seeds are stored in godowns and warehouses in huge and minute quantities and are attacked by different pests. Pulse beetle (PB), *Callosobruchus chinemis* L. is one of the most destructive pests of legumes in store-room. It is broad-based and a severe insect of mung, peas, cowpeas and lentil and has been too reported aggressive legume seed, sorghum and maize (Ahmed et al. 2003). The male PB has pectinate while female has serrate type of antennae (Halstead. 1963). Its damage generally starts in matured pods in the ground from where it is passed to storage space godowns. Both grubs and adults are dependable for causing the harm. They complete their life cycle in 25-34 days during summer whereas 40-50 days in winter (Ghosh and Durbey, 2003). Gujar and Yadav (1978) evaluated that 55-60% loss in the weight of the seed and 45.50-66.30% loss in protein content due to the damage of Dhora and its seeds became unfit for human utilization as well as for further agriculture purpose. The varieties with hard, rough, wrinkled and thick seed coat showed less seed damage and number of holes and were more tolerant than varieties with smooth, soft and thin seed coat.

At current bother manage procedures in storage space on the employ of synthetic pesticides and fumigants. Their unsystematic apply in storage space, though has led to a figure of troubles with pest fight lethal residues in diet legume (Fishwick, 1988), ecological toxic waste (WMO, 1995) and increasing costs of inputs and their appliances (Hashmi, 2016). Moreover, the poor storage facilities of traditional farmers in the developing countries are unsuitable for effective conventional chemical control, as most of the storage types are open to re-infestation by insect pests (Tapondjou *et al* 2002). In the light of mentioned problems together with the up-coming WTO regulations, there is a need to restrict their use globally and implement safe alternatives of conventional insecticides and fumigants to protect stored seed grains from insect damage (Yusof and Ho, 1992; Subramanyam and Hag strum, 1995). At current bother manage procedures in storage space on the employ of synthetic pesticides and fumigants. Their unsystematic apply in storage space, though has led to a figure of troubles with pest fight lethal residues in diet legume, ecological toxic waste. At present merely two fumigants, methyl bromide and phosphine, are commonly old beside stored produce creature vermin. according to the 1997 choice of the 9th Montréal procedure ,methyl bromide, which is a verified ozone depletor in the environment ,will be phased out by 2005 in the superior countries and by 2015 in the upward countries .moreover phosine conflict is appropriate extra ordinary(tyler et al, 1983) and is a material of extensive disquiet. therefore there is critical require to enlarge secure alter sow substance based pesticides are aim precise, safe to human being and helpful organisms, fewer level to pest conflict and revival, recyclable and fewer costly and are capable legume safer. Natives usual pesticides and fumigants to keep stored legumes from creature insect invasion.

Although plant based allelochemicals are widely used herbicide (Arora et al., 2015; Yazlik and Uremis, 2016). In a country like ours; significant work rarely done to recognize the insecticidal potency of plant materials against stored grain pests except neem *{Azadirachta indica),* which is very effective against several insect pests including pulse beetles (Jilani and Malik, 1973; Naqvi, 1987 and Jilani *et al.* 1988; Elzaki and Ali, 2015).

The utilize of sow supplies can direct to recognition of latest bio-pesticides for the help of humid cultivation. Crucial oils have been broadly experienced and have certain hopeful consequences as physically powerful pesticidal action alongside bruchid vermin in laboratory situation (ketoh et al.2005).According to thought (2004), neem work as revolting via disturbing the hunger of creature and retreating their advice to repeat.

According to Bowers (1983), place in the ground pesticides are one occasion some time ago further receiving more and latest notice. There is a much need of knowing the insecticidal potential of indigenous plants against major stored product insect pests as those will be environmentally and socio-economically feasible. Set down income consist of well-off origin of bioactive chemicals (Wink. 1993); that's why they can show to enlargement of new plan of caring annoyance handle agents. A group effort has so been attentive on set down imitative funds for potentially expensive foodstuff as gainful nuisance manage agents. Minute try has been finished to organize stored make vermin throughout with aromatic medical vegetation in spite of their pharmacological measures (Kim et al, 2003). Native lodge equipment are cheaper and risk free of charge in contrast with element pesticides. The occurrence of applications or efficiency of the normal crop useful to stored legumes, though, has not been carefully investigated. These turn into all the further significant since several of the stand crop perform as repellents, gentle pesticides or chemosterilants (saxena et al, 1992). Graing and Ahmad (1988) Reported new than 2400 lodge class possessing insect manage properties. Neem seed oil and neem seed kernel powder individually prevented egg laying of PB for up to 8 months of storage whereas coal ash and soft stone were ineffective. The mortality of PB was dependent on the concentration and duration of exposure to plant materials. High insecticidal activity was complemented by a low adverse effect on grain quality (Singal and Chauhan, 1997). Jacobson critical absent that the the majority capable plants pest manage properties. Agents were in lodge families, of Annonaceae, Asteraceae, Clamiaceae, Meliaceae and Rutaceae. frequently actives next to a partial number pest type as well as exact mark pest and are frequently eco-friendly to harmless food; therefore potentially appropriate for utilize incorporated insect managing.

Powdered spices (flowers of cloves, rhizomes of ginger and turmeric, fruits of black and chili peppers and bulbs of garlic),malathion s(1.5%) and milled stems of *Combretuminmber be* were mixed with grains. Results revealed that Malathion was more effective; however, cloves and black pepper also gave significantly similar effectiveness to that of Malathion against *C. maculatus* in stored seed. (Javaid and Poswal 1995) in another study researchers mixed wheat and lentil seeds with the powdered leaves and seeds of neem tree, *Azadirachta indica* and *A.juss,* at 0, 0.5, 1.0 and 5.0% (w/w) concentrations and tested for a repellent effect against the grain weevil *(Sitophilus granarius* L.), the rice weevil *(S. oryzae*L.) and PB using Nawrot'solfacto-meter. The powdered materials were found to exhibits a strong repellent effect on the grain and rice weevils, while these products were less active against PB. The degree of repellency of plant materials increased with the concentration in the grain product. Ignatowicz *et al.* (1995)

By revelation that grains of legumes can be successfully confined as of PB through addition dried out residue of neem saw dust at the rate of 100-400 mg/50g seeds. It provided good quality result in value of poisonous result; protection and country (Singh *et al,* (1996)) , by testing the efficacy of some botanicals against PB. Five types of botanical dusts were used as insecticides. Black pepper powder, neem leaf dust and mangraila powder were proved to be equally effective in respect of number of eggs laid, number of adults emerged and reduction in damage to grains by the pest. Aslam *et al.* (2002) investigated the insecticidal efficacy of some plant spices against Pulse beetle on chickpea which is sorted. The data were taken on days to complete mortality; days to adult emerged, number of adults emerged and grains weight loss. Least existence to 100% death of PB were experimental in casing of cloves and black pepper treated seed compared to the utmost in be in charge of action. Equivalent figure of existence to mature emerged was counted in all treatments together with manage. Lowest amount figure of mature emerged was counted in casing of cloves and black pepper highest amount was experiential in manage behavior. Power failure was also intended in cloves and black pepper treated grains compared to further treatments.

Experience on the effectiveness of a number of fix foodstuffs adjacent to PB. The management incorporated a variety of lodge foodstuffs similar to folio fine particles of dharek *(Melia azedarach)* and sadabahar *(Ipomoea carnea)* at 10 g/kg of seeds; oils of coconut, mustard and groundnut wherever when neem harvest such the same as achook, nimbicidine and neem-gold at 1 ml/kg of seeds. consequences appeared so as to nimbicidine and achook showed toward exist the the majority effectual in reducing the insect harm awake toward 1.97 and 2.36%, correspondingly, followed by neemgold with 2.61% damage; however, 70.50% damage was recorded in the control. The failure inside power be while elevated since 45.20% into the natural seed, which considerably reduced near a rank of 0.52, 0.93 and 1.07% through the application of nimbicidine. Achook and neemgold correspondingly. (Umrao and Verma 2002)

Neem oil and a number of fix residues going on spawn laying and hatchability of PB spawn resting on stored legume grains. The trial incorporated the action as oil of neem *(A, indica),* residues of grains of golden oleander *(Thevetianeriifolia),* castor *(Ricinuscommunis)* and *Lamanaeamara.* In terms of oviposition, neem oil was the most effective treatment, wherever pest futile on the way to put down progeny resting on the treated grains. But, the smallest amount effectual action was of *L camara.* In conditions of offspring hatchability, golden oleander conduct gave the most excellent consequences with no hatched progeny. Resting on the further supply, action through *R. communis* seed dust was the hatched spawn. On the other hand, treatment with *R. communis* seed residue was the smallest amount effectual, footage a indicate proportion of 90.08 % progeny hatched competent to 93.0% into the natural power. (Angood and Sunaidy 2003). In another study to check the result of neem folio residue lying on swarm of PB in protect seed. Neem leaf powder hence in the direction of exist effectual on the price of 0.5 to 2.0 mg/lOOg of seeds and provided good quality consequences into value of lethal result, protection and country (Singh 2003a).

Customarily old African deposit residues, among a well-known consequence adjacent to (*maculatus,* were extracted with $H_2O$. The extracts, 13 volatile oils, 2 non-volatile oils and 8 slurries were evaluated for their lethal and repulsive special effects aligned with the insect. Appliance of explosive oils led in for the most part gear toward a less amount of progeny lying on treated beans. The explosive oils *Cymbopogon spp.* reasoning greater part of the progeny not toward enlarge interested in mature beetles. Different types of the explosive oils appeared repulsive special effects as non-explosive oils were not repulsive and had no consequence on top of the integer of progeny laid; yet, the enlargement of these spawn was vulnerable, most so through oil of *A. indica.* None of the slurries had a toxic effect on the beetles but some were repellent, whereas the slurry from *A. indica* leaves was attractive (Boeke *et al.* 2004).

The efficacy of vegetable oils next to *C. maculates* violent legume. Groundnut, sesame, soybean, mustard and neem oils (at lOml/kg of seeds) since repulsive, ovipositional deterrent and ovicidal cause beside this beetle were investigated in the laboratory. The entire the oils evaluated were effectual; but, neem oil showed toward be there the most affective whereas every part of the oils except for neem oil showed decreased during efficacy with the wait in action point (Bhatnagar *et al,* 2001).

The period of enlarged stages of PB and known its vulnerable stages on the way to neem and sesame oils. The period for 1-4 instar maggot and pupal condition were 7-12, 12-16, 16-19, 19-22 and 22-27 days living, correspondingly. The consequences appeared to together the oils may possibly manage the maggot within the bean cotyledons indicating so as to oils may penetrate keen on cotyledons and chemically restrain or destroy the young insect within. It was concluded so as to in advance increasing condition were extra liable than the later on ones and the power of maggot site in the bean and its acceptances may exist dependable for the age-dependent property of oils (Ahmad *et al* (2003).

By checking the toxicity of some plant oils of calamus, clove and karanja and its enhancement by combinations of oils against selected insect pests including PB. Calamus oil exhibited the highest toxicity against adults of PB and *Culexquinquefasciatus.* The activity of calamus oil was remarkably enhanced when mixed with inactive oils or other derivatives. The combination of calamus oil and karanja oil (1:1) gave a LC50 of 45.87 p.p.m against *C. quinquefasciatus* larvae, while the combinations of calamus oil and capric acid (1:1) gave a LC50 of 0.51 mg/cm2 against PB (Bhonde *et al.* 2001).

The variant in efficacy of pirimiphos-methyl and oils of neem, sesame, castor and soybean also alone or in combination of four ratios (1:1, 1:2, 1:5, and 1:10) of pirimiphosmethyl and every oil was observed. Each and every type of the oil definitely known preservative, while joint among pirimiphos-methyl except for neem and soybean oils, which exhibit an opposed result on 1:1 percentage not in favor of the beetle. The highest synergism of pirimiphos-methyl was practical on 1:10 percentage having the highest co-toxicity coefficient worth for neem oil (4908.53) obeyed in sesame oil (434.11), castor oil (295.24) and soybean oil (232.93) (Khlequzzaman and Chovvdhury 2003).

The efficacy of different vegetable oils (castor, neem, pongamia, coconut, mustard, sesame, soybean and sunflower) at 2 and 4 ml/kg, eight plant powders including those of datura and neem at 20 and 40 m/kg and three plant extracts at 2.5, 5.0, 7.5 and 10 ml/kg against PB infesting black gram. All the oil treatments were superior in protecting the seeds from PB attack than Malathion treatment (60 p.p.m.) or control. The oils of neem, castor and coconut at both the doses proved most effective in protecting the seeds for about 9 months after treatment. Among the plant powders, *Lantartacamara*and *Tridaxprocumbens* at both concentrations and *Lantartacamara* and *camara* extracts at all concentrations were promising in protecting the seeds. The use of such botanical products did not affect the permeability of the seeds (Tripathy *et al,* 2001).

The botanicals for the management of Pulse beetle violent lentil and legume grains into storage space. The oils of neem, royna and castor at 8 ml/kg wherever when folio dust and rough materials of neem, bankalmi, bishkatali, marigold, castor and mango at 5% (w/w) and at 8 ml/kg, correspondingly, be effectual in reducing egg-laying of PB. The swarm and power failure in lentil and chickpea were prevented through 100% during action by way of oils, folio residues and rough materials of the experienced botanicals up to 9 months of storage space (Bhuiyah *et al.* 2002).

Leaf extracts of *Tabemacmontanalivaricale* and *Tamar'mdusindica*in acetone were found to be an effective anti-feedant and adulticidal agent against PB. The loss in seed weight was minimum (0.46 and 1.37 g) at 100% concentration of *T. livaricale* and *T. indica,* respectively. Percent adult mortality was 72.9% with 7* *livaricale*and 67.6% with *T. indica (*Dwivedi and Venugopalan 2002).

By screening foliar extracts of *Annonasquamosa* for their insecticidal activity against PB. Flavonoids, isolated from aqueous extracts of *A. squamosa,* showed 80% pesticidal action not in favor of PB at a attention of 0.07 mg ml super (-1). Different physico-chemical analysis, chromatographic and spectroscopic study among moderately purify aqueous extort appeared the existence of flavonol sort flavonoids. This could give a helpful start for the growth of botanical insecticides for post-harvest protection of legumes (Kotkar *et al.* 2002). And by studying the pest repulsive action of definite lodge extort next to PB. The extract of *Allium cepa, Eugeniacayophyllifolia, Menthaarvensis, A. indica, Ocimum sanctum* and *Vitexnegundo* showed revolting exploit. In adding up, the revolting motion of delicate take out *of Zingiberofficinale* was as well important not in favor of PB (Kokate and Chintalwar 2003).

The repellent activity of some plant extracts viz.*.Acacia Arabica, Dalbergiasissoo, R. communis* and *Partheniumhysterophorus* against PB on chickpea grains under laboratory conditions. *P.hysterophorus* extract showed the maximum repellency against PB. According to Upasani *et al.* (2003), aqueous folio take out of *R. communis,* a cultivate place in tropic. Upasani *et al.* (2003), aqueous folio take out of *R. communis,* a cultivate plant in hot country, appeared outstanding pesticidal action not in favor of PB. Flavonoids were lonely and experienced as pestiticidal and antimicrobial agents. The remote flavonoids appeared latent pesticidal, ovicidal and oviposition restriction behavior not in favor of PB (Sharma *et al.* 2003).

By Evaluation of insecticidal properties of some plants materials against *Callosobruchus* Sp. On gram. – Thesis (RDVV) Jabalpur. Raina (1970) and Pokharkar and Mehta (2011) reported that egg hatching of *C. chinensis* ranged from 94 to 99 per cent on moong bean at 300C temperature and 70 per cent relative humidity and 92 per cent on chickpea, respectively. However, egg hatching of *C. smaculatus* varied from 57.9 per cent in November and 93.1 per cent in September (Gill and Ramzan, 1998).

The combined average larval and pupal period of *C. chinensis* was observed to be 20 days on cowpea and gram seed at room temperature. Singal and Borah (2001) observed that larval and pupal period of *C.chinensis* was 16.2 ± 0.16 and 7.2 ± 0.18 days, respectively on pods of *Cajanuscajan*. Meghwal and Singh (2005) reported that the developmental period (Larval + Pupal) of *C. chinensis* (L.) on moth bean grains ranged from 20 to 22 days with an average of 20.79 days. Bhargava *et al.* (2008) reported that the mean larval and pupal period of *C. chinensis* ranged from 14.80 to 26.20 and 5.40 to 11.40 days, respectively on different pulses. They also reported that the developmental period of *C. chinensis* varied from 24.60 to 44.80 days on grains of gram cultivar including cowpea, moong bean, gram, pigeonpea, pea and soybean. Pokharkar and Mehta (2011) reported 25.26 ± 3.95 days of larval and pupal period of *C. chinensis* on chickpea. Pathania Mandeep: (RinaKumari 2000).

The worth of several stand crop the same as seed protectants in opposition to the pulse beetle. *Callosobruchus chinensis* (L.) chick pea. Studies were undertaken to evaluate five plant powders, viz., neem (Azadirachta indica A. Juss.) leaf, black pepper (Piper nigrum L.) fruit, aonla (Emblicaofficinalis L.) fruit, five leaved chaste tree (Vitexnegundo L.) leaf, curry leaf tree (Murrayakoenigii (L.) Sprengel) used @ 50 g/kg grains and two oils, viz., mustard (Brassica juncea L.) and neem seed kernel oil used @ 7.5 ml/kg grains for their efficacy as grain protectants against the pulse beetle, *Callosobruchus chinensis* (L.), infesting black gram (Vignamungo (L.) Hepper). The results revealed that black pepper fruit powder, neem oil and mustard oil caused 100 per cent mortality after 7 days'(Thakur A. K. 2009)

The efficiency of different seed protection not in favor of pulse beetle, *Callosobruchus chinensis* (L.) infest green gram, Vignaradiata (L.) Wilczek. Efficacy of 11 seed protection viz., neem grain kernel residues @ 20 g/kg seed, neem oil at the rate of 10 ml/kg, mustard and groundnut oil each @ 07.5 ml/kg, turmeric powder @ 3.5 g/kg, mustard oil +turmeric powder @ 03.75 ml+01.75 g/kg, groundnut oil+turmeric powder @ 03.75 ml+01.75 g/kg, 7 cm cover by means of each one of saw dust, sandy soil, dung cake ash and wheat husk was investigated lying on fully developed death (%) and power failure (%) through *Callosobruchus chinensis* lying on treated and coarse green gram grain (power) on an gap of 1, 35, 70 and 105 time later than storage space (Subhash; *et al.* 2010)..

Consequence of safe to eat and non-safe to eat edible oils on the growth and growth of *Callosobruchus maculates* infesting green gram (vigna radiate). Effect of safe to eat and non- oils of sesame, coconut, groundnut, soybean mustard, mahua, castor, karanja, neem and linseed on *Callosobruchus maculatus* infesting green gram was investigated (KatareSubhash *et al.* 2012).

Suppression of the cow-pea bruchid (*Callosobruchus maculatus* (F.) damaging stored cow-pea (*Vignaunguiculata* (L.)Walp.) Seeds with some edible plant product in powder form. 05 Edible Plant Product Powders (EPPP), West African black pepper (Piper guineense Schum and Thonn), clove (Syzgium aromaticum (L.)Merril and Percy), Ethiopian pepper (Xylopiaaethiopica (Dunn.) A. Rich), Alligator pepper (Aframomummelegueta Schum) and African locust bean (Parkiabiglobosa (Jacq.) R. Br. Ex G. Donf.) were studied for their effectiveness in suppressing oviposition, egg hatch and progeny emergence against *Callosobruchus maculatus* (F.) (Ajayi.F.A; Wintola, H. U. 2006)

## Conclusion

It has been concluded that different plant materials which were used are friendly with the ecosystem have significant results against pulse beetle and serve as an alternative methods for the pest control. They are easy to use for the management of the said pest. This review study opens the door for entomologist to work for design techniques for the management according to the locally available plants materials.

## References

Ahmad (2003) determined the duration of developmental stages of PB and identified its susceptible stages to neem and sesame oils. The duration for 1-4 instar larva and pupal stages were 7-12, 12-16, 16-19, 19-22 days.

Ahmed, K. S., T. Itino and T. Ichikawa. (2003). Duration of developmental Stages of *Callosobruchus chinensis* (Coleoptera: Bruchidae) on Adzuki bean and the effects of neem and sesame oils at different stages of their development. Pak. J. Biol.Sci., 6(10): 932-335.

Ajayi. F.A; Wintola, H. U. (2006). Suppression of the cowpea Bruchid (Callosobruchus maculate (F.) infesting stored cowpea (Vignaunguiculata (L.) Walp.) Seeds with some edible plant product powders.

Angood, S., B. and M. A. Sunaidy. ( 2003). Effect of neem oil and some plant powders on egg laying and hatchability of the cowpea beetle, Callosobruchus chinensis eggs on stored cow pea seeds. Uni. Aden J. Natural App. Sci., 7: 195-202.

Arora, K., D.R. Batish, H.P. Singh and R.K. Kohli. (2015). Allelopathic potential of the essential oil of wild marigold (*Tagetes minuta* L.) against some invasive weeds. J. Environ. Agric. Sci. 3: 56-60.

Aslam, M. (2004).Pest status of stored chickpea beetle, Callosobruchus chinensis L. On chickpea. J. Ento mol. 1 (1): 28-33.

Aslam, M., K. A. Khan and M. Z. H. Bajwa. (2002). Potency of Some Spices against *Callosobruchus chinensis* Linnaeus. OJBS, 2(7): 449-452.

Atwal, A.S. and G.S. Dhaliwal. (2005). Agricultural Pests of South Asia and their Management. 5th edition, Kalyani Publishers, New Delhi, India, 505P.

BBS (Bangladesh Bureau of Statistics). (2008). Statistical Year Book of Bangladesh. Bangladesh Bureau of Statistics, Planning Division, Ministry of Planning, Govt. of the People's Republic of Bangladesh, Dhaka, Bangladesh. Pp. 125-126.

Bhalla, S., K. Gupta, B.Lal, M.L. Kapur and R. K. Khetrapal. (2008). Efficacy of various non-chemical methods against pulse beetle, Callosobruchus maculates (F.). ENDURE International Conference, Diversifying Crop Protection, and France.P. 1-4.

Bhalla, S., K. Gupta, B.Lal, M.L. Kapur and R. K. Khetrapal. (2008). Efficacy of various non-chemical methods against pulse beetle, Callosobruchus maculates (F.). ENDURE International Conference, Diversifying Crop Protection, France. Pp. 1-4.

Bhuiyah, M. I. M., A. N. M. R. Karim, B. N. Islam and Z. Alam. (2002). Test of botanicals for the control of the beetle attacking lentil and chickpea in storage. Bangladesh J. Agri. Res., 27: 349-362.

Boeke, S. J., C. Barnaud, J. J. A. V. Loon, D. K. Kossou, A. V. Huis and M.Dieke. (2004) Efficacy of plant extracts against the cowpea beetle, Callosobruchus maculatus.Int. J. Pest Manag. 50 (4): 251-258.

Bowers, W. S. (1983). The search for fourth-generation pesticides. In "Natural products for innovative pest management" (Eds. Whitehead, D.L and Bowers, W.S.), Pergamon Press Oxford, pp: 47-58.

Dinar musbasalis (Rondani) (Hymenoptera: Pteromalidae).J.Stored Pod. Res. 41: 363-371.

Dwivedi, S. C. and S. Venugopalan.( 2002). Efficacy of Tabernaemontanalivaricate and *Tamarindus indica* as an anti feedant and adulticide against pulse beetle *Callosobruchus chinensis* (L.) (Coleoptera, Bruchidae). Baltic J. Coleop., 2 (2): 159-163.

Egesa, A.O., S.N. Njagi and C.W. Muui. (2016). Effect of facilitative interaction of sorghum-cowpea intercrop on sorghum growth rate and yields. J. Environ. Agric. Sci. 9:50-58.

Elzaki, M.E.A. and A.M. Ali. (2015). Efficacy of neem, basil and chilli powders as antifeedant in the red flour beetle *Tribolium castaneum* (Herbst) (Coleoptera: Tenebrionidae). J. Environ. Agric. Sci. 2: 9.

Fishwick, F. B., (1988). Pesticide residues in grain arising from post harvest Treatments. Aspects of Applied Biol., 17: 37-46.

Ghosh, S. K. and S. L. Durbey. (2003). Integrated Management of Stored Grain Pests. International Book Distributing Co., Luck now, India, pp 263.

Gujar, G.T. and T.D. Yadav. (1978). Feeding of Callosobruchusmaculatus (Fab.) and *Callosobruchus chinensis*. in green gram. Indian J.Entomol. 40: 108-112.

Halstead, D. G. H. (1963). External sex differences in stored products coleoptera. Bull. ENT. Res., 54: 119-134.

Hashmi, M.S. (2016). Agrochemical and agricultural sustainability: A case study of Pakistan. J. Environ. Agric. Sci. 9: 1-9.

Ignatowicz, S., B. Wesolowska and K. Banasik. (1995). Potential of common Plants as grain protectants: Repellent effect of powdered leaves and seeds of the neem tree on stored product pests. Insects: Chemical, Physiological and environmental aspects, Pol and University of Wroclaw, 317-322.

Jacobson, M.( 1989). Botanical pesticides: past, present and future. In: Arnason, J.T., Philogene, B.J.R., Morand, P. (Eds.), Insecticides of Plant Origin. ACS Symposium Series No. 387.American Chemical Society, Washington, pp. 1-10.

Javaid, I. and M. A. T. Poswal. (1995). Evaluation of certain spices for the control of *Callosobruchus maculates* (Fabric us) (Coleoptera: Bruchidae) in cowpea seeds. African.Entomol.3: 87-89.

Jilani, G. and M. M. Malik. (1973). Studies on Neem plants as repellent against stored grain insects. Pak. J. Sci. Ind. Res., 16: 251-254.

Jilani, G. and Saxena, R.C. (1988). Evaluation of neem oil, combination of neem oil and Fumigation and Actellic as paddy / rice protectants against storage insects. Proc.Final Workshop on Botanical Pest Control in Rice-based Cropping Systems.Int. Rice Res.Inst. Manila, 1988. Pp.28.

Ketoh, G. K., H. K. Koumaglo and I A. Glitho. (2005). Inhibition of Callosobruchus maculates (F.) (Coleoptera: Bruchidae) development with essential oil extracted from CymbopogonschoenanthusL Spreng. (Poaceae), and the wasp

Kim, S.I, J. Y. Roh, D. H. Kim, H. S. Lee and Y.J.Ahn. (2003). Insecticidal activities of aromatic plant extracts and essential oils against Sitophilusoryzae and Callosobruchus chinensis. Stored Prod.Res. 39: 293-303.

Kokate, S. D. and G. J. Chintalwar. (2003). Insect repellent activity of certain plant extracts against pulse beetle, *Callosobruchus chinensis*. National Academy Sci. Letters., 26:44-46.

Kotkar, H. M., P. S. Mendki. S. V. G. S. Sadan, S. R. Jha, S. M. Upasani and V. L. Maheshwari. (2002). Antimicrobial and pesticidal activity of partially purified flavonoids of Annonasquamosa. Pest Manage.Sci. 58 (1): 33-37.

Kumari, K and S. N. Singh. (1998). Evaluation of efficacy of botanicals an insecticide against pulse beetle (*Callosobruchus chinensis*.).J.App. Biol., 8:138-140.

Materials on the development of pulse beetle, Callosobruchus chinensis (L.) on stored pigeon pea. Cajanus (L.)Millsp.J.Insect Sci., 10: 196-197.

Mookherjee, P.B., M.G. Jotwani. T.D. Yadav and D. Sircar. (1970). Studies on the Incidence and extent of damage due to insect pests in stored seeds. Leguminous and Vegetable seeds. Indian J. Entomol. 32 (4): 350-355.

Nakakita, H. (1998). Stored Rice and Stored Product Insects. In: Rice Inspection Technology Manual. A. C. E. Corporation, Yokyo, Japan, pp. 49-65.

Rahman R., Huda S. and A. Hussain. (1975). Effect of gamma radiation on different stages Of *Callosobruchus maculates*. Bangladesh J. Zool. 3: 103-109.

Naqvi, S. N. H. (1987). Biological evaluation of fresh neem extracts and some components with reference to abnormalities and esterase activity.Proc. 3rdIntl. Neem Conf (Nairobi 1986), pp. 315-330.

Raina, A.K. (1970).Callosobruchusspp.infesting stored pulses (grain legumes) in India and comparative study of their biology. Indian J.Ent. 32.

Saxena S.C. (1992). Biology of Insects. Published by Raju Primlanifor Oxford and IBH Publishing Co. Pvt. Ltd., 66 Janapath, New Delhi 110001. P.366.

Sharma, S. S., G. S. Yadav, and B. S. Chhillar. (2003). Repellent Activity of Some Plant Extracts against *Callosobruchus chinensis* (L.) in Chickpea Grains. Ann. Biol., 19 (2): 217-218.

Sharma, S.S. 1984. Review of literature of the losses caused by *Callosobruchus chinensis* (L.)(Coleoptera: Bruchidae) during storage of pulses. Bull. Grain Technol.22 (1): 62-.68.

Singal and Borah (2001) observed that larval and pupal period of *C. chinensis* was 16.2 ± 0.16 and 7.2 ± 0.18 days, respectively on pods of Cajanuscajan.

Singal, S. K. and R. Chauhan. (1997). Effect of some plant products and other

Singh, S. C, N. K. Pandey and A. Kumar. (1996). Effect of neem saw dust on mortality of the pulse beetle *(Callosobruchus chinensisL.)* on black gram (Vignamungo).J. Ecotoxicol. Environ. Mon It., 6(1) 69-71.

Singh, S. C. (2003a). Effect of neem leaf powder on infestation of the pulse beetle *Callosobruchus chinensis* in stored khesari. Ind. J. Entomol, 65: 188-192.

Tapondjou, L. A., C. Adler, H. Bouda and D. A. Fontem. (2002). Efficacy of Powder and essential oil from Chenopodiumam brosioides leaves as post-harvest grain protectants against six stored product beetles. J. Stored Prod. Res., 38: 395-402.

Tripathy, M. K., P. Sahoo, B. C. Das and S. Mohanty. (2001). Efficacy of botanical plant powders and extracts against *Callosobruchus chinensis* Linn, attacking black gram (Cv. T9). Legume Res., 24(2):82-86.

Tyler, P. S., R. W. Taylor, D. P. Rees. (1983). Insect resistance to phosphine Fumigation in foodware houses in Bangladesh, Int. Pest Control,25: 10-13.

Umrao, R.S. and R.A. Verma. (2002). Effectiveness of some plant products against pulse beetle on pea. Ind. J. Entomol., 64: 451-453.

Upasani, S. M., H. M. Kotkar, P. S. Mendki and V. L. Maheshwari.(2003).Partial characterization and insecticidal properties of Ricinuscommunis L foliage flavonoids. Pest Manage. Sci., 59 (12): 1349-1354.

Wink, M. (1993). Production and application of phyto-chemicals from an Agricultural perspective. In: T.A. van Beek, H. Breteler, (Eds.), Photochemistry and Agriculture, Vol. 34. Clared on, Oxford, UK, pp.171-21.

Yazlik, A. and İ. Uremis. (2016). Evaluation of autotoxic potential of Johnsongrass and its integrated application with herbicides. J. Environ. Agric. Sci. 9: 44-49.

Yusof, O. and S. H. Ho. (1992). A survey of insecticidal resistance in Sitophilus Zeama is Motsch. In Malaysia and Singapore. J. Plant Protec. in the Tropics, 9: 219-225.

# 7

# Risk assessment of pesticides on honey bee and pollination of agriculture crops in Pakistan

**Saboor Ahmad***, **Muhammad Asif Aziz**
Department of Entomology, PMAS Arid Agriculture University, Rawalpindi, Pakistan

*Corresponding author email: saboorahmad86@yahoo.com

## Abstract

This paper reviews the importance of honey bee as pollinator of agriculture crops, fruit, vegetables, seeds etc. and losses in agriculture, honey bee and honey production due to use of chemical pesticides. Honey bee (Hymenoptera: Apidae) is economically important insect. Honeybees are existing all around world excluding extreme Polar Regions. Honey bees (*Apis mellifera*) add approximately US $200 billion to the global economy every year, mainly through crop pollination. Productivity worth of pollination dependent crops equal to US$ 1590 million in Pakistan where 61 main pollinate crops are considered as foodstuff (26 fruits, 19 vegetable, 7 oilseeds, 4 grain legumes, 2 flavoring crops and 3 nut shrubs) depend on honey bee pollination. Honey bee population is declining with alarming speed all around the world and it is endangering global food security because one third of agriculture production depend upon on pollination particularly by honey bee. Current reduction in honey bee population for agriculture pollination is threatening to nuts, fruit, vegetable and seed production in Pakistan. Several factors responsible for decline of honey bee colonies including climate change, predation by other insects, air pollution, ecosemiotic collapse, the alteration of the bee microbiome, electromagnetic radiation from the sun, expose to initial life stress, nano materials, and biochemical pesticides, diseases, parasites surroundings of honey bees. The toxic chemicals are transported in hive through the pollen and nectar which may destroy the entire honey bee colony. Pesticides pattern applied in agriculture have been changed in last decade. This review research indicates how pesticide is used in agricultural crops for control of pests affecting on honey bee health and decline of bee populations. Then concludes with intensive studies on current management protocols would support in procedure formulation and resolution making polices regarding to shelter honey bees and management of pollination for agriculture purpose, which is lacking.

**Keywords:** Honey bee, Agriculture crops, Pollination, Beekeeper, Pesticides

## Introduction

The domesticated honey bee (Hymenoptera: Apidae) is economically important insect (Michener, 1969). Honey bees are present all around world excluding extreme Polar Regions. In Pakistan, four species of honeybee are present among them three bee species (*Apis dorsata, Apis florae, Apis cerana*) are indigenous while *Apis mellifera* is exotic imported from Russia and Australia during 1979 (Jadran, 2011; Hikmat et al, 2016). Honey bee producing different products including honey, royal jelly, bee venom, pollen, bee wax and propolis that are used in food commodities, medicinal drugs and further marketable products (Wakhal et al., 1999). In Pakistan 400,000 honey bee colonies making 10,000 MT honey annually, almost more than 4000 beekeepers were involved in raising of *Apis mellifera* and 27000 household were getting income from beekeeping industry (PARC, 2010).

Beekeeping remains significant for agricultural and environmental activity. Beekeeping plays a parallel role in cross-pollination of economical farming crops and medicinal plants with the manufacturing of honey. Honey bees increase the production of different fruit crops up to 30-40 % (Jadran, 2011). Pesticides requirement increase over last decade in most developing countries and increased in Pakistan during 1990 (Osteen and Fernandez, 2013; Khan et al., 2002). De A et al. (2014) demonstrated the percentage of pesticides being used 45 % in Europe and 25 % in USA, 4 percent in India and in other parts of world 26%. This value of consumption of pesticide is almost € 36 billion annually (Popp et al., 2013; De A et al., 2014). Pakistan is second largest purchasers of chemical substances in South Asia, 27% pesticides used within fruits, vegetables crops and annually application of pesticides is increasing up to 25 % (Hussain et al., 2004). Honey bees and beekeepers are suffering from the excessive usage of pesticides every year in agriculture sector due to beginning of agriculture modernization (Maini et al., 2010). Honey bee life depends upon flowering plants. Honey bees foragers collect nectar from nectariferous plants, collect pollen only from polliferous plant and some bee foragers collect nectar and pollen on the same plant (Pătruică, 2006). Pollination is an essential process for production of fruit plants, variety of flowering plants and deals fundamental ecosystems facilities to human welfare (Kevan, 1999; Klein et al., 2007). It is confirmed that most of flowering plants are pollinated by insects, animals; including 70 % reproduction occurring in world's flowering plants is due to pollination (Ollerton et al., 2011). Among the pollinators, *Apis mellifera* has been playing crucial role for improving the agricultural crops production throughout the world (Allen-Wardell et al., 1998). Extensive use of pesticides leads to loss of honey bee population and also decrease in pollination process (Chauzat et al., 2006).

## Influence of pesticides on honey bee and honey production

The bee population is reducing all around world since 2006. Average death rate of honey bee colonies are 30 % in USA and some beekeeper claimed even greater injuries (Ellis et al., 2010; Lee et al., 2015). About 25-30 % honey bee population decline in European countries and Canada is due to extreme use of higher concentration of pesticides (Van Engelsdorp and Meixner, 2010; Neumann and Carreck, 2010). There are several factors responsible for decline in honey bee population across the world including weather alteration (Le Conte and Navajas, 2008), air contamination (Girling et al., 2013; Mc Frederick et al., 2008), ecosemiotic collapse (Harries-Jones, 2009), adaptation of the honey bee microbiome (Mattila et al., 2012; Cox-Foster et al., 2007), electromagnetic radiation from the sun (Ferrari, 2014), honey bee colony exposure to primary life pressure (Wray et al., 2011; Clare CR et al., 2015), nano-materials (Milivojevic et al., 2015), and biochemical background of bees (Mullin et al., 2015). The population of honey bee is decreased from 1966-1979 due to excessive use of organophosphorus, organochlorine, carbamate and pyrethroid pesticides contact. The ratio of pesticides usage is comprising 74 percent insecticides, 14 percent weeds killer, 9 percent fungicides, 2 percent acaricides and 1 percent fumigants in Pakistan (Khan, 1998), in which most of pesticides used for cotton (60%), rice (7%), cereals (4%), sugarcane (2%) and rest of pesticides (27%) used for other crops (Kang, 2013). Pesticides are broadly used in Pakistan to controlling the agriculture pest, excessive use of chemical disturb the ecological system and cause pest resistance, kill ecofriendly insects (Predators, Parasitoids and Pollinators) .The consumption of pesticides (Fig 1) as well as the sprayed crops area increased many folds (Fig 2).

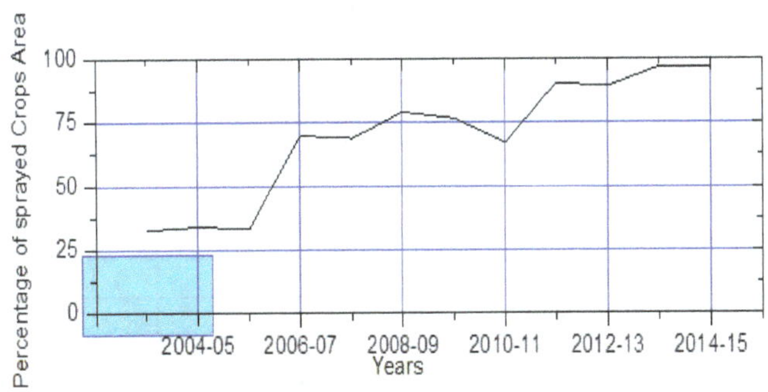


**Fig 1: Consumption of pesticides in Pakistan (Agriculture statistic of Pakistan 2014-15)**

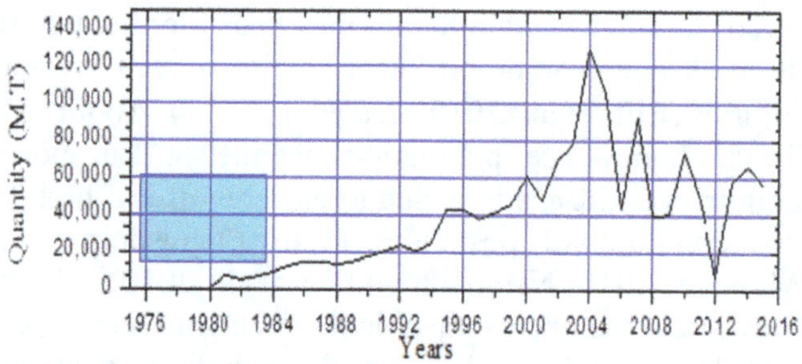


**Fig 2: Percentage of Sprayed Crops Area covered by ground plant protection measures in Pakistan (Agriculture statistic of Pakistan 2014-15)**

In Pakistan most of pesticides used by farmers are insecticides and fungicides. Table 1 shows grouping of pesticides, chemical group, WHO Hazard Category and their target pest. It was observed that farmer also used highly toxic insecticides containing carbofuran and moderately toxic (endosulfan) which are hazardous to human, arthropods including honey bee (Asif et al., 2014).

**Table 1: Characterization of pesticides with toxicity class applied in agricultural crops and residual effect on honey bee health (Jivan and Marian, 2013; Asif et al., 2014; Douglas, 2015).**

| Pesticides common name | Chemicaly Group | Toxicity class | Residual activity | Usess |
|---|---|---|---|---|
| Imidacloprid | Nicotinoid | II | <8 hours | Sucking pests |
| Endosulfan | OC | II | <8 hours | Against lepidopterous larvae |
| Chlorpyrifos | OP | II | 4-6 days | Acaricide |
| Diafenthiuron | Urea | III | 3-4 days | Sucking pests &mite |
| Mancozeb | Dithiocarbmate | U | | Downy mildew |
| Profenophos | OP | II | | Against lepidopterous larvae |
| Cypermethrin | Pyrethroid | II | 3 days | Against lepidopterous larvae |
| Sulfur | Inorganic | III | | |
| Thiamethoxam | Nicotinoid | III | 7-14 days | Sucking pests |
| Lambda-Cyhalothrin | Pyrethroid | II | > 7days | Chewing & sucking pests |
| Dimethoate | OP | II | 3 days | Sucking pests |
| Acetamiprid | Nicotinoid | II | | Sucking pests |
| Permethrin | Pyrethroid | Ib | 1-2 days | Acaricide |
| Emamectin | Avermectin | II | | Lepidopterous Fruit worms |
| Thiophanate methyl | Benzimidazole | U | | Powdery mildew |
| Spinosad | Bacteria | II | <2 hours | Against lepidopterous larvae, thrips |
| Abamectin | Avermectin | II | | Against lepidopterous larvae |
| Metalaxyl | Anilide | II | | Downy mildew, ealry blight of tomatoe |
| Bifenthrin | Pyrethroid | II | | Against lepidopterous larvae |
| Pyridaben | Unclassified | II | <2 hours | Sucking pests |
| Lufenuron | Urea | II | | Against lepidopterous larvae |
| Difenoconazole | Conazole | II | | Early blight of tomato, Powdery mildew diseases |
| Carbofuran | Carbamate | Ib | | Used for lepidopterous larvae |
| Acephate | OP | II | 3 days | Chewing & sucking pests |
| Manzate | Ethylene bisdithiocarbamate | III | | Black spot, flower blight, rust |
| Chloranthraniliprole | | U | | Against lepidopterous larvae |
| Malathion | OP | III | 2-5 days | Chewing & sucking pests |
| Monomehypo | Nereistoxin | NR | | Against lepidopterous larvae |
| Deltamethrin | Pyrethroid | II | 4 hours | Chewing & sucking pests |
| Copper Oxychloride | Inorganic | II | | early blight of tomato |
| Azoxystrobin | Unclassified | III | | Powdery mildew, downy mildew stripe rust, haustorium |
| Indoxacarb | Oxadiazine l | III | | Sucking pests and chewing |

Ib=highly toxic; II=Moderately toxic; III=Slightly toxic; U=Unlikely toxic

Toxicity of insecticides is usually determined by measuring critical contact toxicity standards $LD_{50}$ (It is value which cause 50 % of mortality of population when exposed to pesticides). Honey bee threshold level to toxicity is normally set at extremely lethal (acute $LD_{50}$ < 2 µg /honey bee); moderately lethal (acute $LD_{50}$ equal with 2 - 10.99 µg/honey bee); somewhat lethal (acute LD50 equivalent toward 11 - 100µg/honey bee); non-lethal (acute LD50 > 100µg/honey bee) to developed honey bees (Antonina and Marian, 2013). Honey bee visit in pesticides treated agriculture field and the toxic chemicals are transported in hives through the pollen and nectar, it may kill the entire honey bee colony. The polluted pollen and nectar in bee colony may be impressed due to residual poisonous result of pesticides. When the toxic honey comb introduced into new family of bee brood colony it may also be affected severely (leading to death). The toxic chemicals can entirely destruct the bee population or decrease the population within the bee colony. This weaken colony may enter the winter with short food and may cause the collapse of colony over winter season (Atkins, 1992). There are different situations through which honey bees are exposed to pesticides; exposure through contact when honey bee may be poisoned in flight during application of chemicals or contact with residue on treated plants (Bura et al., 2000); by ingestion when bees collect pesticide contaminated food and it will be used by entire colony which results into colony destruction and storage food (nectar and pollen) in which nectar can be used instantly but pollen need time for fermentation to be digestible for honey bees and it will be used during the period outside the harvest particularly in wintertime, harvested pollen in August can be used in March, April next year (Villa et al., 2000). Insecticides may cause lethal and sub lethal effect on biological system of honey bees. Insecticides with different mode of action effect on molecular level lead to initiation of noticeable and unobservable effect, damage physiological function (thermoregulation and muscle activity), behavior (foraging) and cognitive activities (Belzunces et al., 2012).

**Sub-lethal effect of pesticides**

Sub-lethal influence of pesticides upon honey bee health and former beneficial insect were determined in most publications. For example sub-lethal doses of pesticides (Parathion, Diazinon) disturb the movement of honey bee behavior (familiarize over short distances, familiarize over long distances), harvesting, transport of nectar, development, fertility, fecundity, survival, feeding behavior and thermoregulation (Vandame, 1995; Desneux *et al.*, 2007). Anjum et al. (2001) determined mortality of honey bee in treated cucumber crops with Polo (Diafenthiuron 50 EC) and Endosulfan (35 EC) after 7 days interval. Mortality percentage was 34.64 and 66.64 % respectively after 48 hours .Plant protection products also show negative effect on honey bees on lives, immune capabilities and other behaviors which are important for better development of colony like bee brood feeding, building comb, swarming, and balancing worker and drone population between comb, transferring of information with in new building of colony (Colin et al., 2004).

Dilbar et al. (2014) conducted an experiment on three honey bee species ( *Apis florea, Apis mellifera and Apis dorsata*) to determine the persistence data of confined bee against expose to seven insecticides including bifenthrin, spinosad, indoxacarb, emamectin benzoate, carbosulfan, imidacloprid and chlorpyrifos with concentration (1000, 500, 250,125 and 0 ppm). Carbosulfan and imidacloprid (1000 ppm) were more toxic with $LD_{50}$ after exposure of 4 hrs used for *A. mellifera* while chlorpyrifos and imidacloprid having concentration (1000 ppm) extremely lethal with $LT_{50}$ after expose of 5 hours to *A. florea*. Chlorpyrifos having concentration (1000 ppm) was highly lethal with LT50 after 5 hrs to A. *dorsata*. Therefore spinosad having concentration (125 ppm) with LT50 was enhanced to 15 hours, 18 hours and 20 hours against *A. florea, A. mellifera* and *A. dorsata* respectively.

Ping-Li et al. (2010) examined toxicity of some pyrethroids including deltamethrin and bifenthrin which are extensively applied in agriculture crops (fruits, vegetables, forestry crops). Bifenthrin has highly toxic effect on honey bees health including weight of egg, hatching, weight of larvae, emergence rate, and capping rate in each development stage while deltamethrin influence on larval weight, capping rate, advent rate in uncapped brood stage. Spinosad has also toxic effect on honey bee and causes high mortality when exposure time is increased. Chlorpyrifos and imidacloprid were highly toxic against three species of honey bee (*A. floria, A. mellifera and A. dorsata*) which caused 100 % mortality after 6 hour exposure to these insecticides with high concentration (Johnson et al., 2010). One neonicotinoids pesticide, imidacloprid is broadly used in agricultural crops effecting on the foraging behavior of honey bee and lead to honey bee colony collapse disorder (CCD) (Lu et al. 2014; Karahan *et al.* 2015). Three nonicotinoids pesticides (clothianidin, imidacloprid, thiamethoxam) should not be applied during in oilseed rape and maize during flowering season to save honey bee colonies (Cressey, 2013).

**The synergistic effects**

According to (Colin and Belzunces, 1992) and (Pilling and Jepson, 1993) synergistic effect of mixture of fungicides and pyrethroids applied on honey bee in control condition enhanced toxicity 10 to 100 fold. When different pesticides (pyrethroid and fungicides) applied in agriculture field mixing in the same tank, synergism in their toxicity may happen accidently and show negative effect on insects.

Different insecticides including clothianidin, imidacloprid and thiamethoxam (neonicotinoid) are highly toxic pesticides to honey bees. Krupke et al. (2012) conducted a study to determine exposure of pesticides including neonicotinoids clothianidin, thiomexotham, atrazine (herbicides) and fungicides applied in agriculture during the foraging period. He concluded that propionazole (fungicide) showed synergism with neonicotiniod and pesticide residue becomes a part of soil, plant, pollen and nectar and when bees visit these plants, mortality occur directly (nectar) and indirectly (pollen storage).

Neonicotinoids insecticides are used directly in agriculture crops and also for seed treatment. These neurotoxic chemicals are systemic and become a part of plant including pollens and have been showed negative effect on honey bee health since 1990s (Henry et al., 2012; Kessler et al., 2015). Muhammad et al. (2016) demonstrated the residual effect of neonicotinoids on honey bee in laboratory experiments. Imidacloprid highly lethal than thiamethoxam followed by acetamiprid when use by wet surface method while nitenpyram and thiacloprid least lethal after 3 and 6 hour of exposure. Imidacloprid, acetamiprid extremely toxic than thiamethoxam to worker honey bee when expose to wet surface after 24 and 48 hours. Thiamethoxam, acetamprid and imidacloprid showed highly toxicity whereas thiacloprid and nitenpyram less toxic after 3 to 6 hours when exposure to oral application. Therefore, thiamethoxam was more lethal followed by thiacloprid, imidacloprid while acetamiprid and nitenpyram was less lethal after 24 and 48 hours through oral application. Thiacloprid and nitenpyram should be used nearby blossoming season to protect the honey bee. France and Germany prohibited the use of neonicotinoids during 2008 due to decline of bee colonies. Pesticides which are applied in agriculture field, among them most pesticides are fat soluble and larger quantity of pesticides store in bee wax than honey. Many scientific studies showed that the pollution of bee product in that order; propolis is most polluted followed by wax, pollen and honey (Bogdanov, 2003)

In Pakistan honey bee hives produces an average of 10-15 kg honey in each season but in Australia bee hives produces 25-30 kg honey per season due to good bee breeding facilities (Izhar, 2016). There are several reasons for less production in Pakistan including climatic changes, untrained beekeeper and extensive use of pesticides has severely affected the honey bee colonies and beekeeper annual loss of 5661 MT honey which equal to loss of 9.91 million rupees. In swat valley Wild honey also reduce up to 75 percent due chemical application and now not easily accessible on the marketplace. Farmers keep honeybees with their agricultural crops but pesticides are the main obstacle in beekeeping and also expected 90 percent honey bee deaths every day by bulky application of pesticides (Mohammad *et al.*, 2008). Honey production has declined up to 40% in Pakistan due to major changes in (habitat, biodiversity) and widely application of pesticides lead to reduce honey bees population within the Hindu Kush Himalayan region (Muhammad, 2016).

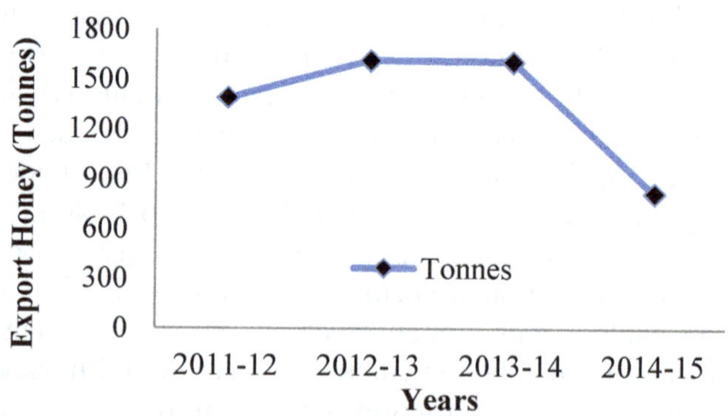


**Fig 3: Export of natural honey from Pakistan (Pakistan Bureau of Statistics)**

**Role of pollination**

Above 25,000 species of bees pollinate almost 70% of crops (87 of the 124 main crops) which are directly used for human consumption in the world. Production increase 5 to 50% in most agriculture crops as a result of pollination done by bees (Klein *et al.*, 2007).

Almost 90 percent of all angiosperms depend on pollinators to survive. Honey bees responsible for third of pollination in agriculture sector. In Pakistan important crops including alfalfa, clover, almond, melon, soybean, sunflower, cucumber, apricot, cherry, peach, citrus, pear, persimmon , prune, apple ,plum, cantaloupe, okra, rape seed carrot onion, avocado, kiwi, cranberry blueberry are mostly require honeybee pollination (Mohammad and Elizabeth, 2014).

In Agriculture worldwide profitable value of pollination service is USA$193 billions annually. Annual economic worth of insects pollination for pulses crop is $2.7 millions, oil seed crops $ 233.1 millions, tree nut crops $ 50.5 millions, for seasonings are $ 5.5 millions and fruit and vegetables are $ 2.3 billions and 78.5 millions respectively ( Gallai et al., 2009). In Europe 84% productions of crops depend upon the insect pollinator (Williams, 1994). Production value of pollinated dependent crop was equal to 1.59 billion US$ in Pakistan, therefore fruit crops are leading with 980 million, vegetable crops 320 million, nuts crop 150 billion, oil-rich seed crops 130 million and spices 4 million US $ (Muhammad and Elizabeth, 2013). Now Pakistan has 61 essential pollinated crops regard as foodstuff containing 26 fruits crop, 7 oil-rich seed crops, 4 grain legumes crop, 19 vegetables crop, 2 spices and 3 nut shrubs which are depend on honey bee pollination (Mohammad and Elizabeth, 2014).

According to (Stephen and Irshad, 2012) production of mustard crops increased 30 % and production of sarsoon (*Brassica rapa trilocularis*) and toria (*Brassica napus*) increased 47 % by increasing the bee hives. The presence of honey bee pollination cucumber and cauliflower yield increased up to 28-32.5 % and 23 % respectively. Different district of Pakistan including Matli, Badin and Golarchi yield of radish, sunflower increased 18.7 and 22 % respectively and seed set of onion increased 62-93 % due to honey bee pollination. Clover yield increased 100 % due to honey bee pollination. Anjum et al. (2001) found the foraging activity of honey bee in cucumber crops. Due to honey bee pollination germination of cucumber crops increased 90.33 % and yield obtained 352.7 kg per hectare. Muhammad et al. (2009) investigated honey bee pollination efficacy in canola crops. Number of pods and number of seed per pod are 815 and 20 with presence of bee pollination while 349 and 15 without pollination, yield of canola crops and weight of seed also increased due to bee pollination. Oz et al. (2009) determined that sunflower hybrid production enhanced seed set ratio, seed weight, number of filled seed per head and seed yield per head due to honey bee pollination.

**Table 2: Economic values of honey bee pollination in different part of world (Uma and Tej, 2012: Mohammad and Elizabeth, 2014).**

| Country | Economic values (billion USA Dollar) | References |
|---|---|---|
| America | 14.6 | (Morse and Calderone, 2001) |
| Canada | 1.2 | (Winston and Scott, 1984) |
| China | 0.7 | (Chen, 1993) |
| New Zealand | 2.3 | (Matheson and Schrader, 1987) |
| European Union | 3 | (Williams, 1992) |
| United Kingdom | 0.15 | (Carreck and Williams, 1998), |
| Brazilian | 8.75 | (Freitas and Imperatriz-Fonesca, 2005) |
| East Africa | 1.2 | (Kasina *et al.*, 2009) |
| Uganda | 0.46 | (Munyuli, 2011) |
| Netherland | 1.25 | (Blacquiere et al., 2010) |
| Himalayan region of Pakistan | 0.95 | (Uma et al., 2012) |

**Deficit of pollination**

According to current population growth rate high level of farming production will be needed to support the world population. The practices which are applied to enhance the yields damaged our ecosystem depict most susceptible to pests. To improve better health, length and quality of human life, food requirement have enhanced, and demanding the efficient pest control by using different pesticides. Approximately two billion individuals involve in agricultural business that mostly used chemical control to save their field crops and farm animals. Pesticides are applied also extensively in garden, throughout home and frame work of public fitness programmed (Oerke, 2006). In USA incomes of some fruit crops, seeds and nuts crop decline more than 90 percent in absence of insect pollinators (Southwick and Southwick, 1992). In USA pesticides have been consider as major factor for loses of honey bee colonies, reduction in honey production, wax yield and pollination to cause losses about $283 millions per year. It is estimated approximately $210 millions losses due to declining of honey bee pollination of crops, residue of pesticides have been found in bee wax, pollen , adult bee and bee pupae (David, 2009; Mullin, 2010). In Pakistan yield of some fruit reduced up to 33.4% due to pollination shortage (Aslam et al., 2004). According to Gallai *et al.* (2008) three main crops categories will be affected due to decline of pollination , in which mainly fruits and vegetables production reduced with losses estimated at 50 billions for each and 39 billions Euros losses by edible seed crops. In Himalayan region of Pakistan the population of pollinator decline due to farmers and institution unawareness about pollination benefits. In mountain region of Azad Kashmir of Pakistan yield of apple decrease every year due to absence of pollinators and farmers replace their apple trees to agriculture crops (Uma and Tej, 2002). Uma et al. (2012) found production of agriculture crops increasing in mountain region of Pakistan by managing the honey bee pollinators. In Peshawar production of loquat decline 64.4% due to absence of honey bee pollinator and apple production also decreases (Khan, 1986; Khan, 2004). Aslam et al. (2004) determined that inflorescence and fruit seed set was in un caged and caged tree 42.5 and 28.3 % respectively, yield decrease 33.4 % due to decline of honey bees pollination. Zafar et al. (2016) examined the residual effect of Malathion (0.002 gm/ml) on pollinator insects including honey bee, bumble bee, syrphid fly and butterflies in *Calendula officinalis* field. After 10 days interval it was observed the forceful decrease in number of pollinator particularly honey bee due to insecticide treatment but significantly higher number of insect pollinator was detected in control field. In current senario colonies of honey bee are declining due to excessive use of chemical pesticides resulting into declining of food production in the absence of pollinators (Chauzat et al. 2006). Therefore excessive application of pesticides not only preserved within fruits and vegetables but also pollute the water and soil then enter into food chain and also enter into human blood through this food and water, pesticides is main cause of environmental pollution, damage the biodiversity and decline of natural habitats (Cerejeira *et al.,* 2003; USAID and I-LED, 2006-2009). It is claimed by many commercial companies of insecticides and fungicides that their products are harmless or only slightly toxic to honey bees but pesticides are dangerous to bees and other pollinators when applied on flowering crops (Abramson *et al.,* 1999). Large amount of insecticides are used to control the pests also having effect on natural enemies (Predators, Parasitoids, Parasites) of agriculture pests, sub-lethal doses of neonicotinoids have shown clearly effect on scrounging activity, natural selection, colony health and new honey bee and bumble bee queen production (Henry *et al.,* 2012; Schneider *et al.,* 2012: Whitehorn *et al.,* 2012) , Herbicides effect directly and indirectly on beneficial arthropods and may indirectly decreases the fitness of pollinating insect (Egan *et al.,* 2014).

## Conclusion

Honey bee is an important pollinator for agriculture crops and increasing the production of fruits, seeds and vegetables crops without any additional investment. Pollination process is helping to improve crops quality, quantity, and genetic diversity and improve over all ecosystems. 15 % of world crops are pollinate by succeeded honey bee colonies and rest of world crops are pollinated by solitary bee and other animals. In developed countries *Apis mellifera* colonies are rented by farmers for pollination in their agricultural crops. It is continuous challenge to excessive use of pesticides in agriculture crops to control the pest. Every year new pesticides are coming to market with new formula and these are hazardous to biodiversity and may cause the animal and livestock death, pesticides residue present in food and water and honey bees severely damaged when exposed to pesticides. Pesticides treatment should avoid in plantation and crops during flowering and pesticides application should ensure protection of honey bee health. If honey bee takes polluted nectar and pollen in hive it may cause mortality of honey bee foragers result in destruction of whole colony. Farmers should use ecofriendly pesticides (derived from plant extracts etc.) like nicotine, quassin, rotenone, ryan, tephorosia, pyrethrins, veratrine, and anabasine which have less toxic effect on honey bee. Farmers must apply integrated pest management practice for control of pests in their agriculture field and should not only rely on chemical control measures.

## References

Asif AJ, Amber RS, Najma M and Shafi MN, 2014. Current scenario of pesticides practice among farmers for vegetables production: A case study in Lower Sindh, Pakistan. Int J. Dev. Sustain. 3(3): 493-504.

Abramson C, Aquino I, Ramalho F and Price J, 1999 - The effect of insecticides on learning in the Africanized honey bee (*Apis mellifera* L.). Arch. Environ.Contam. Toxicol. 37: 529-535.

Allen-Wardell G, Bernhardt P, Bitner R, Burquez A, Buchmann S, Cane J, Cox PA, Dalton V, Feinsinger P, Ingram M, Inouye D, Jones CE, Kennedy K, Kevan P, Koopowitz H, Medellin R, Medellin-Morales S, Nabhan GP, Pavlik B, Tepedino V, Torchio P and Walker S, 1998. The potential consequences of pollinator declines on the conservation of biodiversity and stability of food crop yields. Conserv. Biol. 12: 8-17.

Anjum S, Zain UAD, Javad I, Muhammad I, Umair W, Rab NS and Ikram HQ, 2001. Insecticidal Mortality and Pollination Role of Honeybee (*Apis mellifera* L.) on Cucumber (Cucumis sativus L.) Crop. Int. J. Agri. Biol. 3(4): 501-502.

Antonina J and Marian B, 2013. The Impact of Pesticides on Honey Bees and Hence on Humans. J. Anim. Sci. Biotechnol. 46 (2)

Aslam, M, Razzaq M, Hussain S and Akhter W, 2004. Effect of insect pollinators on fruit setting on Mango (*Mangifera indica* L.). J. Sci. Res. 15:53-58.

Atkins EL, 1992. Injury to honey bee by poisoning, pp. 1153–1208. In: Graham JE (eds),The Hive and the Honey Bee, Dadant and Sons, Hamilton, Canada.

Belzunces LP, Sylvie T and Jean-Luc B, 2012. Neural effects of insecticides in the honey bee. Apidologie, 43:348–370.

Blacquiere T, Van SN and Bitter R, 2010. Bijen-Fascinerend, essentieel enbedreigd. Cachiers Bio-wetenschappen en Maatschappij, kwartaal 4, The Hague, Stiching Bio wetenschappen en Maatschappij.

Bogdanov S, Imdorf A, Charrière JD, Fluri P and Kilchenmann V, 2003. Qualite des produits apicoles et sources de contamination, Rev. Suisse d'Apiculture, 99 (11-12): 8-12.

Bura M, Pătruică S and Bura VAl, 2005. Tehnologieapicolă; (ed.). Solness, Timişoara.

Carreck N and Williams IH, 1998. The economic value of bees in the UK. Bee World. 79: 115–123.

Cerejeira MJ, Viana P, Batista S, Pereira T, Silva E, Valerio MJ, Silva A, Ferreira M and Silva-Fernandes AM, 2003. Pesticides in Portuguese surface and ground waters. Water Res. 37: 1055-1063.

Chauzat MP, Faucon JP, Martel AC, Lachaize J, Cougoule N and Aubert M, 2006. A survey of pesticide residues in pollen loads collected by honey bees in France. J. Econ. Entomol. 99: 253-262.

Chen YC, 1993. Apiculture in China. Agriculture Publishing House Peoples, Beijing, China.

Clare CR, Coombs CB, Frazier M, Grozinger CM and Robinson GE, 2015. Early-life experience affects honey bee aggression and resilience to immune challenge. Sci. Rep. 5: 55-72.

Colin ME and Belzunces LP, 1992. Evidence of synergy between Prochloraz and Deltamethrin in Apis mellifera L. a convenient biological approach. Pest. Sci. 36: 115-119.

Colin ME, Bonmatin JKM, Moineau I, Gaimon C, Brun S and Vermandere JP, 2004. A method to quantify and analyse the foraging activity of honey bees: relevance to the sublethal effects induced by systemic insecticides, Arch. Environ. Contamin. Toxicol. 47: 387- 395.

Cox-Foster DL, Conlan S, Holmes EC, Palacios G, Evans JD, Moran NA, Quan PL, Briese T, Hornig M, Geiser DM, Martinson V, vanEngelsdorp D, Kalkstein AL, Drysdale A, Hui J, Zhai J, Cui L, Hutchison SK, Simons JF, Egholm M, Pettis JS and Lipkin WI, 2007. A metagenomic survey of microbes in honey bee colony collapse disorder. Science, 318: 283–287.

Cressey D, 2013. Europe debates risk to bees: proposed pesticide ban gathers scientific support as some experts call for more field studies. Nature, 496 (7446): pp. 408.

David P, 2009. Environmental and Economic Costs of the Application of Pesticides Primarily in the United States, pp. 89–111. In Peshin R and Dhawan A (eds.), In Integrated Pest Management: Innovation-Development Process. Springer-Verlag, New York, USA.

De A, Bose R, Kumar A and Mozumdar S, 2014. Worldwide pesticide use, pp. 5–6. In: De A, Bose R, Kumar A and Mozumdar S (eds.), Targeted delivery of pesticides using biodegradable polymeric nanoparticles. Springer, New Delhi, India.

Desneux N, Decourtye A and Delpuech J M, 2007. The sublethal effects of pesticides on beneficial arthropods, Annu. Rev. Entomol. 52: 81-106.

Dilbar H, Qasim M, Saleem M, Akhter M and Khan KA, 2014. Bioassay of insecticides against three honey bee species in laboratory conditions. Cercetări Agronomice în Moldova. 2: 158.

Douglas S, 2015. Pesticides – a guide to their effects on honey bees. 2nd ed. New South Wales Department of Primary Industries, Goulburn, Australia.

Egan JF, Bohnenblust E, Goslee S, Mortensen D and Tooker J, 2014. Herbicide drift can affect plant and arthropod communities. Agric. Ecosyst. Environ. 185:77–87

Ellis JD, Evans JD and Pettis J, 2010. Colony losses, managed colony population decline, and colony collapse disorder in the United States. J. Apic. Res. 49(1): 134–136.

Ferrari TE, 2014. Magnets, magnetic field fluctuations and geomagnetic disturbances impair the homing ability of honey bees (Apis mellifera). J. Apic. Res. 53(4): 452–465.

Freitas BM and Imperatriz-Fonseca VL, 2005. A importanica economica da polinizacao Mensagem Doce. 80: 44-46.

Gallai N, Salles JM, Settele J and Vaissiere BE, 2008. Economic valuation of the vulnerability of world agriculture confronted with pollinator decline. Ecological Econom. 68: 810-821.

Gallai, N, Salles JM, Settele J, Vaissière BE, 2009. Economic valuation of the vulnerability of world agriculture confronted with pollinator decline. Ecological Econom. 68: 810–821.

Girling RD, Lusebrink I, Farthing E, Newman TA and Poppy GM, 2013. Diesel exhaust rapidly degrades floral odours used by honeybees. Sci. Rep. 3: 27-79.

Harries-Jones P, 2009. Honeybees, communicative order, and the collapse of ecosystems. Biosemiotics, 2(2): 193–204.

Henry M, Béguin M, Requier F, Rollin O, Odoux JF, Aupinel P, Aptel J, Tchamitchian S and Decourtye A, 2012. A common pesticide decreases foraging success and survival in honey bees. Science. 336: 348–350.

Hikmat UK, Syed IA, Nighat S and Baharullah K, 2016. Honey production potential of the honey bee (*Apis mellifera*) in Karak and Kohat, Pakistan. J. Entomol. Zool. Stud. 4(6): 559-564.

Hussain S, Aslam M, Razzaq M and Akhter W, 2004. Effect of insect pollinators on fruit setting on Mango (*Mangifera indica* L.). J. Sci. Res. 15: 53-58.

Izhar U, 2016. From modest farms to millions: Peshawar's neglected honey industry. (https://www.dawn.com/news/1234013).

Jadran KK, 2011. Measurement of economic efficiency for the projects of bee raising inside and outside towns in Wasit Governorate. Al-Anbar J. Vet. Sci. 9(3): 112-121.

Jivan A and Marian B, 2013. The Impact of Pesticides on Honey Bees and Hence on Humans. Anim. Sci. Technol. 26 (2): 272-277.

Johnson RM, Ellis MD, Mullin CA and Frazier M, 2010. Pesticides and honey bee toxicity–USA. Apidologie. 41(3): 312-331.

Kang SA, 2013. Impacts of pesticide use in Pakistan agriculture: their benefits and hazards. (http://zaraimedia.com/2013/02/13/impact-ofpesticides-use-in-pakistan-agriculture-their-benefitsand- hazards/).

Karahan A, Ibrahim C, John MH, Ismail K and Harrington W, 2015. Sublethal imidacloprid effects on honey bee flower choices when foraging. Ecotoxicol. 2: 15-37.

Kasina JM, Mburu J, Kraemer M and Holm- Mueler K, 2009. Economic benefits of crop pollination by bees: a case study of Kakamega small holder farming in western Kenya. J. Economic. entomol. 102: 467-473.

Kessler SC, Tiedeken E J, Simcock KL, Derveau S, Mitchell J, Softley S, Stout JC and Wright G A, 2015. Bees prefer foods containing neonicotinoid pesticides. Nature, 521(7550): 74-76.

Kevan PG, 1999. Pollinators as bio indicators of the state of the environment: species, activity and diversity. Agric. Ecosyst. Environ. 74: 373-393.

Khan BM, Shahid M and Chaudhry MI, 1986. Effect of honey bee pollination on the fruit setting and yield of loquat, *Eriobotrya japonica*. Pak. J. Forestry. 36(2): 73-77.

Khan MA, Iqbal M, Ahmad I, Soomro MH and Chaudhary MA, 2002 Economic evaluation of pesticide use externalities in the cotton zones of Punjab, Pakistan. Pak. Dev. Rev. 41:683–698

Khan MR and Khan MR, 2004. The role of honey bees (Apis mellifera L.) (Hymenoptera: Apidae) in pollination of apple. Pak J. Biol. Sci. 7(3):359-362.

Klein AM, Vaissiere BE, Cane JH, Steffan-Dewenter I, Cunningham SA, Kremen C and Tscharntke T, 2007. Importance of pollinators in changing landscapes for world crops. Proc. R. Soc. L. 274: 303-313.

Krupke C, Hunt G, Eitzer B, Andino G and Given K, 2012. Multiple routes of pesticide exposure for honey bees living near agricultural fields. PLoS ONE. 7(1): 68-92.

Le Conte Y and Navajas M, 2008. Climate change: impact on honey bee populations and diseases. Revue Scientifique Et Technique-Office International Des Epizooties. 27(2): 499 –510.

Lee KV, Steinhauer N, Rennich K, Wilson ME, Tarpy DR, Caron DM, Rose R, Delaplane KS, Baylis K, Lengerich EJ, Pettis J, Skinner JA, Wilkes JT, Sagili R and vanEngelsdorp D, 2015. A national survey of managed honey bee 2013–2014 annual colony losses in the USA. Apidologie. 46(3): 292–305.

Lu C, Warchol KM and Callahan RA, 2014. Sub-lethal exposure to neonicotinoids impaired honey bees winterization before proceeding to colony collapse disorder. Bull.Insectol. 67(1): 125-130.

Maini S, Medrzycki P and Porrini C, 2010. The puzzle of honey bee losses: a brief review. Bull. Insectology. 63(1): 153-160.

Matheson AG and Schrader M, 1987. The value of honeybees to New Zealand's primary production. Nelson, New Zealand: Ministry of Agriculture and Fisheries.

Mattila HR, Rios D, Walker-Sperling VE, Roeselers G and Newton ILG, 2012. Characterization of the active microbiotas associated with honey bees reveals healthier and broader communities when colonies are genetically diverse. PLoS One. 7(3): 29-62.

McFrederick QS, Kathilankal JC and Fuentes JD, 2008. Air pollution modifies floral scent trails. Atmospheric Environ.. 42(10): 2336 –2348.

Michener C, 1969. Comparative social behavior of bees. Annu. Rev. Entomol. 14: 299-342.

Milivojevic T, Glavan G, Bozic J, Sepcic K, Mesaric T and Drobne D, 2015. Neurotoxic potential of ingested ZnO nanomaterials on bees. Chemosphere. 120: 547–554.

Mohammad I and Elizabeth S, 2013. Value of insect pollinators to agriculture of Pakistan Int. J. Agron. and Agric. Res. 3(5): 14-21.

Mohammad I and Elizabeth S, 2014. Review: pollination, pollinated and pollinators interaction in Pakistan. J. Bioresource Manage. 1(1) 19-25.

Mohammad N, Hizbullah K and Mohammad RJ, 2008. Pesticide Use in Swat Valley, Pakistan Exploring Remedial Measures to Mitigate Environmental and Socioeconomic Impacts. Mt. Res. Dev. 28(3): 201-204.

Morse RA and Calderone NW, 2001. The value of honey bees as pollinators of US crops in 2000. Bee Cult. 128: 1–15.

Mohammad ZK, 2015. Pakistan is losing its honey bees to climate change. (https://www.thethirdpole.net/2015/11/24/pakistan-is-losing-its-honey-bees-to-climate-change/).

Muhammad W, Munir A, Muhammad AA and Muhammad N, 2016. Lethal and sublethal effects of new chemistry insecticides against honeybees, *Apis mellifera.* 13th Asian Apicultural Association Conference, Jeddah, Kingdom of Saudi Arabia. Page numbers?

Muhammad SM, Shazia R, Mahjabeen S, Shahid N and Muhammad A, 2009. The pollination by honeybee (*Apis mellifera* L.) increases field of canola (*Brassica napus* L.) Pak. Entomol. 31(2).

Mullin CA, Chen J, Fine JD, Frazier MT and Frazier JL, 2015. The formulation makes the honey bee poison. Pest. Biochem. Physiol. 120: 27–35.

Mullin CA, Frazier M, Frazier JL, Ashcraft S, Simonds R and Pettis JS, 2010. High levels of miticides and agrochemicals in North American apiaries: Implications for honey bee health. PLoS ONE. 5: 54-97.

Munyuli MBT, 2011. Pollinator biodiversity in Uganda and in Sub Sahara Africa: landscape and habitat management strategies for its conservation. Int. J. Biodivers. Conserv. 3: 551-609.

Neumann P and Carreck NL, 2010. Honey bee colony losses. J. Apic. Res. 49(1): 1–6.

Oerke EC, 2006. Crop losses to pests. J. Agric. Sci. 144:31–43

Ollerton J, Winfree R and Tarrant S, 2011. How many flowering plants are pollinated by animals? Oikos. 120: 321–326.

Osteen CD and Fernandez-Cornejo J, 2013. Economic and policy issues of US agricultural pesticide use trends. Pest. Manag. Sci. 69:1001–1025

Oz M, Karasu A, Cakmak I, Goksoy AT and Turan ZM, 2009. Effects of honeybee (*Apis mellifera*) pollination on seed set in hybrid sunflower (*Helianthus annuus* L.). African J. Biotechnol. 8(6): 1037-1043.

PARC. 2010. Honeybee, Annual Report of PARC. 2010-11. pp. 98.

Pătruică S, 2006. Tehnologia creşterii albinelor, Curs universitar, Edit. Eurobit. Timişoara.

Pilling ED and Jepson PC, 1993. Synergism between EBI fungicides and a pyrethroid insecticide in the honeybee (*Apis mellifera* L.). Pest. Sci. 39: 293-297.

Ping-Li D, Wang Q, Sun JH, Liu F, Wang X, Wu YY and Zhou T, 2010.Effects of sublethal concentrations of bifenthrin and deltamethrin on fecundity, growth and development of the honeybee *Apis mellifera* ligustica. Environ.Toxicol. Chem. 29: 644-649.

Popp J, Peto K and Nagy J, 2013. Pesticide productivity and food security: A review. Agron. Sustain. Dev. 33:243–255.

Schneider CW, Tautz J, Grünewald B and Fuchs S, 2012. RFID tracking of sublethal effects of two neonicotinoid insecticides on the foraging behavior of *Apis mellifera*. PLoS One. 7: 23-30.

Simeria G, Popescu G and Giuchici C, 2001. Combaterea biologică în sens strict" a patogenilor şi dăunătorilor plantelor, Ed. Mirton. Timişoara.

Southwick EE and Southwick JL, 1992. Estimating the economic value of honey bees (Hymenoptera: Apidae) as agricultural pollinators in the United States. J. Econ. Entomol. 85: 621–633.

Stephen E and Irshad M, 2012. Economics impact of pollinators in crop production of Pakistan. UNEP-GEFFAO Project. Islamabad, pp. 10.

Uma P and Tej P, 2002. Warning signals from the apple valleys of Hindu Kush- Himalayas: Productivity concerns and pollination problems. International Centre for Integrated Mountain Development, Katmandu, Nepal. Page numbers?

Uma P, Tej P, Harish KS, Pushkin P, Aungsathwi M, Nar BT, Tan K and Muhammad SM, 2012. Value of insect pollinators to Himalayan agricultural economies. International Centre for Integrated Mountain Development, Kathmandu, Nepal.

USAID and I-LED, 2006-2009. Pesticide Evaluation Report and Safer Use Action Plan. Pakistan.

Vandame R, Meled M, Colin ME and Belzunces LP, 1995. Alteration of the homing-flight in the honey bee *Apis mellifera* L. exposed to sublethal dose of deltamethrin. Environ. Toxicol. Chem. 14: 855-860.

Van Engelsdorp D and Meixner MD, 2010. A historical review of managed honey bee populations in Europe and the United States and the factors that may affect them. J. Invert. Pathol. 103: 80–95.

Villa S Vighi M Finizio A and Serini GB, 2000. Risk assessment for honeybees from pesticide-exposed pollen. Ecotoxicol. 9: 287–297.

Wakhal MD, Bhujbal M and Pais EVD, 1999. Analysis of Honey, pollen and royal jelly by high performance liquid chromatography-A review. Apiacta. 30(4): 6-11.

Whitehorn PR, Oconnor S, Wackers FL and Goulson D, 2012. Neonicotinoid pesticide reduces bumble bee colony growth and queen production. Science. 336: 351-352.

Williams IH, 1994. The dependences of crop production within the European Union on pollination by honey bees. Agric. Zool. Rev. 6: 229–257.

Williams IH, 1992. Apiculture for agriculture. pp. 14–18. In British Beekeepers' Association, Spring Convention, Stone Leigh, UK.

Winston ML and Scott CD, 1984. The value of bee pollination to Canadian agriculture. Canadian Beekeeping, 11: 134.

Wray MK, Mattila R and Seeley TD, 2011. Collective personalities in honeybee colonies are linked to colony fitness. Anim. Behav. 81(3): 559-568.

Zafar IK, Hafiz MT, Salma B, Kafeel A, Saira B, Rabia Y and Ijaz RN, 2016. Toxic effect of Malathion on insect pollinators visiting marigold flowers. Biologia. 62 (1): 1-7.

# 8

# Efficacy of Biomax and Fiprokill against *Heterotermes indicola* (Wasmann)

**Ayesha Aihetasham[1]*, Muhammad Saeed Akhtar[1], Tayyaba Bibi[1], Imran Bodlah[3]**
[1]Department of Zoology, University of the Punjab, Lahore-54590, Pakistan
[2]Department of Entomology, PMAS-Arid Agriculture University, Rawalpindi, Pakistan

*Corresponding author email: misswaqar@yahoo.com

## Abstract

Termites cause profound economic losses worldwide. *Heterotermes indicola* causes tremendous damages to wood work in the buildings and forest trees. Different types of insecticides have been used to control this subterranean termite. In the present study, efficacy of Biomax (Chlorpyrifos) and Fiprokill (Fipronil) against *Heterotermes indicola* (Wasmann) was studied. Toxicities and repellencies of these chemicals were compared at same concentrations. Mortality during 8 hours exposure for Biomax at different concentrations 480, 240, 120, 60, 30, 15, 7.5 and 3.75 ppm was 93.3%, 80%, 73.3%, 63.3%, 56.6%, 50%, 45.6% and 43.3%, respectively. Fiprokill at the same concentration (480, 240, 120, 60, 30, 15, 7.5 and 3.75 ppm) caused 66.6%, 56.6%, 50%, 43.3%, 33.3%, 30%, 26.6% and 20% mortality, respectively. Biomax was found to be more toxic as compared to Fiprokill. $LC_{50}$ values for *H. indicola* exposed to soil treated with Biomax and Fiprokill were 5.735 and 221.676 ppm, whereas $LC_{90}$ values for *H. indicola* exposed to soil treated with Biomax and Fiprokill were 359.290 and 813.722 ppm, respectively. Biomax was repellent at 480, 240, 120 and 60 ppm while Fiprokill was found non-repellent at all the concentrations used.

**Keywords:** Toxicity, *H. indicola*, Repellency, $LC_{50}$, $LC_{90}$

## Introduction

Termites are the most important eusocial insect pests of human structure, causing significant damages by feeding on all types of cellulose containing materials (Hickey, 2006; Ahmed and French, 2008; Khan et al., 2015). They cause significant damage to plants, agricultural crops like rice, barley, sugarcane, millet and household materials (Sattar et al., 2014; Aihetasham et al., 2015). Termites have also been frequently considered as the decomposers of different lignocellulosic materials (Lenz, M. et al., 2011). Lingocellulose waste degrading capability of termites attributes them a central place in carbon cycle (Brune, 2014; Manzoor et al., 2015).

Different methods such as baiting system, wood treatment, physical barriers and biological control agents have been used for termite control, but the most effective and extensively used method is the chemical control. Chemicals including imidacloprid, silafluofen, chloropyrifos, cypermethrin, spinosad, bifenthrin and fipronil were found effective in controlling different species of termites (Manzoor et al., 2014). The chemicals are differed in their functions, Biomax (Chlorpyrifos) belongs to organophosphorus insecticides, acaracide and miticide. It acts on the nervous system of insects inhibiting acetylcholinesterase (nplc.orst.edu/factsheets/chlorpgen.html). On contrast, Fiprokill (Fipnonil), a slow acting toxicant, belongs to chemical family of phenypyrazole. It is a widely used insecticide, and a strong disrupter of insect central nervous. It inhibits GABA- gated and glutamate-gated chloride channels chloride (GluCI), causing toxicity of central nervous system (npic.orst.edu/factsheets/fipronil.html).

Now a days, non-repelllent insecticides are preferred over repellent insecticides as these chemicals mainly depend on foraging of termites in treated areas to attain maximum lethal contact, in addition, they cause delayed mortality and trophallaxis (Su and Scheffrahn, 2000; Thorne and Breisch, 2001; Misbah-ul-Haq et al., 2016).Fifty species of termites have been described from Pakistan. Among these, 11 species have been observed destroying wooden structure of buildings (Akhtar, 1983; Aihetasham et al., 2015). Amongst subterranean termites, *Heterotermes indicola* (Wasmann) is a serious pest known to urban and rural areas (Aihetasham et al., 2015). It has ability to damage standing trees by hollowing them out from the inside, without producing any external signs of injury (Balachander et al., 2013; Misbah-ul-Haq et al., 2016).As maximum damage to wood in Pakistan is caused by subterranean termites therefore it is necessary to investigate different methods for their control. Present study involve efficacy of two insecticides, Biomax (Chlorpyrifos) and Fiprokill (Fipronil) against *Heterotermes indicola*.

## Materials and Methods

### Collection of termites

Termites (*Heterotermes indicola)* were collected by hand picking using fine forceps and camel hair brush from infested old trees of *Populus euramericana* from botanical garden, University of the Punjab, New Campus, Lahore. Termites were kept in Laboratory at 26±2°C and 80% relative humidity. Active and healthy, termites workers were selected for experiment. They were placed in Petri-plates (70mm x 10mm) having moist filter paper (soaked in distilled water) for further experiments.

### Soil for Bioassay

The soil was collected from garden of the Department of Zoology, University of the Punjab, New Campus, Lahore, Pakistan. The collected soil samples were with sandy and loam texture and pesticide contamination free as there had been no known chemical application. After sieving soil with the help of thin net, oven dried for 24 hours at 70°C.

### Insecticides

Two insecticides Biomax (chlorpyriphos) 48% and Fiprokill (Slow acting) 25 TC were selected for the experiments because of their repute as an effective pesticide against different crops pest. To compare their effectiveness on *Heterotermes indicola*, a particular amount of Biomax was dissolved in distilled water. Stock solution of Biomax (480 ppm) was prepared with eight different concentrations i.e., 480, 240, 120, 60, 30, 15, 7.5 and 3.75 ppm.Similarly, a particular amount of Fiprokill was also dissolved in distilled water. Stock solution of fiprokill (480 ppm) was prepared with eight different concentrations i.e., 480, 240, 120, 60, 30, 15, 7.5 and 3.75 ppm. For control, soil was treated only with distilled water.

### Toxicity test

Procedure used by Smith (1979) was adopted for soil toxicity. Bottom of each petri plate was evenly covered with 10 grams of weighted oven dried soil. The procedure involved the application of Eight different dilutions 480, 240, 120, 60, 30, 15, 7.5 and 3.75ppm (6ml each) of Biomax were applied to soil samples. Similarly, these eight different dilutions of Fiprokill were used for treatment of soil. These Petri-plates were sterilized at 70°C for 24 hours prior to the experiment. In each petri plate ten termites were released. Three replicates of each concentration of insecticides were used with control. Mortality data of termites was observed after every half hour up to eight hours.

### Repellency test

Similarly, for repellency test, procedure of Smith (1979) was adopted using the same. In this regard, Twenty-seven Petri-plates were washed and then oven dried at 70°C for 24 hours. Ten grams of oven dried soil was weighted in Petri-plates and with the help of glass slide the soil was evenly divided in two halves so that one half of the bottom of each petri plate was covered with 5 gm of treated soil and the other half with 5 gm of untreated soil. Ten termites were released in the centre of each petri plate to record their orientation on untreated soil. Three replicates of each concentration were used. For minimizing effect of light, plates were completely covered with black cloth, providing natural environment and maintaining room temperature at 25-26°C and 80% RH. After 15 minutes intervals, the numbers of termites were recorded on untreated soil to take 5 observations. A treatment concentration at which 21 or more out of 30 termites i.e. the aggregate of three replicates of termites found on untreated soil was considered as repellent at all five observation time.

**Calculation of $LC_{50}$ and $LC_{90}$ values and Statistical analysis**

Data collected for $LC_{50}$ and $LC_{90}$ of Biomax and Fiprokill for *H. indicola* after 8 hours exposure was statistically analyzed by Probit analysis using MINITAB version 17.0 software. Probability plots for Response of *H. indicola* to each chemical were plotted and determined.

## Results and Discussion

For *H. indicola*, the percentage mortality in soil treated with Biomax, after 8 hours was 93.3%, 80%, 73.3%, 63.3%, 56.6%, 50%, 46.6% and 43.3% in 480, 240, 120, 60, 30, 15, 7.5 and 3.75 ppm respectively, whereas, at the same concentrations, the percentage mortality in soil treated with Fiprokill, after 8 hours was 66.6%, 56.6%, 50%, 43.3%, 33.3%, 30%, 26.6% and 20% respectively (Figure. 1)

$LC_{50}$ of Biomax and Fiprokill for *H. indicola* was 5.735 ppm and 221.676 ppm, respectively (Table. 1). It was noticed that at higher $LC_{50}$ insecticide required greater time to kill a population. SimilarlyLC$_{90}$ of Biomax and Fiprokill for *H. indicola* was 359.290 and 813.722 ppm, respectively (Table. 1). Probability plots for response of *H. indicola* to Biomax and Fiprokill at eight different concentrations (Fig.2-3).

**Table 1. $LC_{50}$ and $LC_{90}$ values ofInsecticides (Biomax and Fiprokill) against *H. indicola***

| Serial No. | Insecticides | $LC_{50}$(ppm) | $LC_{90}$(ppm) |
|---|---|---|---|
| 1 | Biomax | 5.735 | 359.290 |
| 2 | Fiprokill | 221.676 | 813.722 |

Soil toxicity test showed that mortality increased with increase in concentration of Insecticide. The comparison of toxicity of Biomax and Fiprokill showed that Biomax was found to be more toxic as compared to Fiprokill.

Our results are in confirmation with Ahmed and Farhan (2006) who studied the effectiveness of chlorpyrifos, imidacloprid, bifenthrin and thiamethoxam against *Microtermesobesi* and found chlorpyrifos as the most effective in controlling termites. Organophosphates such as chlorpyrifos, degrades quickly, whereas chloronicotinyls and pyrethroid degrade slowly (Baker and Bellamy, 2006). Rao *et al.* (2005) studied the effect of chlorpyrifos and monocrotophos on the locomotorbehaviour (velocity) and acetylcholinesterase activity of *Odontotermes obesus* at intervals of 4, 8, 12, 16, 20 and 24 h and found $LC_{50}$ values for chlorpyrifos and monocrotophos as 0.046 and 0.148 μg $cm^{-2}$, respectively. After 24 h termites showed 97 and 88% reduction in locomotorbehaviour.

Results showed that chlorpyrifos was 3.22-fold more toxic than monocrotophos. Sheikh (2015) also evaluated the efficacy of Tenekil (poly chlorinated petroleum hydrocarbon), Terminus (chlorpyrifos) and Termidor (fipronil) on *M. obesi*. Their $LC_{50}$ values were 114.55, 3.15, 7.816 and ppm, respectively in 8h exposure.The results showed that terminus caused maximum mortality (93.3%) at higher doses i.e.,500 ppm after 8 hour exposure.

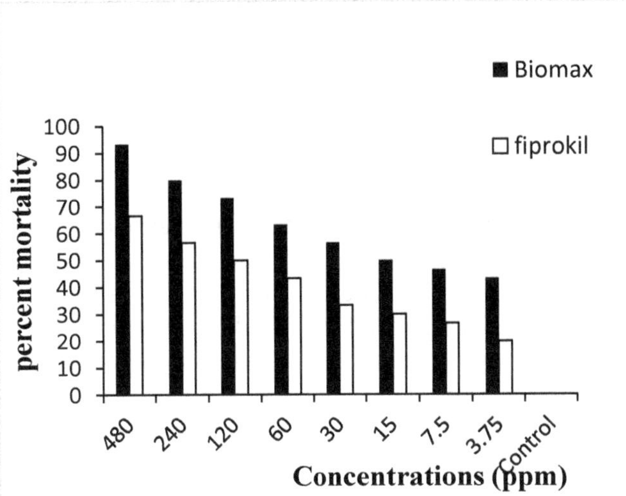


**Fig. 1. A comparison of Mortality (%) of *H. indicola* exposed to eight different concentrations ofBiomax and Fiprokill**

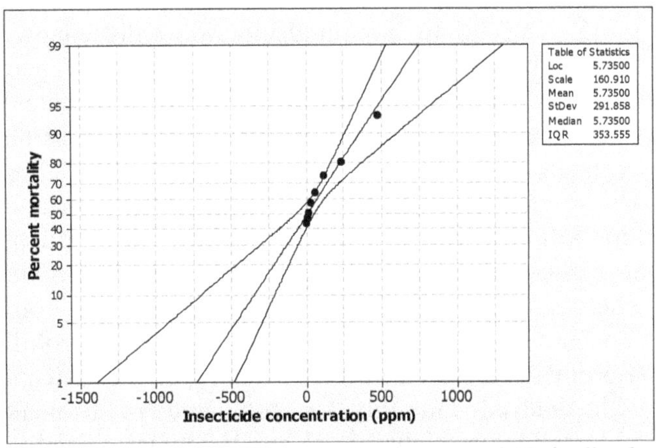


**Fig. 2: Probability Plot for Response of *H. indicola* to Biomax at different concentrations**

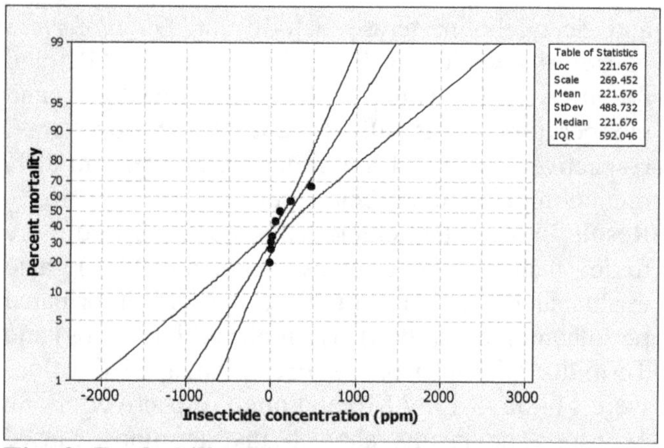


**Fig. 3: Probability Plot for Response of *H. indicola* to Fiprokill at different concentrations**

Our findings are also in accordance with Ahmed *et al.* (2006) who reported toxicity of chlorpyrifos, imidacloprid and monomehypo against termites on sugarcane. After 15, 45, 60, and 75 days of application on setts chlorpyrifos gave significant control of termites population and reduced the number of termites as compared to imidacloprid and monomehypo. Iqbal *et al.* (2013) also evaluated toxicity of six new chemical insecticides (Chlorfenapyr, spinosad, thiamethoxam, fipronil and imidacloprid) against *Microtermes mycophagus* D. $LC_{50}$ values for *M. mycophagus* D. exposed to soil treated with these insecticides were alculated and found in ranges of 3.72-5.88 ppm and 2.80-3.45 ppm, for chlorfenapyr and fipronil, respectively. Manzoor *et al.* (2012) also explained toxicity and repellency of four insecticides cadusafos, chlorfenapyr, imidacloprid, bifenthrin and fipronil at different concentrations against *Heterotermes indicola*.

Their $LC_{50}$ values for *H. indicola* were 346.75, 75.86, 14.45, 1.05, 0.46 for imidacloprid, chlorfenapyr, fipronil, bifenthrin and cadusafos, respectively. Their results confirmed that cadusafos (organophosphates) was more toxic with $LC_{50}$ significantly lower in 8h treatment.

The results regarding the repellency of Biomax and Fiprokill also showed that Biomax became repellent at higher doses i.e, 480, 240, 120, and 60 ppm, whereas, Fiprokill was non-repellent at all concentrations (Figure 4). This proved that Fiprokill (fipronil) is a slow acting and non-repellent toxicant with delayed moratlity as compared to Biomax (chlorpyrifos) which caused quick contact mortality and its repellency effects hence it can be used for long term control programme. Similar results were obtained by Hu (2005) who studied the effect of fipronil and indoxacarb against two subterranean termites i.e., *C. formosanus* (Shiraki) and *R. flavipes* (Kollar) as both were toxic, non-repellent and had delayed mode of activity.

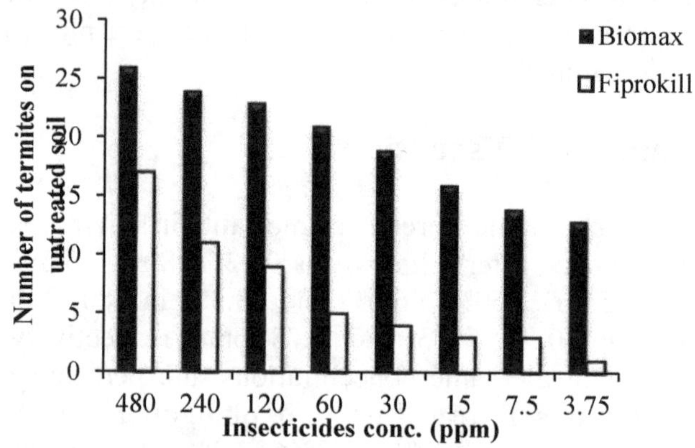


**Fig. 4: A comparison of Repellent effect of Biomax and Fiprokill to *H. indicola* at different concentrations**

## Conclusion

Both insecticides were toxic to *H. indicola* however Biomax with $LC_{50}$ 5.735 ppm and $LC_{90}$ 221.676 was found to be more toxic as compared to Fiprokill with $LC_{50}$ 359.290ppm and $LC_{90}$ 813.722 against *H. indicola*. Fiprokill was non repellent at all concentrations as compared to Biomax. Therefore Fiprokill being slow acting toxicant and non-repellent termiticide is the best solution for effective and long term termite control

## References

Ahmad S, Mustafa T, Riaz MA and Abid H, 2006. Efficacy of insecticides against subterranean termites in sugarcane. Int. J. Agric. Biol. 8(4): 508-510.

Ahmed BM and French JRJ, 2008. An overview of termites control methods in Australia and there links to aspects of termite biology and ecology. Pak. Entomol. 30(2): 101-118.

Ahmed and Farhan M, 2006. Laboratory evaluation of chlorpyrifos, bifentthrin, imidacloprid, thiomethoxam and flufenoxuron against *Microtermes obesi* (Isoptera: Termitidae). Pak. Entomol. 28(2): 45-50.

Aihetasham A, Umer M, Akhtar MS, Din MI and Rasib KZ, 2015. Bioactivity of medicinal plants *Menthaarvensis* and *Peganum harmala* extracts against *Heterotermes indicola* (Wasmann) (Isoptera). Int. J. Biosci. 7(5): 116-126.

Akhtar MS, 1983. Wood destroying termites (Isoptera: Rhinotermitidae) of Pakistan, Mater. Organismen. 18(4): 277-291.

Baker PB and Bellamy DE, 2006. Field and Laboratory Evaluation of Persistence and Bioavailability of Soil Termiticides to Desert subterranean Termite *Heterotermes aureus* (Isoptera: Rhinotermitidae). J. Econ. Entomol. 99(4): 1345-1353.

Balachander M, Remadevi O and Sasidharan T, 2013. Dissemination of *Metarhizium anisopliae* infection among the population of *Odontotermes obesus*(Isoptera: Termitidae) by augmenting the fungal conidia with attractants. J. Asia Pac. Entomol. 16: 199-208.

Brune A, 2014. Symbiotic digestion of lignocellulose in termite guts. Nat. Rev. Microbiol.12: 168-180.

Hickey CD, 2006. Effects of Disodium OctaborateTetrahydrate in Ethylene Glycol on Consumption and Mortality of the Eastern Subterranean Termite. M.Sc. Thesis Florida: http://dx.doi.org/10.14411/eje.2008.025.

Iqbal N and Saeed S, 2013. Toxicity of six new chemical insecticides against the termite, *Microtermes mycophagus* D. (Isoptera: Termitidae: Macrotermitinae). Pakistan J. Zool. 45(3): 709-713.

Khan IA, Zaman M, Akbar R, Ali I, Alam M, Saeed M and Farid A, 2015. Efficacy of single and mixed particles sand size as physical barrier against *Heterotermes indicola* under laboratory conditions. J. Entomol. Zool. Stud. 3(4): 106-109.

Lenz M, Lee CY, Lacey MJ, Yoshimura T and Tsunoda K, 2011. The potential and limits of termites (Isoptera) as decomposers of waste paper products. J. Encon. Entomol. 104: 232-242.

Manzoor F, Abbas M and Latif MU, 2015. Comparative Study of Resistance and Feeding Preference of 24 Wood Species to Attack by *Heterotermes indicola* (Wasmann) and *Coptotermes heimi* (Isoptera: Rhinotermitidae,Termitidae) in Pakistan. Sociobiol. 62(3): 417-425.

Manzoor F, Saleem S and Abbas M, 2014. Laboratory Evaluation of Imidacloprid against *Microtermes obesi* (Holmgren) (Isoptera: Macrotermitinae). *Proceedings of the Pakistan Academy of Sciences*. 51 (1): 43–48.

Manzoor F, Sayyed AH, Rafique T and Malik, SA, 2012. Toxicity and repellency of different insecticides against *Heterotermes indicola* (Isoptera: Rhinotermitidae). J. Anim. Plant Sci. 22(1): 65-71.

Misbah-Ul-Haq M, Khan IA, Farid A, Ullah M, Gouge DH and Baker PB, 2016. Efficacy of indoxacarb and chlorfenapyr against Subterranean termite *Heterotermes indicola* (Wasmann) (Isoptera: Rhinotermitidae) in the laboratory. Turk. Entomol. 40 (3): 227-241.

Rao JV, Parvathi K, Kavitha P, Jakka NM and Pallela R 2005. Effect of chlorpyrifos and monocrotophos on locomotor behavior and acetylcholinesterase activity of subterranean termites, *Odontotermes obesus*. Pest Manage.Sci. 61(4): 417-421.

Sattar A, Naeem M and Ehsan-Ul-Haq, 2014. Efficacy of Plant Extracts Against Subterranean Termites i.e. *Microtermes obesi* and *Odontotermes lokanandi* (Blattodea:Termitidae). J. Biodivers. Biopros. Dev. 1: 122.

Smith VK, 1979. Improved techniques designed for screening candidate termiticides on soil in the laboratory. J. Econ. Entomol. 72(6): 877-879.

Su NY and Scheffrahn RH, 2000. Termites as pest of buildings, In: Abe, T., Bignell, D.E. & Higashi, M., (eds.), Termites: evolution, sociality, symbiosis, ecology. Kluwer Academic Publishers, Dordrecht, Boston, London. pp. 437-453.

Thorne BL and Breisch NL, 2001. Effects of sub lethal exposure to imidacloprid on subsequent behavior of subterranean termite *Reticulitermes virginicus* (Isoptera: Rhinotermitidae). J. Econ. Entomol. 94: 492-498.

9

# Improving the growth of *Centella asiatica* using surfactant modified natural zeolite loaded with NPK nutrients

**Noor Asyraf Noor Azman, Nik Ahmad Nizam Nik Malek*, Norfarah Syuhaidah Mohd Noor, Muhammad Arshad Javed**

Department of Biotechnology and Medical Engineering, Faculty of Biosciences and Medical Engineering, Universiti Teknologi Malaysia, 81310 UTM, Skudai, Johor, Malaysia

## Abstract

There is great interest in using sustainable fertilizer to enhance medicinal plant quality such as *Centella asiatica*, which is known for its various medicinal properties. Here, we examined the performance of NPK-Organo-Zeolite (NPKOZ) as a controlled release fertilizer for the growth of *C. asiatica*. Natural zeolite that has been modified with surfactant hexadecyltrimethyl ammonium (HDTMA) and addition with ammonium (N), phosphate (P) and potassium (K) was characterized with X-ray diffraction and Fourier transform infrared spectroscopy. In a leaching study, it was found the NPKOZ showed fewer losses of nutrients. For the plant growth study, a comparison was made with eight different treatments including chemical fertilizers monoammonium phosphate and monopotassium phosphate for 70 days. Morphological (number of flowers, branches and leaves, specific leaf area and plant biomass) and biochemical growth parameters (N, P, K and chlorophyll contents) of the plant were analysed. Second cultivation was also conducted to compare potted, hydroponic and verticulture systems with or without NPKOZ. The NPKOZs treatments resulted in the highest number of leaves, branches and flowers among all treatments. In addition, the chlorophyll contents were also the highest with NPKOZs application. The NPK contents in the *C. asiatica* plants were comparable with that of chemical fertilizers tested. This study showed that NPKOZ fertilizer with verticulture system is a suitable sustainable controlled release fertilizer for the improvement of the growth of *C. asiatica*.

*Corresponding author email: niknizam@fbb.utm.my

**Keywords:** *Centella asiatica,* Surfactant modified zeolite, Controlled release fertilizer

## Introduction

Vegetables contribute an important part in malnutrition problems and resulted in a higher rate of mortality such as accident happened in Mali and Nigeria (Keatinge et al., 2011). Plant herbs especially the leafy or parts of soft flowering offered valuable fibre and medicinal properties. One of many potential vegetable that needs to be improved is *Centella asiatica (Pennywort)* that belongs to the family of Apiaceae. *C. asiatica* which flourishes in damp places could be eaten as a raw vegetable in Malaysia and Indonesia and mostly found in tropical and subtropical countries such as India, Southeast Asia and South Africa (Huda-Faujan et al., 2009; Gohil et al., 2010). Different types of phytochemical compounds such as triterpene acid, flavonoids and alkaloids are useful in medicinal problems. Pennywort represents one of the herbs that can treat numerous ethnomedical and pharmacological problems, for example as a wound dressing, antioxidant properties as well as reducing Parkinson's disease symptoms (Jin et al., 2015; Nataraj et al., 2017). Thus, studying the improvement of *C. asiatica* yield is very important because of its high valuable properties and application.

The usage of excess fertilizers in agriculture could harm the environment such as leaching of nutrients from the fertilizers (Lubkowski, 2016). This nutrients leaching contributes significantly to the eutrophication and groundwater contamination (Penuelas et al., 2009). For example in China and Taiwan, the problem with nitrogen leaching from agricultural activities that contaminates groundwater becoming a national major concern (Xu et al., 2010, Adhikari and Chen, 2013). Besides, Penuelas *et al.* (2009) documented the effect of algal bloom caused by the excess of phosphorus in fertilizer in Catalonia, Spain. Furthermore, Ah Tung *et al.* (2009) also reported that N and K fertilizers implementation in oil palm showed a concern for groundwater quality in Malaysia. There are various methods in combating nutrients leaching and one of them is the application of a slow release fertilizer (SRF) that can reduce nutrients losses, lowering fertilization expenditure and increased crop yields (Sempeho, 2014).

In this part, a 3-dimensional aluminosilicates mineral material called zeolite is a candidate of choice as a controlled release fertilizer for cultivation because it has high adsorption efficiency (Polat et al., 2004). Raw zeolite is inadequate to help plant growth since it lacks in major plant nutrients. Macronutrient like N is needed to improve the stimulate the green growth production of the plant, P as an important phosphate source for nucleic acid and K provides a vital role in photosynthesis activity (Prajapati and Modi, 2012). The zeolite can only adsorb and retain cationic elements such as ammonium and potassium because of the negative charge of the zeolite framework. Fortunately, zeolite efficiency can be improved by modifying its external surface using cationic surfactant molecules (Reháková et al., 2004; Bansiwal et al., 2006; Bhardwaj et al., 2012). The surfactant modified zeolite can hold P element and the combination of zeolite itself could hold N and K nutrients. One of the major characteristics of this fertilizer is the ability of slowly released the nutrients. Nevertheless, the related particular reports only mentioned single N-loaded or P-loaded surfactant modified zeolite (SMZ) as a fertilizer (Li, 2003; Bansiwal et al., 2006; Bhardwaj et al., 2012). The introduction of N, P and K onto zeolite is a better way of increasing nutrients retention capacity in medicinal plant production especially for the *C. asiatica* growth. Furthermore, there are diverse methods on how to grow a plant, including the outdoor and indoor techniques. Basically, improvements from year to year are made to culture the plant in a way of lessening the fertilizer, water and spaces used. Common ways of planting a plant include potted technique, which have different container types that can be made from ceramic, plastic or fibre waste. In addition, a liquid based culture known as a hydroponic method excels in regulating humidity, nutrients percentage and quality of water (Rajpurkar, 2016). Vertical farming or verticulture is another form of cultivating the plant but in a vertical condition (Sitawati et al., 2016). These three cultivation methods also showed a possibility of decreasing nutrient leaching when incorporated with suitable fertilizer. The basic concept of this study was to compare the system used to grow *C. asiatica* with NPK-Organo-Zeolite (NPKOZ) and conventional fertilizers with the development of new technique in growing the herb by using NPKOZ. Thus, it is hoped to bring a valuable solution to increase the desired yield of *C. asiatica* and reducing environmental concern for leaching problems by applying slow release fertilizer (SRF) (Sempeho, 2014).

## Materials and Methods

### Preparation and Characterization of NPK-Organo-Zeolite

Zeolite from Indonesia was obtained from Provet Group of Companies Sdn Bhd, Serdang, Selangor. Monoammonium phosphate (MAP) and monopotassium phosphate (MKP) fertilizers were supplied by Greentrade Sdn Bhd while Ferti 45 (F45) fertilizer was obtained from Fertiland Trading Co.. The surfactant hexadecyltrimethyl ammonium bromide (HDTMA-Br) and potassium chloride (KCl) were supplied by Qrec (Asia) Sdn Bhd. The preparation process of NPK-Organo-Zeolite was referred to Malek et al. (2014) which was begin with the addition of 40 g zeolite (zeo) with KCl and stirred overnight by using a magnetic stirrer. Then, the mixture was filtered with 185 mm Macherey-Nagel filter paper and dried in an oven at 80°C overnight. Next, the solid mesh was ground with mortar and pestle into powder form and sieved. The steps were repeated three times to ensure most K ions are located inside the zeolite framework forming K-zeo. Next, 10 g of K-cli were mixed with 4 mM of HDTMA and Organo-K-zeo (OKZ) was obtained after the grinding process. Lastly, 40 g of OKZ was added into MAP to form the final product which is NPKOZ. Each product formed was characterized by X-ray diffraction (XRD) and Fourier transform infrared spectroscopy (FTIR) with ATR (Attenuated Total Reflectance) sampling technique (Malek, 2011; Malek et al., 2014).

**Elemental Analysis of Fertilizers**

All fertilizers were analyzed by inductively coupled plasma optical emission spectrometry (ICP-OES) Perkin Elmer Optima at Ayer Hitam Agriculture Institute (IPAH), Johor for K and the content of trace elements (%) is shown in Table - 1. The determination of N content (%) in the materials was conducted by using Kjeldahl method and P content (%) with atomic absorption spectroscopy (AAS), AAnalyst 300, Perkin Elmer at Institute of Veterinary Malaysia, Johor.

**Column Leaching Study**

The release patterns of nutrients from 8 different treatments in 15 days were determined by column leaching study (Li, 2003; Malek et al., 2015). The treatments were applied by top dressing on top of the sand and it was done in triplicates and the experimental setup includes retort stand, transparent plastic cup, 185 mm filter paper and 200 ml collection plastic cup. Each cup (top) was filled in with approximately 200 g of sand and fixed amount of fertilizers (Table - 2). Then, 100 ml of distilled water was poured into each cup for every 24 hours and leachate was then collected and stored in 50 ml plastic bottle. NANOCOLOR® Test 1-05 (Ref 918 05) and NANOCOLOR® Test 1-77 (Ref 918 77), Macherey-Nagel were used to measure ammonium ($NH_4^+$) and phosphate in the leachate solution, respectively. Whereas, K nutrient ($K^+$) in the leachates was analysed using ICP-OES.

**Table -1: Elemental analysis of the materials used in the study**

| No. | Materials | N (%) | P (%) | K (%) | Zn (%) | Mn (%) | Fe (%) |
|---|---|---|---|---|---|---|---|
| 1 | Zeo | 0.00 ± 0.00 | 0.02 ± 0.00 | 0.20 ± 0.00 | 27.33 ± 0.88 | 128.67 ± 3.71 | 0.33 ± 0.03 |
| 2 | MAP | 14.38 ± 0.01 | 1.11 ± 0.01 | 0.00 ± 0.00 | 0.00 ± 0.00 | 0.00 ± 0.00 | 0.00 ± 0.00 |
| 3 | MKP | 0.00 ± 0.00 | 1.11 ± 0.03 | 31.37 ± 0.85 | 0.00 ± 0.00 | 0.00 ± 0.00 | 0.00 ± 0.00 |
| 4 | F45 | 18.29 ± 0.38 | 0.37 ± 0.01 | 14.57 ± 0.20 | 15.00 ± 0.58 | 102.67 ± 0.88 | 0.10 ± 0.00 |
| 5 | NPKOZ | 0.69 ± 0.00 | 0.14 ± 0.00 | 2.83 ± 0.03 | 15.67 ± 1.76 | 110.00 ± 4.73 | 0.53 ± 0.07 |

**Table - 2: List of treatments for column leaching study**

| No. | Treatments | Description |
|---|---|---|
| 1 | Control | Sand (200 g) (No treatment) |
| 2 | Zeo | Sand (200 g) + Zeolite (5 g) |
| 3 | MAP | Sand (200 g) + Monoammonium phosphate (5 g) |
| 4 | MKP | Sand (200 g) + Monopotassium phosphate (5 g) |
| 5 | F45 | Sand (200 g) + F45 (5 g) |
| 6 | NPKOZ5 | Sand (200 g ) + NPK-Organo-Zeolite (5 g) |
| 7 | NPKOZ10 | Sand (200 g) + NPK-Organo-Zeolite (10 g) |
| 8 | NPKOZ15 | Sand (200 g) + NPK-Organo-Zeolite (15 g) |

**Cultivation Process of *Centella asiatica* for Fertilizers Comparison**

Firstly, the soil pH was maintained at pH 7 and sandy loamy soil was used throughout the experiment (Devkota and Jha, 2009). *C. asiatica* was obtained from Institute of Veterinary Malaysia, Johor. The *C. asiatica* was cultivated and grouped for eight different treatments as in Table - 2 in a semi-greenhouse for 35 days. At least 3 leaves per plantlet were planted into a mixture of sand and soil (40%:60%) that was evenly mixed (Devkota and Jha, 2009). The harvest period was done after 70 days of cultivation (Zainal, 2005). Morphological growth parameters like number of leaves, branches, flowers, leaf area (LA) ($cm^2$) and biggest specific leaf area (BSLA) ($cm^2/g$) were measured. Meanwhile, the total fresh biomass (TFB) and total dried biomass (TDB) of *C. asiatica* were weighed and then dried in an oven at 60°C for 48 hours (Devkota and Jha, 2009). Determination of the biochemical growth parameters was conducted after harvesting the plants including the estimation of total chlorophyll content in leaf using acetone which were determined their absorbance values at 662, 646, and 470 nm (Pompelli et al., 2013). Lastly, the leaves samples were analysed for N, P and K contents in *C. asiatica*.

**Cultivation Process of *Centella asiatica* for Systems Comparison**

For systems comparison, three different cultivation techniques with or without NPKOZ was compared with similar propagation, soil pH and location study. The morphological and biochemical growth parameters used in the second comparison were also similar with previous procedure. However, the system used in this study is the potted technique with fixed rectangular shape made from plastic material (48.5 cm × 19.5 cm × 14.5 cm). The second system was based on hydroponic method by the Deep Water Culture (DWC) system with little modification (57.0 cm × 43.0 cm × 13.5 cm) (Okemwa, 2015). The last method which is verticulture system required the usage of vertical gardening concept by using polyvinyl chloride (PVC) with the height of 72.5 cm and 8.6 cm in diameter. A total of eight holes were made and the PVC was then submerged in small pot filled with 2 kg of soil for stability.

**Statistical Analysis**

The data from each parameter were analysed by computing software, Statistical Package for the Social Sciences (SPSS) software version 16. A one-way analysis of variance (ANOVA) test involving one species (*C. asiatica*), morphological growth parameters and eight treatments was interpreted. The post hoc Duncan Multiple Range Test (DMRT) was conducted to compare the means and find the existence of variation. Lastly, Pearson Correlation test was carried out to identify the correlation between the parameters with the significance level of ($p \leq 0.05$).

## Results and Discussion

**Characterization of NPK-Organo-Zeolite**

XRD patterns of modified zeolite (K-zeo, OKZ and NPKOZ) were compared with raw zeolite shown in Figure 1(i). Most of the detected materials consist of natural zeolite clinoptilolite (C) and some quartz and mordenite as its impurities (Treacy and Higgins, 2007). Characterization of the prepared materials by FTIR spectroscopy was performed to detect the presence of HDTMA molecules in the sample (Figure 1(ii)). There are important bands were observed (in circle) in the FTIR spectra of the surfactant modified zeolites (Organo-K-zeo (2921 and 2850 $cm^{-1}$) and NPKOZ (2923 and 2851 $cm^{-1}$) samples). Both bands were in line with the finding from previous study (Malek, 2011) where these bands proved the presence of $CH_3$ asymmetric and $CH_2$ stretching of the HDTMA molecules, respectively. The FTIR peaks below 1000 $cm^{-1}$ represent Si-O-Al and Si-O-Si bonding properties of the zeolite framework with a weak peak at 793 $cm^{-1}$ for the symmetric bonding. In addition, the strong intensity of the wide peak at 1008 $cm^{-1}$ is referred to asymmetric bonding which similarly observed at bands around 1000 $cm^{-1}$ and 500 $cm^{-1}$ (Sharuddin *et al.* 2014). The presence of clinoptilolite can also be seen in many specific bands ranging from 3400 and 3700 $cm^{-1}$ bands (Mansouri et al., 2013).

The negative charge of the zeolite surfaces made it possible to attract other cations especially ammonium and potassium cations and also cationic surfactant HDTMA-Br. The attachment of HDTMA-Br which has polar head surrounded with a charge on the zeolite surfaces creates surfactant modified zeolite (SMZ) and eventually, the SMZ has the ability in adsorbing anionic compounds such as phosphate or nitrate (Haggerty and Bowman, 1994). From the XRD, the crystalline structure of the modified zeolite unchanged and stable after undergoes several reaction processes which are heating and nutrients loaded processes (Malek *et al.* 2015). Thus, the stable ionic nature of the zeolite framework helped attracting the cationic charge for instance, ammonium and potassium cations as well as amphiphilic compounds of HDTMA-Br surfactant in attaching phosphate ions towards the formation of NPKOZ. It is possible that the zeolite will remain intact as a medium in the soil (Reháková et al., 2004; De Smedt et al., 2015).

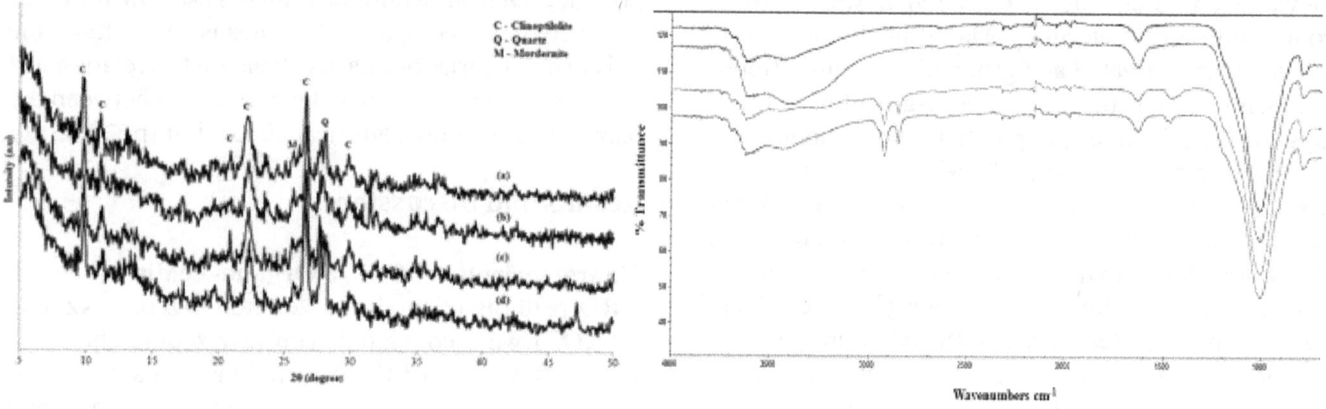


**Figure 1**

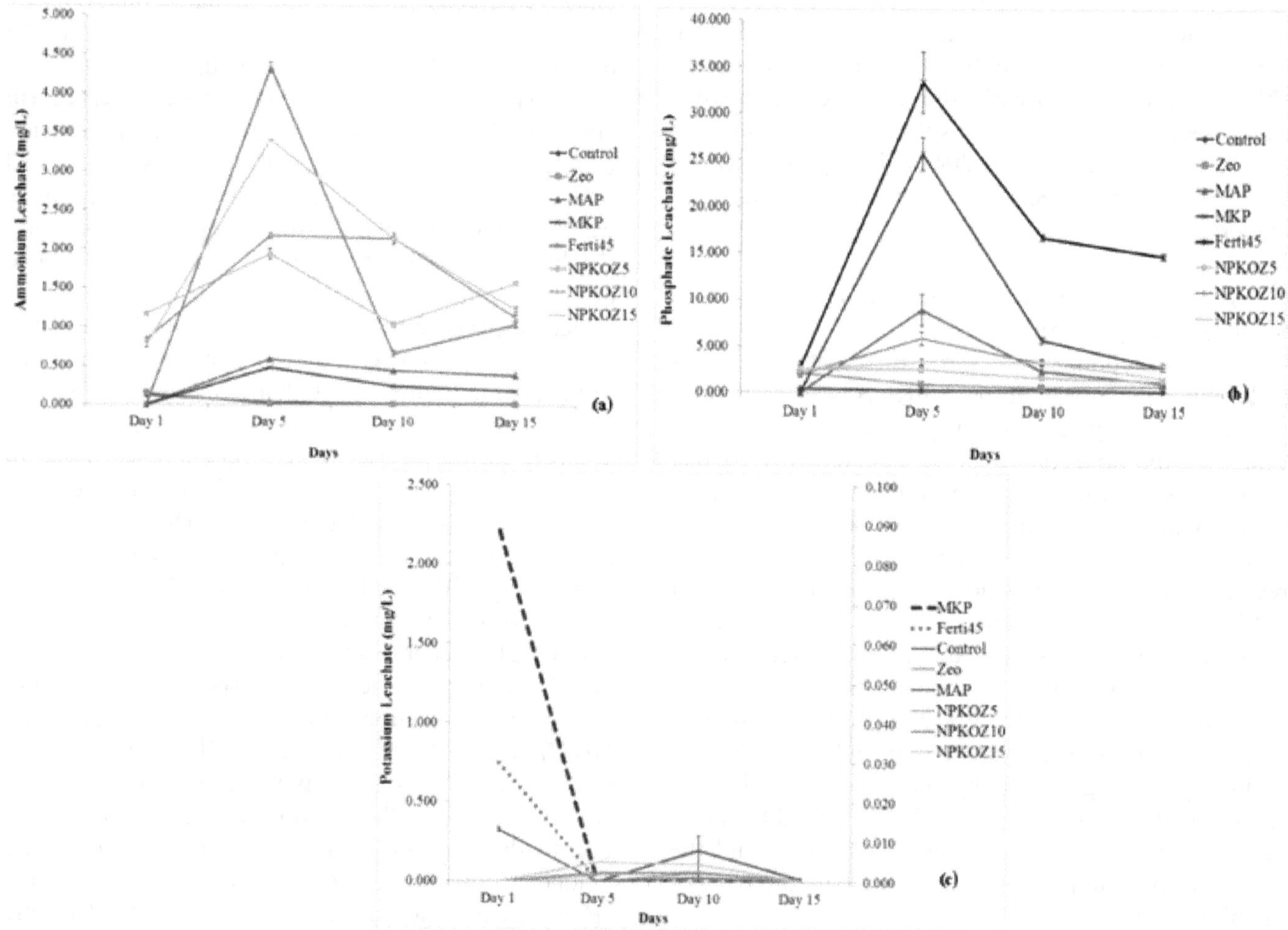


**Figure 2**

### Column Leaching Study

Based on Figure 2, all fertilizer samples released nutrients at maximum level at day 5 and gradually decreased until day 15. A high phosphate ion was found to release from Ferti 45 fertilizer on day 5 as compared to other samples (Figure 2b). The phosphate released percentage by Ferti 45 as compared to NPKOZ exceeded more than 500%. Furthermore, the result of ammonium leaching was similarly recorded by Ferti 45 fertilizer where it had a dramatic loss from day 5 to 10. For the potassium released ($K^+$) in the column leaching study, monopotassium phosphate fertilizer released the highest $K^+$ on day 1, which is linked to its highest K content (Figure 2c). Ferti 45 fertilizer also marked second in rapid losses of $K^+$ ion while the NPKOZ only released $K^+$ on day 5 afterward in a small quantity. The absence of physical nutrients carrier such as zeolite in the chemical fertilizer resulted in a higher loss of respective nutrients from chemical fertilizers.

In column leaching study, the initial detection of nutrients in the first week was considered normal and leached out until the value is detected over the course (Lou et al., 2015). The chemical fertilizer such as F45 was observed to release more nutrients than their counterpart, NPKOZ fertilizer. At higher concentration of P in fertilizer, it can be indicated that the increasing P will also increase its leaching content. The slow release of phosphate ion from NPKOZ happens because the SMZ has capabilities in adsorbing phosphate ions from aqueous solution (Bansiwal, 2006; Hamzah et al., 2014). Meanwhile, the strong attraction of $K^+$ towards zeolite anionic framework is achieved by cation exchange property of the zeolite (Li et al., 2013). The incorporation of N and K inside the zeolite is not only contributes to nutrients loss but also a better performance for plant growth (Li et al., 2013).

### Comparison between NPKOZ and Conventional Fertilizers towards the *C. asiatica* Growth

The morphological growth parameters in Table - 3 showed that NPKOZ15 has the highest number of leaves after 70 days of cultivation. In contrast, MAP with a lot of ammonium element for leaves development did not resulted in a high number of leaves. Moreover, the branches numbers and number of flowers were higher in NPKOZ10 and NPKOZ15 than other treatments. Interestingly, even though F45 had the highest TFB, the biggest reduction of its weight after heating treatment into the TDB was notifiable. On the other hand, the NPKOZs treatments increased steadily for its TFB and TDB than other systems except for the Ferti 45 system. Based on the result of leaves number, six out of eight treatments mean score were considered rejecting the null hypothesis except for control (M = 2.33, SE = 0.33) and NPKOZ15 (M = 12.67, SE = 1.45) that did not differ significantly from other treatments. In term of the biggest leaf area ($cm^2$) (BLA) and the biggest specific leaf area ($cm^2/g$) (BSLA) selected during harvesting time, F45 application contributed to the highest value compared to other systems. However, NPKOZ5 application resulted as good as MAP and MKP fertilizers performance. In overall, the increased in the amount of NPKOZ used (5, 10 and 15 g) resulted in the increased morphological growth parameters. There are two major biochemical growth parameters presented in this study; chlorophyll contents and N, P, and K contents for the growth of *C. asiatica* (Figure 3). In this study, the application of NPKOZ5, NPKOZ10 and NPKOZ15 dominated the total chlorophyll content (TCC) while F45 resulted in the lowest TCC value. In term of N and K contents, F45 showed the highest percentage while NPKOZs treatments were comparable with that of MAP and MKP. On the other hand, the number of branches, leaves and flowers along with the changes of TFB to TDB showed that the NPKOZs application was better than other treatments which are known for its slow release properties. There was a strong, positive correlation between number of branches ($r = 0.948$, $n = 24$, $P \leq 0.01$), number of flowers ($r = 0.817$, $n = 24$, $P \leq 0.01$), TFB ($r = 0.908$, $n = 24$, $P \leq 0.01$), TDB ($r = 0.866$, $n = 24$, $P \leq 0.01$), and BLA ($r = 0.522$, $n = 24$, $P \leq 0.01$) towards the number of leaves in Table - 3. A similar results pattern can be seen in parameter number of branches and chlorophylls. Reháková *et al.* (2004) suggested that the zeo loaded with certain nutrients might thrive longer during vegetation, as the nutrients were released slowly. This finding might be contributed in the making of NPKOZ that was added with the MAP at the last stage of the preparation. By the supplement of MAP onto the surfactant layer, NPKOZ was able to discharge the ammonium element that helped in leaf formation and phosphate element for the cell nucleic development for the growth of the plant stem. Besides that, the last nutrient to be released is K element inside NPKOZ that regularly known to induce the growth of flowers (Hitchmough, 2008). The presence of flower as an essential organ for the plant reproduction could have been related to the previous numbers of branches formed, where the flower arose from beneath. In term of biochemical parameters, chlorophyll is one of the main component in the photosynthesis process. Chlorophyll also served as parts of compounds in antioxidant property, namely the carotenoid. This outcome showed a valuable property of NPKOZ as a single solution to increase the chlorophylls content in the leafy plant. Although the uptakes of NPK are fewer by the whole plant for NPKOZs application compared to F45, the initial deduction was that the NPKOZs are slightly better than its counterpart. The ability of NPKOZ to release the nutrients as discussed in the characterization section might provide a sufficient source of nutrients even with fewer amounts.

**Comparison between Cultivation System towards the *C. asiatica* Growth**

In morphological growth parameters based on Table - 5, plant treated by both NPKOZ Pot (M = 20.00, SE = 2.97) and NPKOZ Verti (M = 21.20, SE = 1.88) exceeded more than 20 leaves, which is higher than other systems. The results are not the same for the hydroponic system where the observed number of leaves was found similarly lower in both control and NPKOZ. Furthermore, petiole length for the plant after treated with NPKOZ Pot and Verti systems showed the highest length as compared to other systems. Moreover, the measured leaf area signified the highest value for NPKOZ Verti and Pot systems, where it is almost doubled compared to the control system. The root length of the plant did not differ in the verticulture and pot systems whereby it is higher in the hydroponic system due to its liquid nature based or Deep Water Culture (DWC) bed (Okemwa, 2015). Besides that, the TFB and TDB results for NPKOZ systems clearly outnumbered the control systems. For instance, NPKOZ Verti (M = 11.29, SE = 1.96) weighted in almost doubled for its fresh biomass than the control Verti (M = 7.41, SE = 0.96). Apart from the morphological parameters, the biochemical properties for the second cultivation were also recorded as positive results in chlorophyll contents and nutrient contents in *C. asiatica*. Among all NPKOZ systems, NPKOZ Verti is observed as the highest recorded data for the combined chlorophyll a and b. Besides that, it is noticeable that phosphate content in the three systems containing NPKOZ contributed to the highest value of P in *C. asiatica*. Initial conclusion showed that the verticulture systems contributed more positive parameters than other cultivation systems, based on either morphological or biochemical growth parameters.

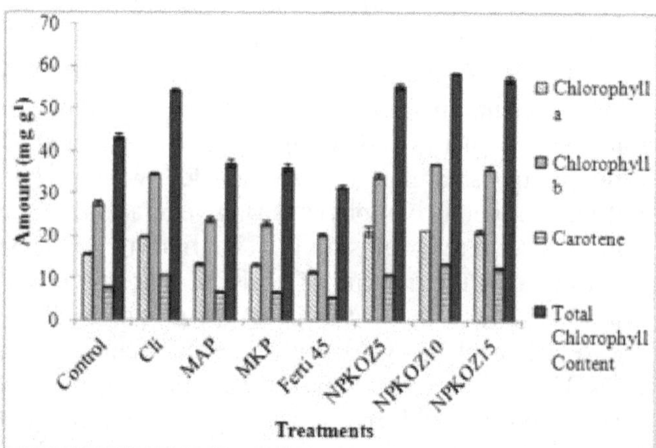


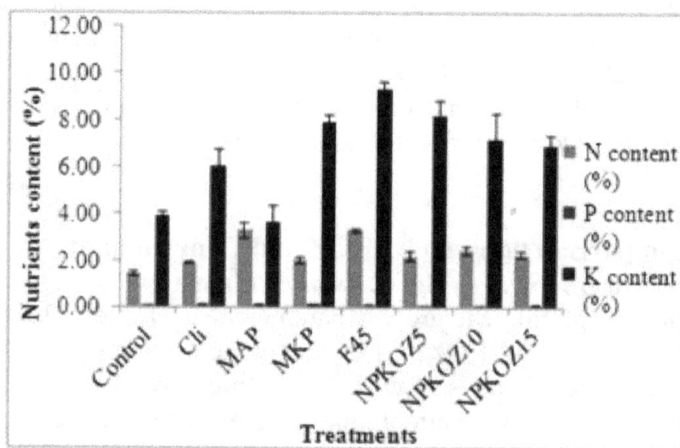


**Figure 3:**

**Table - 3: Means comparisons with morphological growth parameters in eight treatments towards the growth of *C. asiatica* after 70 days. For each parameter, significant difference between means among different treatments are indicated by different letters (Duncan test, α= 0.05).**

| Treatments | No. of leaves | No. of branches | No. of flowers | TFB (g) | TDB (g) | BLA ($cm^2$) | BSLA ($cm^2/g$) |
|---|---|---|---|---|---|---|---|
| **Control** | $2.33^{a} \pm 0.33$ | $1.00^{a} \pm 0.00$ | $0.00^{a} \pm 0.00$ | $1.28^{a} \pm 0.45$ | $0.10^{a} \pm 0.03$ | $7.33^{a} \pm 0.33$ | $383.33^{a} \pm 16.67$ |
| **Zeo** | $4.33^{ab} \pm 0.88$ | $1.33^{ab} \pm 0.33$ | $0.33^{a} \pm 0.33$ | $3.64^{ab} \pm 0.24$ | $0.47^{ab} \pm 0.05$ | $12.33a^{b} \pm 1.09$ | $316.11^{a} \pm 33.89$ |
| **MAP** | $6.00^{abc} \pm 1.53$ | $2.33^{ab} \pm 0.67$ | $0.67^{a} \pm 0.33$ | $4.58^{ab} \pm 1.88$ | $0.51^{ab} \pm 0.20$ | $14.67a^{bc} \pm 3.72$ | $330.24^{a} \pm 25.78$ |
| **MKP** | $5.00^{abc} \pm 1.00$ | $2.33^{ab} \pm 1.33$ | $1.33^{a} \pm 1.33$ | $3.35^{ab} \pm 0.60$ | $0.37^{ab} \pm 0.08$ | $13.83^{abc} \pm 0.73$ | $422.22^{a} \pm 43.39$ |
| **F45** | $11.00^{bc} \pm 4.04$ | $4.33^{ab} \pm 1.76$ | $1.67^{a} \pm 1.67$ | $8.91^{b} \pm 2.80$ | $0.77^{b} \pm 0.28$ | $21.33^{c} \pm 3.81$ | $452.50^{a} \pm 58.86$ |
| **NPKOZ5** | $5.33^{abc} \pm 1.45$ | $2.00^{ab} \pm 1.00$ | $1.33^{a} \pm 0.88$ | $3.67^{ab} \pm 0.46$ | $0.39^{ab} \pm 0.46$ | $15.67^{abc} \pm 2.09$ | $428.33^{a} \pm 73.39$ |
| **NPKOZ10** | $12.00^{bc} \pm 4.36$ | $5.00^{ab} \pm 2.31$ | $3.33^{a} \pm 1.76$ | $6.40^{ab} \pm 2.30$ | $0.69^{ab} \pm 0.24$ | $14.67^{abc} \pm 1.96$ | $346.67^{a} \pm 39.30$ |
| **NPKOZ15** | $12.67^{c} \pm 1.45$ | $5.33^{b} \pm 0.88$ | $3.00^{a} \pm 1.53$ | $8.13^{b} \pm 2.12$ | $0.91^{b} \pm 0.30$ | $19.33^{bc} \pm 4.10$ | $348.33^{a} \pm 53.26$ |

Notes: (Included Standard Error Mean (SEM), Total Fresh Biomass (TFB), Total Dry Biomass (TDB), Biggest Leaf Area (BLA), and Biggest Specific Leaf Area (BSLA))

**Table - 4: Pearson correlation coefficients comparisons with morphological growth parameters in eight different systems towards *C. asiatica* growth at significance level Sig (2-tailed). Noted that * represent significant ($p<0.05$) and ** represent highly significant correlation ($p<0.01$).**

| Parameters | No. of leaves | No. of branches | No. of flowers | TFB (g) | TDB (g) | BLA ($cm^2$) | BSLA ($cm^2/g$) |
|---|---|---|---|---|---|---|---|
| **No. of leaves** | 1 | | | | | | |
| **No. of branches** | .948** | 1 | | | | | |
| **No. of flowers** | .817** | .874** | 1 | | | | |
| **TFB (g)** | .908** | .814** | .759** | 1 | | | |
| **TDB (g)** | .866** | .768** | .785** | .969** | 1 | | |
| **BLA ($cm^2$)** | .522** | .398 | .443* | .688** | .663** | 1 | |
| **BSLA ($cm^2/g$)** | -.095 | -.105 | -.108 | .018 | -.105 | .070 | 1 |

Notes: (Total Fresh Biomass (TFB), Total Dry Biomass (TDB), Biggest Leaf Area (BLA), Biggest Specific Leaf Area (BSLA))

**Table - 5: Means comparisons with morphological growth parameters in three systems towards the growth of *C. asiatica* after 70 days. For each parameter, significant difference between means among different treatments are indicated by different letters (Duncan test, α= 0.05).**

| Systems | No. of leaves | TPL (cm) | Root length (cm) | TFB (g) | TDB (g) | Leaf area ($cm^2$) | BSLA ($cm^2/g$) |
|---|---|---|---|---|---|---|---|
| **Control P** | $18.80^{bc} \pm 2.35$ | $8.38^{a} \pm 0.77$ | $14.59^{a} \pm 0.82$ | $7.28^{ab} \pm 0.61$ | $0.83^{b} \pm 0.11$ | $10.56^{bc} \pm 0.89$ | $655.93^{bc} \pm 86.00$ |
| **NPKOZ P** | $20.00^{c} \pm 2.97$ | $12.18^{bc} \pm 1.68$ | $14.53^{a} \pm 4.49$ | $8.85^{bc} \pm 1.27$ | $1.15^{c} \pm 0.12$ | $19.53^{c} \pm 2.14$ | $491.25^{ab} \pm 20.12$ |
| **Control H** | $10.60^{a} \pm 0.93$ | $8.24^{a} \pm 1.24$ | $16.21^{a} \pm 2.41$ | $4.47^{a} \pm 0.38$ | $0.53^{a} \pm 0.04$ | $8.21^{a} \pm 1.00$ | $623.00^{bc} \pm 86.54$ |
| **NPKOZ H** | $10.60^{a} \pm 1.21$ | $8.02^{a} \pm 0.41$ | $11.66^{a} \pm 1.45$ | $4.27^{a} \pm 0.31$ | $0.51^{a} \pm 0.04$ | $9.21^{bc} \pm 1.37$ | $780.60^{c} \pm 69.79$ |
| **Control V** | $13.40^{ab} \pm 0.93$ | $10.60^{ab} \pm 0.97$ | $12.09^{a} \pm 0.95$ | $7.41^{ab} \pm 0.65$ | $0.98^{bc} \pm 0.09$ | $13.05^{b} \pm 0.83$ | $401.43^{a} \pm 61.09$ |
| **NPKOZ V** | $21.20^{c} \pm 1.88$ | $15.04^{c} \pm 1.24$ | $14.95^{a} \pm 1.27$ | $11.29^{c} \pm 1.96$ | $1.50^{d} \pm 0.14$ | $21.51^{c} \pm 1.83$ | $486.38^{ab} \pm 46.47$ |

Notes: (Included Standard Error Mean (SEM), Total Fresh Biomass (TFB), Total Dry Biomass (TDB), Tallest Petiole Length (TPL), and Biggest Specific Leaf Area (BSLA))

**Table - 6: Pearson correlation coefficients comparisons with morphological growth parameters in three systems towards *C. asiatica* growth at significance level Sig (2-tailed). Noted that * represent significant ($p<0.05$) and ** represent highly significant correlation ($p<0.01$).**

| Parameters | No. of leaves | TPL (cm) | Leaf area ($cm^2$) | Root length (cm) | TFB (g) | TDB (g) | BSLA ($cm^2/g$) |
|---|---|---|---|---|---|---|---|
| No. of leaves | 1 | | | | | | |
| TPL (cm) | .493* | 1 | | | | | |
| Leaf area ($cm^2$) | .685** | .884** | 1 | | | | |
| Root length (cm) | .595** | .429 | .404 | 1 | | | |
| TFB (g) | .729** | .773** | .917** | .524* | 1 | | |
| TDB (g) | .697** | .827** | .930** | .506* | .901** | 1 | |
| BSLA ($cm^2/g$) | -.049 | -.605** | -.505* | -.261 | -.501* | -.479* | 1 |

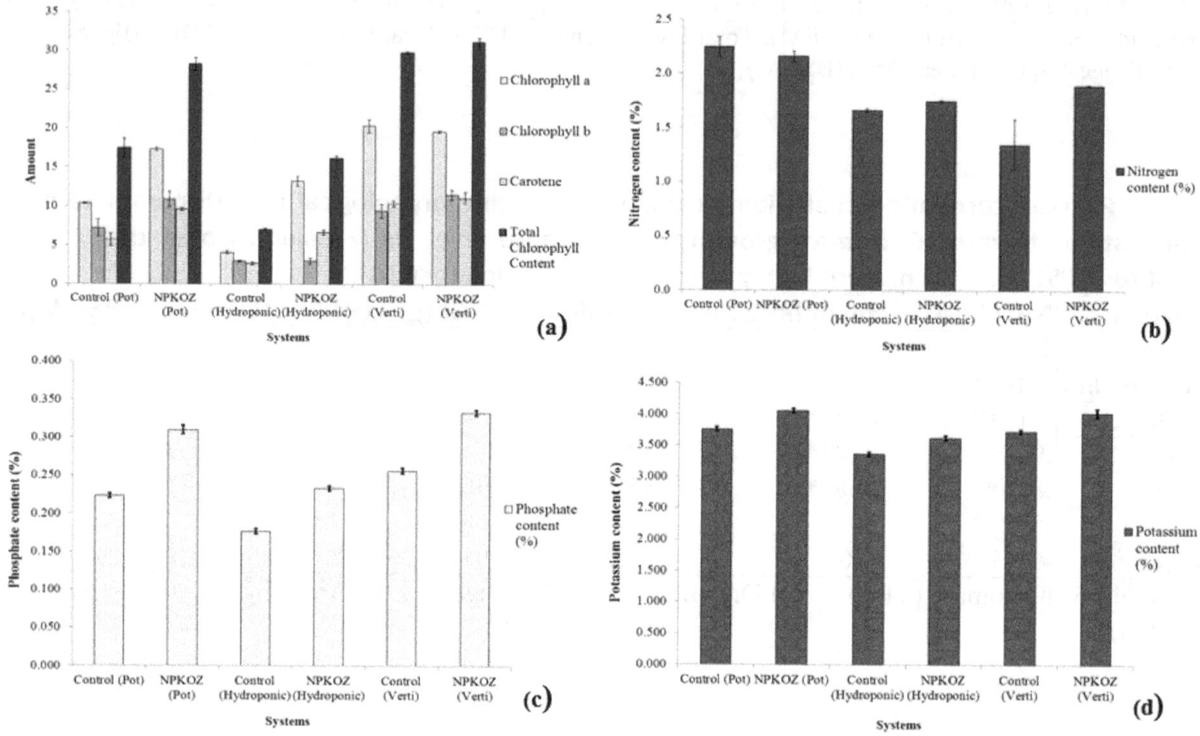


**Figure 4**

The second cultivation of *C. asiatica* by using three systems applied with NPKOZ also recorded significant results, especially the high number of leaves. Nitrogen source plays an important role in the development of plant leaves, which is the main reason of the highest number of leaves. Ohyama (2010) briefly described that nitrogen (N) element not only provide a shoot growth in excess, but also preferred the young leaves if the N source is depleted. In contrast, NPKOZ application in the hydroponic system provided only slightly positive results in overall parameters compared to control. The hydroponic system used in this study was based on Deep Water Culture (DWC) bed which needs small plant pots with coco peat media (Okemwa, 2015). The little pots keep afloat in water based bed where it is usually filled or replace with liquid organic fertilizers that provide nutrients in liquid formed. In this experiment, no solution fertilizer based was used and furthermore, NPKOZ is insoluble in water. For this reason, NPKOZ in teabag method is subjected to the loss of nutrients before being taken by the plant. For the chlorophyll parameter part, NPKOZ application on each system of the cultivation showed higher values as compared to the control system in term of TCC. The greener the tone colour of the plant itself, the higher the chances of chlorophyll content piling up. Limantara *et al* (2015) studied the comparison between ten green leafy vegetables where they found that the darker green colour contributed to the total peak of chlorophyll content using *Brassica rapa L.* (Chinese flowering cabbage). In this study, almost all morphological and biochemical growth parameters recorded either significant or highly significant correlation. For example in signification correlation, there was a positive correlation between tallest petiole length ($r = .493$, $n = 18$, $P<0.05$), total chlorophyll content ($r = .518$, $n = 18$, $P<0.05$) and N content ($r = .572$, $n = 18$, $P<0.05$) towards the number of leaves. In broad, NPKOZ Verti took two out of four morphological growth parameters highest assessments. Based on the positive results of the verticulture system in BGP (chlorophyll and phosphate content) and high in four MGP parameters (NOL, TPL, leaf area, TFB and TDB), the initial conclusion clarified that the verticulture is the best system among other systems.

## Conclusion

From characterization of the prepared materials by XRD, FTIR and leaching study, it was proven that the cationic surfactant HDTMA molecules were attached on the zeolite framework without changing the framework structure of the zeolite. In addition, these surfactant molecules assist zeolite in adsorbing phosphate anions. Chemical fertilizers resulted in a dramatic loss of major nutrients in leachate, especially F45 but, NPKOZ fertilizers able to attain the highest number of leaves, flowers and branches as well as overall chlorophyll contents because of the slowly release of the plant nutrients. Furthermore, the verticulture method applied with NPKOZ is selected as the best cultivation system and it is a potential sustainable fertilizer for the high yield of *C. asiatica*.

## Acknowledgments

Authors would like to thank staff in Faculty of Biosciences and Medical Engineering (FBME), UTM and Provet Group of Companies Sdn. Bhd. This project was funded by Research Grant University (RUG) (Vot 13H53) provided by UTM and Ministry of Higher Education, Malaysia.

## References

Adhikari KR and Chen ZS, 2013. Reducing nitrogen leaching from agriculture: an essential step toward protecting the environment in high rainfall areas of Taiwan. Compost. Sci. Util. 21: 2-15.

Ah Tung PG, Yusoff MK, Majid NM, Goh Kah J and Gan Huang H, 2009. Effect of N and K fertilizers on nutrient leaching and groundwater quality under mature oil palm in Sabah during the monsoon period. Am. J. Appl. Sci. 6: 1788-1799.

Bansiwal AK, Rayalu SS, Labhasetwar NK, Juwarkar AA and Devotta S, 2006. Surfactant-modified zeolite as a slow release fertilizer for phosphorus. J. Agr. Food Chem. 54: 4773-4779.

Bhardwaj D, Sharma M, Sharma P and Tomar R 2012. Synthesis and surfactant modification of clinoptilolite and montmorillonite for the removal of nitrate and preparation of slow release nitrogen fertilizer. J. Hazard. Mater. 227: 292-300.

De Smedt C, Someus E and Spanoghe P, 2015. Potential and actual uses of zeolites in crop protection. Pest Manage. Sci. 71: 1355-1367.

Devkota A and Jha PK, 2009. Variation in growth of *Centella asiatica* along different soil composition. Bot. Res. Int. 2: 55-60.

Gohil KJ, Patel JA, Gajjar AK, 2010. Pharmacological review on *Centella asiatica*: a potential herbal cure-all. Indian J. Pharm. Sci. 72: 546–556.

Haggerty GM and Bowman RS, 1994. Sorption of chromate and other inorganic anions by organo-zeolite. Environ. Sci. Technol. 28: 452-458.

Hitchmough J and Fieldhouse K, 2008. Plant user handbook: A guide to effective specifying. 1st ed. Blackwell Science Ltd., Oxford.

Huda-Faujan N, Noriham A, Norrakiah AS and Babji AS, 2009. Antioxidant activity of plants methonolic extracts containing phenolic compounds. African J. Biotechnol. 8: 484-489.

Jin SG, Kim KS, Yousaf AM, Kim DW, Jang SW, Son MW, Kim YH, Yong CS, Kim JO and Choi HG, 2015. Mechanical properties and in vivo healing evaluation of a novel *Centella asiatica*-loaded hydrocolloid wound dressing. Int. J. Pharm. 490: 240-247.

Keatinge JDH, Yang RY, Hughes JDA, Easdown WJ and Holmer R, 2011. The importance of vegetables in ensuring both food and nutritional security in attainment of the Millennium Development Goals. Food Secur. 3: 491-501.

Li J, Wee C and Sohn B, 2013. Effect of ammonium-and potassium-loaded zeolite on kale (Brassica alboglabra) growth and soil property. Am. J. Plant Sci. 4: 1976-1982.

Limantara L, Dettling M, Indrawati R and Brotosudarmo THP, 2015. Analysis on the chlorophyll content of commercial green leafy vegetables. Procedia Chem. 14: 225-231.

Li Z, 2003. Use of surfactant-modified zeolite as fertilizer carriers to control nitrate release. Microporous Mesoporous Material. 61: 181-188.

Lou Z, Zhu J, Wang Z, Baig SA, Fang L, Hu B and Xu X, 2015. Release characteristics and control of nitrogen, phosphate, organic matter from spent mushroom compost amended soil in a column experiment. Process Saf. Environ. Prot. 98: 417-423.

Lubkowski K, 2016. Environmental impact of fertilizer use and slow release of mineral nutrients as a response to this challenge. Pol. J. Chem. Technol. 18: 72-79.

Malek NANN, 2011. Surfactant modified zeolites as drug support system and antibacterial agent, PhD dissertation, Universiti Teknologi Malaysia, Malaysia.

Malek NANN, Hamzah NS, Dzkulfli NH, Abdullah WMMW and Hamdan S, 2014. Effect of natural zeolite (clinoptilolite) and urea on the growth of *Amaranthus gangeticus*, *Clinachantus nutans* and *Capsicum annuum*. Jurnal Teknologi. 68: 2180-3722.

Malek, NANN, Siang TW, Hamzah NS and Dzkulfli NH, 2015. Evaluation of NPK-organo-zeolite as controlled release fertilizer for the growth of *Clinacanthus nutans*. 5th International Conference on Environment 2015.

Mansouri N, Rikhtegar N, Panahi HA, Atabi F and Shahraki BK, 2013. Porosity, Characterization and structural properties of natural zeolite-clinoptilolite-as a sorbent. Environ. Prot. Eng. 39: 139-152.

Nataraj J, Manivasagam T, Thenmozhi AJ and Essa MM, 2017. Neurotrophic effect of asiatic acid, a triterpene of *Centella asiatica* against chronic 1-methyl 4-phenyl 1, 2, 3, 6-tetrahydropyridine hydrochloride/probenecid mouse model of Parkinson's disease: The role of MAPK, PI3K-Akt-GSK3β and mTOR signalling pathways. Neurochem. Res. 42: 1354-1365.

Ohyama T, 2010. Nitrogen assimilation in plants. Research Signpost, Kerala, India.

Okemwa E, 2015. Challenges and opportunities to sustainability in aquaponic and hydroponics systems. Int. J. Sci. Res. Innov. Technol. 2: 54-76

Penuelas J, Sardans J, Alcaniz JM and Poch JM, 2009. Increased eutrophication and nutrient imbalances in the agricultural soil of NE Catalonia, Spain. J. Environ. Biol. 30: 841-846.

Polat E, Karaca M, Demir H and Naci-Onus A, 2004. Use of natural zeolite (clinoptilolite) in agriculture. J. Fruit Ornam. Plant Res. 12: 183-189.

Pompelli MF, França SC, Tigre RC, De Oliveira MT, Sacilot M and Pereira EC, 2013. Spectrophotometric determinations of chloroplastidic pigments in acetone, ethanol and dimethylsulphoxide. Braz. J. Biosci. 11: 52-58.

Prajapati K and Modi HA, 2012. The importance of potassium in plant growth—a review. Indian J. Plant Sci. 1: 177-186.

Rajpurkar Y, 2016. Hydroponics. Building greenhouses and hydroponic farms to secure and flourish food in Qatar. Grin Publishing, Munich, Germany.

Reháková M, Čuvanová S, Dzivák M, Rimár J and Gavaľová Z, 2004. Agricultural and agrochemical uses of natural zeolite of the clinoptilolite type. Curr. Opin. Solid State Mater. Sci. 8: 397–404.

Sempeho SI, Kim HT, Mubofu E and Hilonga A, 2014. Meticulous overview on the controlled release fertilizers. Adv. Chem. ID 363071.

Sharuddin SSM, Ping JLW, Othaman NS, Mohtar SS, Saman N, Johari K, Tian SX, Malek NAN and Mat H, 2014. Natural zeolite characterization for adsorptive coagulation flocculation (ACF) removal of ammonium in drinking water treatment process, In Proceedings, 8th SEATUC Symposium. 4-5 March 2014. Universiti Teknologi Malaysia.

Sitawati S, Suryanto A and Nurlaelih EE, 2016. Optimization of plant growth and yield through innovation of the materials and medium verticulture. Res. J. Life Sci. 3: 55-64.

Treacy MM and Higgins JB, 2007. Collection of simulated XRD powder patterns for zeolites, 5th ed. Elsevier Science, Amsterdam, The Netherlands.

Xu L, Zhang Q and Huang L, 2010. Nitrogen leaching in a typical agricultural extensively cropped catchment, China: experiments and modelling. Water Environ. J. 24: 97-106.

Wang S and Peng Y, 2010. Natural zeolites as effective adsorbents in water and wastewater treatment. Chem. Eng. J. 156: 11-24.

# 10

# Bioefficacy of plant powders against *Callosobruchus chinensis* L (Coleoptera: Bruchidae) in infested chickpea grains

**Asma Tariq[1], Sumera Afsheen[1*], Mobushir Hussain[1], Ahmed Zia[3], Syed Shakeel Shah[2*], Sabila Afzal[2], Inamullah Khan[4], Yousaf Hayat[5]**

[1]Department of Zoology, University of Gujrat, Gujrat, Punjab, Pakistan
[2]Department of Zoology, University of Gujrat, Sub-Campus Narowal, Pakistan
[3]National Insect Museum, National Agriculture Research Centre, Islamabad, Pakistan
[4]Department of Plant Protection, Agricultural University Peshawar, KPK, Pakistan
[5]Department of Statistics, Mathematics and Computer Science, University of Agriculture, Peshawar, KPK, Pakistan

*Corresponding author email: sumera.afsheen@uog.edu.pk, shakeel.shah@uog.edu.pk

## Abstract

In this study we used dried leaf powders of five plants i.e., *Syzygium cumini*, *Citrus limon*, *Momordica charantia*, *Eucalyptus globulus* and *Piper nigrum* against *Callosobruchus chinensis* infesting chickpea seeds during storage. Free choice and no choice tests were conducted. Plant powders were applied at dose of 2% (w/w). Efficacy of all plant powders were compared by orientation, oviposition, adult mortality and adult emergence of the pulse beetle. Effect of plant powder on germination of seed was also observed by germination test. Leaf powder of *M. charantia* was found to be more effective while leaf powder of *S. cumini* was found to be least effective in adult mortality and oviposition deterrence. Furthermore, the results showed that leaf powders of all plants had no effects on germination of seeds.

**Keywords:** Leaf Powders, Chickpea seeds, Seed viability, Pulse beetle

## Introduction

Pulses are among the major food sources in developing countries and vegetarian societies ((Aslam et al., 2006)). Chickpea (*Cicerarietinum* L.) contains 38-59% carbohydrates and 25.3-28.9% proteins (Khan et al. 1978 a; Khan 1981 b; Hulse 1991. Chickpea is widely cultivated crop in Pakistan, mostly after monsoon season.

Stored chickpea is heavily infested by pulse beetle (*Callosobruchus chinensis)*. Pulse beetles causes great economic losses as it reduces the dry weight and makes nutritional values poor. (Ahmed et al., 2003; Ahmed et al., 2009; Kumar et al., 2009; Righi-Assia et al., 2010). Pulse beetle, commonly known as Dhora, causes heavy loss about 90%, in stored grams (Qayyum et al., 1978; Kakepoter et al., 1985; Upadhyay et al., 2006). The use of commercially available insecticides and fumigants has adverse effects on environment as well as humans (Talukder (2006). Synthetic chemicals also induces the resistance in insects pests, toxic effects to grain handlers, and toxic residues affect the entire food chain after utilization (Schoonhoven (1982); Sharaby (1988); Shaaya et al. (1997); Upadhyay et al. (2011)). A number of plant products have been successfully tried as protectants against various stored grain insect pests (Gill et al., 1971; Pandey et al., 1986; Jilani et al., 1988; Ayyanger et al., 1989; Dixit et al., 1990; Kumari et al., 1998; Verma et al., 1999; Kim et al., 2003).These plant products included ash (Ajayi et al., 1987) vegetable oil (Schoovoven 1978;

Kazi et al., 1999) and plant powders and extracts (Abdullahi et al., 2004; Partil et al., 2006; Gupta et al., 2008; Shukla et al., 2009).The powders of certain plant parts i.e. leaves, bark, seeds as well as oil extracts suppressed the oviposition and adult emergence and increased mortality rate of bruchids and maintained the seed quality as well, when mixed with stored grains (Tapondjou et al., 2002; Adedire et al., 2004; Rahman et al., 2006; Bamaryi et al., 2007; Shukla et al.,2007). Botanical powders are safe and interesting alternatives of chemicals insecticides especially in developing tropical countries where plants are abundantly present throughout the year (Khalequzzaman et al., 2009). Keeping in view the usefulness of plant products, the present research is performed to evaluate the repellent and deterrent effects of botanical powders from five angiospermic plants on adult mortality, oviposition and adult emergence of pulse beetle in stored chickpea. The weight loss and germination of treated seeds were also checked. In this research, leaf powders of *Syzygiumcumini* (Jaman), *Citrus limon* (Lemon), *Momordica charantia* (Bitter gourd), *Eucalyptus globulus* (Eucalyptus) and *Piper nigrum* (Black pepper) are used and according to best of my knowledge, leaf powders of bitter gourd, eucalyptus and black pepper are used for the first time against *Callosobruchus chinensis*.

## Materials and Methods

**Rearing of *Callosobruchus chinensis***

Chickpea seeds were taken from a local market in Gujranwala and naturally infested adult pulse beetles (*Callosobruchus chinensis*) were obtained. From these infested seeds hundred adult beetles were separated in order to obtain newly emerged pulse beetles (Halstead, 1963). Newly emerged pulse beetles were released in 1 kg clean and un-infested chickpea seeds in a plastic jar covered with cloth and capped with muslin cloth. The jar was kept in lab at temperature 25-35 C. Adult beetles were left for about seven days for the purpose of mating and oviposition. After seven days, the parent beetles were separated from chickpea seeds and eggs. Chickpea seeds with eggs were left in jar for about 25 days in order to obtain adult beetles which were used in the entire study.

**Stored product**

Five kilogram of chickpea seeds was bought and placed in refrigerator for 4 days in order to prevent prestorage infestation.

**Preparation of botanical powders**

i. *Piper nigrum* (Black pepper)
ii. *Citrus limon* (Lemon)
iii. *Eucalyptus globules* (Eucalyptus)
iv. *Momordica charantia* (Bitter gourd)
v. *Syzygium cumini* (Jaman)

Cleaned leaves of these plants were dried in shade. Then crush into powdered form by using electric grinder (Nisar et al., 2012; Udoh et al., 2010). Finely ground powder was collected by passing through 25-mesh sieve.

**Free choice test**

For this test, free choice chamber used which was circular in shape and having 6 transparent plastic jars. Side walls of jars were wrapped with black paper. Then jars were positioned at equal distance from each other and a transparent jar was placed in center. All jars connected with central jar through plastic tube. The orientation of adult pulse beetles i.e. treated chickpeas and untreated chickpeas were studied by performing free choice test. Fifty grams of chickpea seeds were treated separately with 5 botanical powders at dose of 2% (w/w). The treated seeds were placed in 5 boxes of the device. The control set of chickpea seeds were also placed in the remaining box without any treatment. In the central box of chamber 20 adult pulse beetles were placed and the chamber was placed in a dark room. The number of adults oriented in each treated and control set of chickpea seeds were counted daily for 3 days. The experiment was repeated 3 times.

**No choice test**

Fifty grams of chickpea seeds were treated with each botanical powder separately at dose of 2% (w/w) and then placed in 5 plastic jars separately. Control set was also placed in one container without any treatment. About 0-2 days old five pairs of pulse beetles were introduced into each jar.

Jars were placed in lab shelf after covering with muslin cloth. Weight loss in the seeds, mortality of adult beetles, adult emergence and oviposition was observed. Treated seeds were then subjected to germination test at end of experiment. The experiment was repeated three times.

**Statistical analysis**

Data was analyzed statistically by using one way ANOVA and for significant means; LSD was applied (Steel et al., 1980). For statistical analysis a software package Statistics was used. To check % oviposition deterrence, % adult emergence deterrence and % feeding deterrent index (FDI), Abbott's formula was applied (Abbott (1925) as follows.

$$\textit{Per cent control} = X - Y / X \times 100$$

## Results and Discussion

In order to observe effects of botanical powders free choice and no choice tests were performed. Weight loss and germination of chickpea seeds, Adult orientation, adult mortality, oviposition and adult emergence of *Callosobruchus chinensis,* were observed. The obtained results are as follows:

**Effect of powders on orientation of *C. chinensis***

In this study the plant powders of *Syzygium cumini*, Citrus limon, *Momordica charantia*, Eucalyptus globulus and *Piper nigrum* were used. Powders found effective over control in repelling the *C. chinensis* adults, when their orientation was observed through free choice test (Table 1). Leaf powders of *Momordica charantia* and *Eucalyptus globulus* significantly ($P<0.05$) hindered the orientation of adult beetles as compared to other three. Among the treatments, *Citrus limon* leaf powder showed the highest mean orientation of beetles followed by *Syzygium cumini* and *Piper nigrum* and these three powders were not significantly different from each other. Maximum orientation of bruchids was recorded in control set of chickpeas.

In this study it was found that leaf powders of these plants contains repellent properties against *C. chinensis*. In case of each treatment diverse results found that pest repelling factors are not evenly present in the plants. The leaf powder of *Momordica charantia* was most effective then *Eucalyptus globulus* in repellency and these results are in accordance with the previous studies. Oil of *E. globulus* exposed to contain highest insecticidal potential among other treatments i.e. neem oil, castor oil and sunflower oil (Lal et al., 2012). From different studies it was found that fruit powder of P. nigrum at different concentration is proved effective post-harvest grain protectant (Aslam et al., 2002; Mahdi et al., 2008). In this study P. nigrum was also found effective against *C. chinensis* which confirmed the previous findings. But here for the first time, powder of leaves of *P. nigrum* was used.

**Table 1: Effects of leaf powders on orientation of *C. chinensis***

| Treatments 2% (w/w) | Insects with respect to days | | | Grand Mean |
|---|---|---|---|---|
| | 1DAT | 2DAT | 3DAT | |
| *Syzygiumcumini* | 3.00 ± 1.00 | 2.66 ± 1.2 | 3.00 ± 0.00 | 2.89 [b] ± 0.45 |
| *Citrus limon* | 3.33 ± 0.33 | 3.33 ± 0.33 | 3.00 ± 0.58 | 3.22 [b] ± 0.22 |
| *Momordica charantia* | 1.00 ± 0.58 | 1.00 ± 1.00 | 1.00 ± 0.58 | 1.00 [c] ± 0.37 |
| *Eucalyptus globules* | 1.00 ± 0.58 | 1.33 ± 0.66 | 1.33 ± 0.33 | 1.22 [c] ± 0.28 |
| *Piper nigrum* | 3.33 ± 0.66 | 2.66 ± 0.33 | 2.33 ± 0.33 | 2.78 [b] ± 0.28 |
| Control | 6.00 ± 0.58 | *7.33* ± 0.33 | 7.00 ± 1.15 | 6.78 [a] ± 0.43 |
| Grand Mean | 2.94 ± 0.47 | *3.06* ± *0.56* | 2.94 ± 0.52 | 2.98 ± 0.29 |

DAT= Days after treatment

Each value represents the mean of number of insects oriented in each treatment. The mean followed by same letter are not different significantly according to LSD.

The mean difference is significant at the level of 0.05.

**Effect of powders on adult mortality of *C. chinensis***

In table 2 mortality of adult *C. chinensis* described which were recorded after 2nd, 4th and 6th day of treatment and revealed the effects of plant powders. Adult mortality was observed during 6 days of exposure to treatments. The greatest mortality was presented by *Momordica charantia* and *Citrus limon* followed by Eucalyptus globulus and *Piper nigrum* while slightest mortality was observed in *Syzygium cumini*, still higher than control set of chickpeas.

The powders play important role in killing insects by occlusion of spiracles, by causing tracheal resistance in respiration and by obstructing movement of insects, eventually causing death (Shukla et al., 2007). To the best of our knowledge, the dried leaf powders of *M. charantia, E. globulus and P. nigrum* are tested for the first time for the control of *C. chinensis* in chickpea seeds. It was hypothesized that powder leaves of *M. charantia* would be efficient in grain protection after harvesting and results of this study supported our hypothesis. *M. charantia* leaf powder caused greatest mortality, followed by *C. limon*. All the Citrus spp. possesses insecticidal factors. A high mortality rate was reported in *C. chinensis* and other insect pests from vital oils of Citrus spp. such as *C. reticulata, C. sinensis, C. paradise* and *C. grandis* (Zia et al., 2013). *C. medica* reported to be highly active as compared to *Syzygium cumini, Aegle marmelos* and *Ammomum subulatum* in controlling of pulse beetle (Shukla et al., 2007).

**Table 2: Effects of leaf powders on life span of *C. chinensis***

| Treatments 2% (w/w) | Insects with respect to days | | | Grand Mean |
|---|---|---|---|---|
| | 1DAT | 2DAT | 3DAT | |
| *Syzygium cumini* | 1.67 ± 0.67 | 2.67 ± 0.33 | 2.67 ± 0.67 | 2.33 [a] ± 0.33 |
| *Citrus limon* | 1.67 ± 1.2 | 4.00 ± 0.58 | 3.33 ± 1.20 | 3.00 [a] ± 0.62 |
| *Momordica charantia* | 3.33 ± 0.33 | 4.33 ± 0.33 | 1.67 ± 1.20 | 3.11 [a] ± 0.54 |
| *Eucalyptus globulus* | 2.67 ± 0.67 | 3.33 ± 0.88 | 2.33 ± 1.86 | 2.78 [a] ± 0.64 |
| *Piper nigrum* | 4.33 ± 1.2 | 2.33 ± 0.33 | 1.67 ± 0.33 | 2.78 [a] ± 0.6 |
| Control | 1.33 ± 0.88 | *1.67 ± 0.33* | 0.33 ± 0.33 | 1.11 [b] ± 0.35 |
| Grand Mean | 2.5 ± 0.4 | *3.06 ± 0.31* | 2.00 ± 0.44 | 2.52 ± 0.23 |

DAT= Days after treatment

Each value represents the mean of number of insects died in each treatment. The mean followed by same letter are not different significantly according to LSD. The mean difference is significant at the level of 0.05.

## Effect of powders on oviposition of *C. chinensis*

In this experiment all the plant powders used found to be effective in reducing the oviposition of *C. chinensis*. The results in treated and untreated chickpea seeds regarding oviposition are shown in Table 3. Chickpea seeds treated with *Momordica charantia* shows maximum prevention in oviposition i.e. 93.86% followed by *Eucalyptus globulus* and *Citrus limon* causing 63.60% and 61.84 % prevention in the oviposition of bruchids on chickpea seeds respectively. *Citrus limon* powder effective at very small rate on oviposition of bruchids. Powder of *Syzygium cumini* found least effective in oviposition deterrence as shown in adult mortality. *C. limon* leaf powder causes high mortality of bruchids as compared to *E. globulus* and *P. nigrum* while *E. globulus* and *P. nigrum* reduced oviposition as compared to *C. limon*. The reason behind this may be due to semi chemical nature of *E. globulus* and *P. nigrum* which affects behaviour and interrupt physiology of the insects' thus disturbing oviposition and adult emergence. *S. cumini* found least effective in oviposition deterrence and adult mortality as found in another study in which it was used along with other treatments i.e. *Aeglemarmelos, Eupatorium cannabinum, Murraya koenigii, Ammomum subulatum* and *Citrus medica against C. chinensis* in stored chickpea (Shukla et al., 2007).

**Table 3: Effects of leaf powders on oviposition of *C. chinensis***

| Treatments 2% (w/w) | Mean number of eggs laid | % deterrency |
|---|---|---|
| *Syzygium cumini* | 121.67 [ab] ± 23.6 | 19.95% |
| *Citrus limon* | 112.00 [ab] ± 32.6 | 26.32% |
| *Momordica charantia* | 9.33 [c] ± 6.01 | 93.86% |
| *Eucalyptus globules* | 55.33 [bc] ± 31.54 | 63.60% |
| *Piper nigrum* | 58.00 [bc] ± 16.92 | 61.84% |
| Control | 152.00 [a] ± 27.57 | |
| Grand Mean | 84.72 ± 14.45 | |

The mean followed by the same letter are not significantly different according to LSD. The mean difference is significant at the 0.05 level.

Percent deterrency = (control – mean of each treatment / control x 100).

## Effect of powders on adult emergence of *C. chinensis*

The adult emergence of *C. chinensis* was found to be reduced in treatment sets which are given in Table 4. Reduction in emergence of adults from eggs was found 47.24%-97.12% in all treatments. In case of adult emergence *Momordica charantia* showed highest i.e. 97.12% reduction and the lowest of *Citrus limon* i.e., 47.24% was observed in treated chickpea seeds. *Syzygium cumini* powder found more effective giving 64.51% reduction in emergence of adults as compared to its results shown in adult mortality and oviposition of beetles.

Leaf powder of P. nigrum had shown 88.73% deterrence in adult emergence. In order to control pulse beetle in stored green and black gram insecticidal potency of some spices i.e. clove, cinnamon, black cardamom, black cumin, nutmeg, turmeric, black pepper, red pepper, cumin, green cardamom, bay leaf, aonla and ginger was tested. Black pepper and clove were observed highly effective as compared to all other spices (Gautam et al., 2000; Aslam et al., 2002; Mahdi et al., 2008).

**Table 4: Effects of leaf powders on adult emergence of *C. chinensis***

| Treatments 2% (w/w) | Mean number of adults emerged | % Deterrency |
|---|---|---|
| *Syzygiumcumini* | 49.33 $^{bc}$ ± 24.99 | 64.51% |
| *Citrus limon* | 73.33 $^{b}$ ± 20.61 | 47.24% |
| *Momordica charantia* | 4.00 $^{c}$ ± 3.06 | 97.12% |
| *Eucalyptus globules* | 23.00 $^{bc}$ ± 21.50 | 83.45% |
| *Piper nigrum* | 15.67 $^{bc}$ ± 10.73 | 88.73% |
| Control | 139.00 $^{a}$ ± 24.70 | |
| Grand Mean | 50.72 ± 12.90 | |

The mean followed by the same letter are not significantly different according to LSD. The mean difference is significant at the 0.05 level.
Percent deterrency = (control – mean of each treatment / control x 100).

### Effect of powders on weight loss and germination of chickpea seeds

The weight loss of chickpea seeds is due to the feeding activities of larvae of pulse beetle. So, more the emerged beetles, more the weight of seed will be lost as depicted in our findings as well. Weight loss in high amount recorded in the seeds treated with *Citrus limon* and seeds of control set as well and the same results were obtained in case of adult emergence. Least weight loss observed in *Momordica charantia* and *Eucalyptus globulus* treated seeds. In case of each treatment feeding deterrent index (FDI) was calculated by using Abotts formula. *Momordica charantia* (98.03%) was observed having maximum feeding deterrent index followed by *Piper nigrum* (88.24%) and minimum in *Citrus limon* (56.85%) shown in (Table5). The seeds, attained from each treatment, were germinated envisaged no effects of plant powders on the germination of seeds i.e., 100% germination result was obtained in case of each treatment.

**Table 5: Percentage seed germination and weight loss in treated and untreated seeds**

| Treatments 2% (w/w) | % Seed Germination | Mean Weight loss | % FDI |
|---|---|---|---|
| *Syzygiumcumini* | 100% | 50.03 $^{ab}$ ± 0.52 | 71.56% |
| *Citrus limon* | 100% | 49.53 $^{b}$ ± 0.47 | 56.85% |
| *Momordica charantia* | 100% | 50.93 $^{a}$ ± 0.07 | 98.03% |
| *Eucalyptus globulus* | 100% | 50.60 $^{ab}$ ± 0.40 | 88.24% |
| *Piper nigrum* | 100% | 50.57 $^{ab}$ ± 0.28 | 87.26% |
| Control | 100% | 47.60 $^{c}$ ± 0.47 | |
| Grand Mean | | 49.88 | |

The mean followed by the same letter are not significantly different according to LSD. The mean difference is significant at the level of 0.05.
FDI= Feeding Deterrent Index
% FDI = (weight loss in control – weight loss in treatment / weight loss in control x 100)

## Conclusion

All the results proved that *Momordica charantia* (Bitter gourd) is best protectant among all the plants tested in this study. The leaf powders of this plant gave maximum mortality and ovipositional and adult emergence detergency against *C. chinensis*.
Leaf powder of *C. limon* when compared with the leaf powder of *E. globulus* was found more effective in reduction of life span than egg laying ability and adult emergence, whereas the leaf powder of *E. globulus* was found highly effective in reduction of egg laying abilities and adult emergence. So, mixtures of both powders can be applied for effective control in already infested chickpeas.

It is concluded that botanical powders used in this study can be suggested as economical and easily available post-harvest grain protectants due to their non-toxic and eco-friendly nature. Insecticidal potential of other plants can be find by further studies on them. Moreover, different concentrations of these plants powders can also be tested.

## Acknowledgements

The authors are thankful to the Competent Authority at University of Gujrat, Gujrat, Pakistan and all others for all the possible help extended for the accomplishment of this project.

## References


Abbott WS, 1925. A method of computing the effectiveness of an insecticide. J. Economic Entomol. 18: 265-267.

Abdullahi YM and Muhammad S, 2004. Assessment of the toxic potentials of some plants powders on survival and development of *Callosobruchus maculates*. African J. Biotech. 3: 60- 62.

Adedire CO and Akinneye JO, 2004. Biological activity of tree marigold, *Tithonia diversifolia*, on cowpea seed bruchid, *Callosobruchus maculatus* (Coleoptera: Bruchi-dae). Ann. Appl. Biol. 144:185-189.

Ahmed K, Itino T and Ichikawa T, 2003. Duration of developmental stages of *Callosobruchu schinensis* (Coleoptera: Bruchidae) on Adzuki bean and the effects of Neem and Sesame oils at different stages of their development. Pak. J. Biol. Sci. 6: 932-335.

Ahmed S and Din N, 2009. Leaf Powders of basil (*Ocimum basilicum* L.), lantana (*Lantana camara* L.) and gardenia (*Gardenia jasminoides* Ellis) affect biology of *Callosobruchu schinensis* L. (Coleoptera: Bruchidae). Pak. Entomol.31: 5-9.

Ajayi OJT, Arokoyo JT, Nesan OO, Olaniyan M, Ndire- Mbula M and Kannike OA, 1987. Laboratory assessment of the efficacy of some local materials for the control of storage insect pests. Samaru. J. Agric. Res. 5(8):1-85.

Aslam M, Khan KA and Bajwa MJH, 2002. Potency of some spices against *C. chinensis.* L. Online J. Biol. Sci.2: 449–452.

Aslam M, Shaheen FA and Ayyaz A, 2006. Management of *Callosobruchus chinensis* Linnaeus in Stored Chickpea through Interspecific and Intraspecific Predation by Ants. World J. Agri. Sci. 2(1): 85- 89.

Ayyanger GSG and Rao UJ, 1989. Neem extract as larval repellent and oviposition deterrents to *Spodopteralitura* (Fabr.). Indian J. Entomol.51: 121-124.

Bamaryi LJ, Ndams IS, Toro WA and Odekrna S, 2007. Laboratory evaluation of Mahogany (Khaya senegalensis Desv.) Seed oil and seed powder for the control of *Callosobruchus maculatus* (Fab.) (Coleóptera: Bruchidae) on stored Cowpea. J. Entomol. 4: 237-242.

Dixit OP and Saxena RC, 1990. Insecticidal action of *Premnaintegri folia* against *Callosobruchus chinensis*. Pesticides. 24: 29-31.

Gautam P, Vaidya DN and Mehta PK, 2000. Evaluation of some edible plant products against pulse beetle, *Callosobruchus analis* (Fabr.) infesting green gram. Pest Management and Economic Zool. 8: 145-150.

Gill JS and Lewis CT, 1971. Systemic action of an insect feeding deterrent. Nature, 232: 402-403.

Gupta L and Srivastava M, 2008. Effect of *Withania somnifera* extracts on the mortality of *Callosobruchus chinensis.* L. J. Biopesticides. 1(2):190-192.

Halstead DGH, 1963. External sex differences in stored products Coleoptera. Bull. Ent. Res. 54: 119–134.

Hulse JH, 1991. Nature, composition and utilization of grain legumes, In: Uses of tropical Legumes. Proceedings of a Consultant Meetings, 27-30 March 1989, ICRISAT Centre, Patancheru, A.P. 502324, India.

Jilani G, Khan MJ and Ghiasuddin AL, 1988. Studies on insecticidal activity of some indigenous plant materials against the pulse weevil *Callosobruchus analis* F. Pakistan. J. Entomol. 3: 21-32.

Kakepoter TM, Naqvi SMSH, Rustami MA and Munshi GM, 1985. Relative Resistance of Different Pulses to *Callosobruchus chinensis* Linnaeus and its Population Fluctuations, pp. 101-106. In Proceedings, Entomological Society, Karachi, Sindh Agric. Univ. Tandojam, Pakistan.

Kazi SA, Takao I and Toshihide I, 1999. Effects of plant oils on oviposition preference and larval survivorship of *Callosobruchus chinensis* (Coleoptera: Bruchidae) on azuki bean. J. App. Entomol. and Zool. 34(4): 547-550.

Khalequzzaman M and Goni SHMO, 2009. Toxic potentials of some plant powders on survival and development of *Callosobruchu smaculatus*(F.) and *Callosobruchu schinensis.* L. J. Life Earth Sci. Vols. 3-4: 1-6.

Khan MA and Eggum BO, 1978. The nutritive value of some Pakistani diets. J. Sci. Food Agric.28: 1023-1029.

Khan MA, 1981. Nutritive value of cereals in relation to human needs. Pakistan. J. Agric. Res. 2: 141-146.

Kim SI, Roh JY, Kim DH, Lee HS and Ahn YJ. 2003. Insecticidal activities of aromatic plant extracts and essential oils against *Sitophilus oryzae* and *Callosobruchus chinensis*. J. Stored Prod. 39: 293-303.

Kumar A, Shukla R, Singh P, Singh A and Dubeya N, 2009. Use of essential oil from *Mentha arvensis* L. to control storage moulds and insects in stored chickpea. J. Sci. Food Agric. 89: 2643–2649.

Kumari K and Singh SN, 1998. Evaluation of efficacy of botanicals as insecticide against pulse beetle, *Callosobruchus chinensis*. L. J. App. Biol. 8:138-140.

Lal D and Raj DV, 2012. Efficacy of application of four vegetable oils as grain protectant against the growth and development of *Callosobruchus maculatus* and on its damage. Advances in Bioresearch. 3 (2): 55- 59.

Mahdi SHA and Rahman MK, 2008. Insecticidal effect of some spices on *Callosobruchus maculatus* (Fabricius) in black gram seeds: Univ. J. Zool. 27: 47-50.

Nisar MS, Ahmed S, Ashfaq M and Sahi ST, 2012. Effect of leaf and seed extracts of Jatrophacurcas Linn. on mortality and tunneling of subterranean termites, Odontotermesobesus (Ramb.) (Termitidae: Isoptera). Pak. J. Life Soc. Sci. 10(1): 33-38.

Pandey ND, Mathur KK, Pandey S and Tripathi RA, 1986. Effect of some plant extracts against pulse beetle *Callosobruchus chinensis*. Indian J. Entomol. 38: 110-113.

Partil PB, Holihosur SN and Kallapur VL, 2006. Efficacy of natural product, *Clerodendro nineme* against mosquito vector *Aedes aegypti*. Current Sci. 90(8):1064-1066.

Qayyum HA and Zafar MA, 1978. Research on Stored Grain Pests in Pakistan.pp. 67-112. Final Rep. PL480 Project, Department of Entomol., Univ. Agric. Faisalabad.

Rahman A and Talukder FA, 2006.Bio efficacy of some plant derivatives that protect grain against the pulse beetle, *Callosobruchus maculates*. J. Insect Sci. 6(3): 19-25.

Righi-Assia A, Khelil M, Medjdoub-Bensaad F and Righi K, (2010).Efficacy of oils and powders of some medicinal plants in biological control of the pea weevil (*Callosobruchus chinensis*L.). Afr. J. Agri. Res.5: 1474-1481.

Schoonhoven LM, 1982. Biological Aspects of Antifeedants. Entom. Exp. Appl. 31: 57-69.

Schoovoven AV, 1978.Use of vegetable oils to protect stored beans from bruchid attack. J. Economic Entomol. 71:254-256.

Shaaya E, Kostijukovski M, Eilberg J and Sukprakaran C, 1997.Plants oils as fumigants and contact insecticides for the control of stored product Insects. J. Stored Prod. Res. 33(1):7-15.

Sharaby A, 1988. Evaluation of some Myrtaceae Plant Leaves as Protectants against the Infestation by *Sitophilus oryzae* L. and *Sitophilus granaries* L. Insect Sci. Appl. 9(4): 465-468.

Shukla R, Kumar A, Prasad CS and Srivastava B, 2009. Efficacy of *Acoruscalamus* L. leaves and rhizome on mortality and reproduction of *Callosobruchus chinensis* L. (Coleoptera: Bruchidae). J. Appl. Entomol. Zool. 44(2): 241-247.

Shukla R, Srivastava B, Kumar R and Dubey NK, 2007. Potential of some botanical powders in reducing infestation of chickpea by Callosobruchus chinensis L. (Coleoptera:Bruchidae). J. Agri. Tech. 3(1): 11-19.

Steel RGD and Torrie JH, (1980). Principles and Procedures of Statistics: A Biometrical Approach, 2nd Ed. McGraw-Hill, New York.

Talukder, 2006. Plant Products as Potential Stored-product Insect Management Agents. Emir. J. Agric. Sci.18 (1):17-32.

Tapondjou LA, Adler C, Bouda H and Fontem DA, 2002. Efficacy of powder and essential oil from *Chenopodium ambrosioides* leaves as post- harvest grain protectants against six stored product beetles. J. Stored Prod. Res. 38: 395-402.

Udoh PB, Udoh FV, Umo IS and Ikpeme EV, 2010.Effect of Ipomoea batatas leaf extract on the thyroid-gonadal axis of male Wistar rats. J. Physio. Pathophysio. 1(2): 18-21.

Upadhyay RK and Ahmad S, 2011. Management Strategies for Control of Stored Grain Insects Pests in Farmer Stores and Public Warehouses. World J. Agric. Sci. 7(5): 527-549.

Upadhyay RK, Rohatgi L, Chaubey MK and Jain SC, 2006. Ovipositional responses of the pulse Beetle, *Bruchus chinensis* (Coleoptera: Bruchidae) to extracts and compounds of *Capparis decidua*. J. Agri. Food Chem. 54: 9747-9751.

Verma J and Dubey NK, 1999. Prospectives of botanical and microbial products as pesticides of tomorrow. Current Sci. 76(2): 172-179.

Zia S, Sagheer M, Abdul R, Mahboob A, Mehmood K and Haider Z, 2013. Comparative bioefficacy of different Citrus peel extracts as grain protectant against *Callosobruchus chinensis*, *Trogoder magranarium* and *Tribolium castaneum*. World Appl. Sci. J. 21(12): 1760-1769.

11

# Effect of seed soaking with bacillus sp and organic fertilizer on growth of mustard green (*Brassica juncea* L)

**I Ketut Widnyana*, I Wayan Seputra Kuspianta, Putu Lasmi Yulianthi Sapanca**
Faculty of Agriculture, Mahasaraswati University of Denpasar Bali, Indonesia

**Abstract**
Bacillus sp. is one type of bacteria that functions as PGPR (Plant Growth Promoting Rhizobacteria). This study is designed to determine the effect of treatment duration time immersion of mustard seeds with Bacillus sp and the doses of organic fertilizer on growth and yield of mustard plants. The research was designed using Randomized Block Design (RBD) Factorial with two treatments i.e. time soaking of mustard seed with Bacillus sp. and organic fertilizer dose. The results showed that there was a very significant interaction (P <0.01) between the treatment of soil seed immersion with Bacillus sp. and the dosage of organic fertilizer on most parameters observed, except for the root length of the plant. The highest fresh weight of the mustard plant was found in the combination of the treatment of soaking of mustard seeds with Bacillus sp. for 20 minutes with a dosage of 1.50 kg organic fertilizer per plant that is 131,900 g, and the lowest was found in the combination treatment of soaking of mustard seeds for 30 minutes with a dosage of 0 kg organic fertilizer i.e. 34.333 g, so an increase of 384.178%. The highest drying weight of the mustard oven was found in a combination of treatment of soaking of mustard seeds with Bacillus sp. for 30 minutes with a dosage of 1.50 kg organic fertilizer per plant of 8,000 g, and the lowest was found in a combination treatment of immersion of mustard seed for 10 minutes with a dose of organic 0 kg fertilizer per plant that is 4.617 g.

*Corresponding author email: widnyanaketut@gmail.com

**Keywords**: Soaking of Seed, Bacillus sp., PGPR, Organic Fertilizer

## Introduction

Mustard green (Brassica juncea L.) called sawi hijau (Indonesia); sawi bunga (Malaysia); phakkwangtung or phakkatkheokwangtung (Thailand); False pakchoi, Mock pakchoi (English); Cai xin (China) is a vegetable that is quite popular, easily cultivated and can be eaten fresh (usually withered with hot water) or processed into pickles. Mustard green has a wide environmental adaptation, both in the highlands and in the lowlands and widely consumed by people, short-lived 20-45 days. When these vegetables are left to age 60-65 days after planting, it will usually be harvested seeds (Poespodarsono, 1998).

Mustard green is very important role for the health of the human body because these vegetables contain vitamins, minerals and vegetable proteins that are needed to improve the nutritional value of food. Each adult is estimated to require 150 grams of vegetables daily (Sunaryono, 1994). In addition, the development of tourism sector in the area of Bali provides development opportunities for the agricultural sector, especially vegetable cultivation for there are 33 types of vegetables needed by hotels and restaurants. Some vegetables are still imported from outside the island (Antara and Susrusa, 1991).

The last decades of mustard green planting completely use chemical fertilizers as well as synthetic chemical pesticides. This type of cultivation is not good for consumer health, soil ecosystem damage physically, chemically and biologically. Technical improvements of mustard green cultivation should be pursued by utilizing environmentally friendly local resources, such as the utilization of PGPR bacteria such as Bacillus sp. derived from bamboo rhizosphere and combined with organic fertilizers sourced from the surrounding environment such as crop wastes, cattle dung which are abundant but not well utilized.

Rhizobacteria is a group of bacteria with plant roots habitat area (rhizosphere) that has been researched and proven to improve soil fertility, increase plant resistance and can suppress plant pathogens. Rhizobacteria act directly as biological fertilizer and stimulants biologically by producing hormones to grow crops such as IAA (indole acetic acid), gibberellin, cytokine, ethylene, dissolving minerals, and indirectly also prevents pathogenic microorganisms through formation of siderophore and an antibiotic (McMilan, 2007 ; Sarma et al., 2009). Bacillus sp. is a gram-positive bacteria are often used as the biological control of the root disease. The results have shown the potential of using selected strains of B. subtilis in the biological control of seed pathogens, as well as in promoting soybean growth (Araújo et al., 2005).

Bacillus sp. produce phytohormones which is useful directly or indirectly in the farm system. The benefits of phytohormones directly are able to increase the growth of plants and act as facilitators in nutrient uptake, and indirectly as an inhibitor of pathogenic activity in plants (Dwipayana, 2010).

Several studies have been done related to the utilization of PGPR in the cultivation of plants. Several researchers proved that treatments with Pseudomonas spp. can improve the plant growth of wheat and protect them from infection of Phytium spp. through seed treatment (Weller and Cook, 1986). Pseudomonas spp. can increase the number of leaves, height, and yield of tomato plants up to 81,7% through soaking of tomato seeds in suspension of the bacteria (Widnyana et al., 2013). Bacillus sp. tends to give a tomato seedling growing faster than the Pseudomonas spp. isolates until 2-3 days, while a mixture of both isolate the bacteria does not give a different impression with a single treatment (Widnyana and Javandira, 2015).

The study is conducted to obtain information on the benefits of Bacillus sp. rizobacteria combined with organic fertilizer for the growth of mustard green plants in the hope of suppressing the use of synthetic chemical fertilizers.

## Material and Methods

The research was conducted during August and September 2016 in a greenhouse in Badung regency Bali Indonesia. The materials used are green mustard seed, PGPR Bacillus sp. suspension from bamboo rizosphere, organic fertilizer, and sterile soil. The research was designed using Factorial Random Block Design consisting of two factors: the time of soaking of mustard green seed with Bacillus sp. (T) and the dose of organic fertilizer (D). There are 3 levels of immersion time that is 10 (T1), 20 (T2), and 30 minutes (T3), with 4 levels of fertilizer dosage : without organic fertilizer (D0), 0.50 kg organic plant$^{-1}$ (D1), 1.00 kg organic plant$^{-1}$ (D2) and 1.50 kg organic plant$^{-1}$ (D3). Thus there are 12 combinations of treatments with 3 replications so that there are 36 experimental units.

Green mustard seeds were soaked with Bacillus sp. suspension with a concentration of 10 ml in 1 liter of sterile water according to the treatment. Furthermore in the seedling nursery media. After the green mustard seedlings 4 leaves (age 14 days) are transferred to polybags that already contain soil mixed with organic fertilizer (2: 1) that has been roasted to minimize contaminants. Harvesting is done at the time of the growth of mustard green 21 days that has been marked by the emergence of flowers in most plants.

## Results and Discussion

The results showed the significant interaction effect ($P < 0.05$) to very significant ($P < 0.01$) between PGPR Bacillus sp. soaking time with a dose of organic fertilizer on parameters plant height, number of leaves, plant fresh weight, root fresh weight, oven dry weight, and fresh weight of green mustard greens ha$^{-1}$; but did not interact significantly ($P > 0.01$) to the roots length of mustard green. Effect of soaking time interaction seed treatment with Bacillus sp. and dose of organic fertilizer to green mustard plants are presented in Table 1.

**Table 1. Effect of interaction between time of soaking with Bacillus sp. and organic fertilizer dosage to mustard green height (cm).**

| Treatment | Plant height (cm) Dose of organic fertilizer (D) (kg plant$^{-1}$) | | | |
|---|---|---|---|---|
| | 0 ($D_0$) | 0,50 ($D_1$) | 1,00 ($D_2$) | 1,50 ($D_3$) |
| Soaking seed (T) | | | | |
| 10 minute ($T_1$) | 29,633 fg | 40,400 b | 46,167 a | 44,167 a |
| 20 minute ($T_2$) | 29,733 ef | 42,933 b | 44,000 ab | 46,233 a |
| 30 minute ($T_3$) | 27,033 g | 35,033 d | 40.400 c | 40,967 b |

Note: values on rows and columns followed by the same letter, are not significantly different in DMRT 5%

The result of statistical analysis in Table 1. Shows that the combination of T2D3 treatment reached maximum plant height of 46,233 cm and not significantly different with T2D2, T1D3 and T1D2 treatment. Whereas minimum plant height reached at 27.033 sm in T3D0. This suggests that soaking mustard greens for 10 minutes or 20 minutes combined with a 1 kg or 1.5 kg fertilizer dose gives the same effect to the height of mustard green plants.

Effect of interaction soaking time seed treatment with Bacillus sp. and organic fertilizers on the number of green mustard leaves are showed in Table 2.

**Table 2. Effect of interaction between the time soaking seeds with Bacillus sp. and a dose of organic fertilizer to the number of green cabbage leaves (leaf).**

| Treatment | Number of leaves (leaf) Dose of organic fertilizer (D) (kg plant$^{-1}$) | | | |
|---|---|---|---|---|
| | 0 ($D_0$) | 0,50 ($D_1$) | 1,00 ($D_2$) | 1,50 ($D_3$) |
| Soaking seed (T) | | | | |
| 10 minute ($T_1$) | 6.933 cd | 7,067 b | 8,267 a | 7,800 a |
| 20 minute ($T_2$) | 6,067 d | 7,267 a | 7,333 a | 7,533 a |
| 30 minute ($T_3$) | 7,467 a | 8,000 a | 7,800 a | 8,133 a |

Note: values on rows and columns followed by the same letter, are not significantly different in DMRT 5%

The result of statistical analysis in Table 2. Showed that the highest number of leaves was found in T3D3 with mean of 8,133 strands, but not significantly different with T3D2, T3D1, T3DO, T2D3, T2D2, T2D1, T1D3 and T1D2 treatment. The lowest number of mustard green leaves found in T2D0 is 6,067 and not different with T1D0. These results indicate that the doses of organic fertilizers give almost the same effect to the number of mustard green leaves as long as the green mustard seeds are soaked 20 to 30 minutes.

In the root mustard green parameter, there was no interaction between the treatment of soaking time with Bacillus sp. and the dosage of organic fertilizer, so that the result of the research was analyzed to obtain the effect of single factor, as presented in Table 3.

**Table 3. Effect of soaking treatment of mustard green seed with Bacillus sp. and dosage of organic fertilizer on root length of plant$^{-1}$ (cm).**

| Treatment | root length of mustard green (cm) |
|---|---|
| Soaking seed (T) | |
| 10 minute ($T_1$) | 10,19 c |
| 20 minute ($T_2$) | 13,01 b |
| 30 minute ($T_3$) | 15,01 a |
| BNT 5% | 0,087 |
| Dose of organic fertilizer (D) (kg plant$^{-1}$) | |
| 0 kg ($D_0$) | 11,29 a |
| 0,50 kg ($D_1$) | 12,50 a |
| 1,00 kg ($D_2$) | 13,52 a |
| 1,50 kg ($D_3$) | 13,64 a |
| BNT 5% | 2,60 |

Note: The values followed by the same letter on the same treatment and column are not significantly different on LSD 5%.

The statistical analysis of the effect of the treatment on length of root mustard as presented in Table 3 shows that soaking with Bacillus sp. for 30 minutes gives the best effect to the root of the green mustard with an average of 15.01 cm, in contrast with 20 minutes (13.01 cm) and soaking 10 minutes (10.19 cm). But the treatment of organic fertilizers with different doses did not give a real different effect on root mustard green length.

The effect of the interaction of time soaking of the mustard seed with Bacillus sp. and the dosage of organic fertilizer to the fresh weight of the mustard green plant is presented in Table 4.

**Table 4. Effect of the interaction between green mustard seed soaking treatment with Bacillus sp. and a dose of organic fertilizer to the plant fresh weight (g).**

| Treatment | The fresh weight of the plant$^{-1}$ (g) Dose of organic fertilizer (D) (kg plant$^{-1}$) | | | |
|---|---|---|---|---|
| | 0 ($D_0$) | 0,50 ($D_1$) | 1,00 ($D_2$) | 1,50 ($D_3$) |
| Soaking seed (T) | | | | |
| 10 minute ($T_1$) | 36,333 ef | 63,333 cd | 96,000 a | 104,000 a |
| 20 minute ($T_2$) | 27,667 f | 74,000 bc | 85,333 abc | 109,333 a |
| 30 minute ($T_3$) | 29,667 f | 62,667 d | 87,000 abc | 104,000 a |

Note: values on rows and columns followed by the same letter, are not significantly different in DMRT 5%

The result of statistical analysis of parameter to fresh mustard green weight in Table 4, indicated that the highest fresh green mustard weight found in T2D3 is 109.333 g is not significantly different from T1D3, T3D3, T2D2 and T3D2 treatment. The lowest fresh mustard green weight found in T2D0 treatment of 27,667 g was not significantly different from T1DO and T1D3. These results indicate that the soaking time of 10, 20, and 30 minutes combined with a dose of 1 kg or 1.5 kg per plant gives no significant effect on the mean fresh weight of the green mustard.

The effect of the interaction of soaking mustard seeds with Bacillus sp. and the dosage of organic fertilizer to oven dry weight of mustard green plant is presented in Table 5.

The statistical analysis of dry oven mustard green weight in Table 5 shows that the highest dry mustard green oven weight found in a combination of T3D3 treatment of 8,000 g was not significantly different from T3D2, T3 D1, T3D2, T2D2 and T1D2. The lowest dry oven mustard green weight found in T1D0 treatment of 4,100 g was significantly different from all other treatments except T1D3, T2D3, and T3D3.

The effect of the treatment of soaking time of the mustard seed with Bacillus sp. and the dosage of organic fertilizer to the fresh weight of the mustard green ha$^{-1}$ is presented in Table 6.

**Table 5. Effect of interaction between soaking treatment of mustard green seed with Bacillus sp. and dose of organic fertilizer on dry weight of oven crop$^{-1}$ (g).**

| Treatment | Oven dry weight of plant$^{-1}$ (g) Dose of organic fertilizer (D) (kg plant$^{-1}$) | | | |
|---|---|---|---|---|
| | 0 ($D_0$) | 0,50 ($D_1$) | 1,00 ($D_2$) | 1,50 ($D_3$) |
| Soaking seed (T) | | | | |
| 10 minute ($T_1$) | 4,100 c | 4,617 c | 6,100 abc | 7,347 ab |
| 20 minute ($T_2$) | 4,533 c | 5,310 bc | 5,867 abc | 7,833 ab |
| 30 minute ($T_3$) | 4,617 c | 5,333 bc | 6,367 abc | 8,000 a |

Note: values on rows and columns followed by the same letter, are not significantly different in DMRT 5%

**Table 6. Effect of interaction between soaking time treatment of mustard green seed with Bacillus sp. and dose of organic fertilizer on fresh weight of mustard green ha$^{-1}$ (t)**

| Treatment | Plant fresh weight ha$^{-1}$ (t) Dose of organic fertilizer (D) (kg plant$^{-1}$) | | | |
|---|---|---|---|---|
| | 0 ($D_0$) | 0,50 ($D_1$) | 1,00 ($D_2$) | 1,50 ($D_3$) |
| Soaking seed (T) | | | | |
| 10 minute ($T_1$) | 9,184 cd | 21,088 ab | 23,980 a | 23,810 a |
| 20 minute ($T_2$) | 8,163 d | 24,490 a | 24,150 a | 25,850 a |
| 30 minute ($T_3$) | 7,313 d | 17,007 b | 23,469 a | 21,088 ab |

Note: values on rows and columns followed by the same letter, are not significantly different in DMRT 5%

The result of statistical analysis on the fresh weight of mustard green ha$^{-1}$ shows that after conversion in the hectare planting area (ha$^{-1}$) the highest production is found in T2D3 treatment (soaking seed 20 minutes and 1.5 kg of organic fertilizer per plant) of 25,850 tons ha$^{-1}$. This result is the highest yield potential of the type of green mustard Tosakan variety in accordance with the product description of 20 - 25 tons ha$^{-1}$. Production of green mustard on T2D3 is not different from T2D1

(24,490 ton $ha^{-1}$), T2D2 (24,150 ton $ha^{-1}$), T1D2 (23,980 ton $ha^{-1}$), T1D3 (23,810 ton $ha^{-1}$), T3D3 (21,088 ton $ha^{-1}$) and T1D1 (21,088 tons $ha^{-1}$).The lowest production of mustard green was found in a combination of T3D0 treatment of 7,313 tons $ha^{-1}$ was not significantly different from T2D0 (8,163 ton $ha^{-1}$) and T1D0 (9,184 ton $ha^{1}$).

Based on the understanding of previously presented data it can be stated generally that the immersion of green mustard seed with Bacillus sp. PGPR rizobakteri for 10, 20 and 30 minutes combined with 1 kg organic fertilizer dose or 1.5 kg per plant gives no significant effect to plant growth parameters such as plant height, leaf number, fresh weight of plant, oven dry weight of plant and fresh weight of mustard green $ha^{-1}$. Plant growth promoting bacteri (PGPB) are naturally soil bacteria that aggressively colonize plant roots and benefit plants by providing growth promotion. This is in line with opinion Antoun and Kloepper (2001), inoculation of crop plants with certain of PGPB at an early stage of development improves biomass production through direct effects on root and shoot growth. The results of this study according to opinion Widnyana and Javandira (2015) that tomato seed treatment by soaking period of Bacillus sp. isolates between 10 to 30 minutes does not give a different effect on the time of germination and growth of tomato seedlings. Similar research that has been done by Widnyana, et al. (2018) showed that the soaking treatment of Swamp Cabbage (Ipomoea reptans Poir) seeds with suspense P. alcaligenes as PGPR encouraged 25% faster germination, increased yield by 24.4%, increased leaf count to 23.15%, extended stem until 25%, extend the roots to 46.90%, and increase the weight of fresh stems up to 67.07% and dry oven bar stems up to 84.21% compared to soaking the seeds in sterile water.

Rhizobacteria Bacillus sp. can produce phytohormones that have the potential to develop sustainable agricultural systems. The phytohormones produced by these soil bacteria can affect plant growth, either directly or indirectly. The phytohormone indirectly inhibits pathogenic activity in plants, whereas the immediate effect of fitohormon is to increase plant growth and can act as a facilitator in the absorption of some nutrients from the environment.

However, if considered in terms of efficiency of organic fertilizer use and based on the extent of planting ha-1 it is important to know that to obtain maximum potential yield with the lowest use of fertilizer, the soaking of mustard green seed with Bacillus sp. is done for 20 minutes and the dose of organic fertilizer used 0.5 kg of plant -1 will provide the highest yield potential that is 24,490 ton ha-1 or lower 1,360 tonnes compared to 20 minutes immersion and the use of 1.5 kg ha-1 organic fertilizer is 25,850 ton ha-1. When converted to the needs of cost, then the addition of 1 kg of organic fertilizer plant-1, or about 40 tons ha-1 only able to add production of about 1,360 tons of mustard green. By increasing the cost of purchasing organic fertilizer 40 million rupiah can only get the results of 40 million and 480 thousand rupiah excluding freight costs and application fees. This suggests that mustard green plants require 0.5 kg of organic fertilizer plant-1, and if given more than that dose does not result in a significant increase in fresh weight.

Organic materials in addition to the effect on soil nutrient supply also to the physical, biological and chemical properties of the soil. Good soil physical condition can guarantee the growth of plant roots and able as a place of aeration and soil moisture. Organic matter plays an important role in the formation of good soil aggregates and structures, thus improving soil physics conditions, and ultimately facilitating water penetration, water absorption, root development, and increased resistance to erosion (Thompson and Goyne, 2012).

Soil organic matter determines the level of soil fertility, both chemical fertility, physics, and soil biology. This can be due to organic matter capable of contributing nutrients after decomposition, improving soil drainage, increasing infiltration, retention and transmission of water in the soil, and loosening soil and stabilizing soil aggregates, so that plant roots can thrive for nutrients and water for growth. The process of decomposition or mineralization, in addition to being influenced by the quality of organic matter, is also influenced by the frequency of adding organic matter, the particle size of the material, the drought, and the way it is used (mixed or dispersed on the surface) (Vanlauwe et al., 1997).

## Conclusion

Treatment of green mustard seed soaking time and dose of organic fertilizer give real effect to the very real interaction on plant height, leaf number, fresh weight, oven dry weight, and plant fresh weight per hectare; but not significantly different for the root length. The soaking of green mustard seed with Bacillus sp.as PGPR rhizobacteria for 10, 20 and 30 minutes combined with 1 kg or 1.5 kg organic fertilizer dose per plant gives no significant effect to plant growth parameters. The soaking of mustard green seed with Bacillus sp is done for 20 minutes and the dose of organic fertilizer used 0.5 kg of plant $^{-1}$ will provide the highest yield potential that is 24,490 ton ha$^{-1}$ a combination treatment of the most effective, efficient and profitable.

## Acknowledgement

I would like to extend my gratitude to the Director of Directorate of Research and Community Service Kemenristekdikti, Rector of Mahasaraswati University of Denpasar, and Dean of Faculty of Agriculture Mahasaraswati Denpasar University for the facilities provided.

## References

Antara M and Susrusa KB, 1991. Intensification of Quality of Vegetable Demand by Hotels and Restaurants of Badung Regency of Bali Province. Denpasar: Faculty of Agriculture Universitas Udayana Denpasar (in bahasa)

Antoun H and Kloepper JW, 2001. Plant growth–promoting rhizobacteria (PGPR). In: Brenner S, Miller JH (eds.) Encyclopedia of genetics. Academic, New York, pp 1477–1480

Araújo FF, Henning AA and Hungria M, 2005. Phytohormones and Antibiotics Produced by Bacillus subtilis and their effects on Seed Pathogenic Fungi and on Soybean Root Development. World J. Microbiol. Biotech. 21(8): 1639-1645.

Dwipayana HD, 2010. Identification of Bacterial Diversity in Waste Recycling Paint Sludge by Conventional Microbiological Technique. Environmental Engineering Study Program. Bandung.

McMilan S, 2007. Promoting Growth with PGPR. The Canadian Organic Grower. http//:www.cog.ca/

Poespodarsono S, 1998. Mustard plant. Malang: Faculty of Agriculture Universitas Brawijaya. (in bahasa)

Sarma MV, Saharan RK, Prakash K, Bisaria A and Sahai V, 2009. Application of Fluorescent Pseudomonads Inoculant Formulations on Vigna mungo through Field Trial. Int. J. Biol. Life Sci. 1(1): 25-29.

Sunaryono, 1994. Planting Vegetables Grow key important in Indonesia. Jakarta: Horticultural Research Institute Pasar Minggu Jakarta.(in bahasa)

Thompson A and Goyne KW, 2012. Introduction to the sorption of chemical constituents in soils. Nature Edu. Knowledge. 3(6): 15.

Van Peer R and Schippers B, 1988. Plant Growth Responses to Bacterization with Selected Pseudomonas Spp. Strains and Rhizosphere Microbial Development in Hydroponic Cultures. Can. J. Microbiol. 35: 456–463

Vanlauwe B, Diel J, Sanginga N and Merckx R, 1997. Residue quality and decomposition: An unsteady relationship. In Dirven by Nature Plant Litter Quality and Decomposition, (Eds. Cadisch, G. and K.E. Giller), pp. 157-166. Department of Biological Sciences, Wey College, University of London, UK.

Weller DM and Cook RJ, 1986. Increased Growth of Wheat by Seed Treatments with Fluorescens Pseudomonads, and Implications of Phytium Control. Can. J. Plant Pathol. 8:328-334.

Widnyana IK, Suprapta DN, Sudana IM and Rai Maya Temaja IG, 2013. *Pseudomonas alcaligenes* Potential Antagonist against *Fusarium oxysporum* f. sp. lycopersicum the Cause of Fusarium Wilt Disease on Tomato. J. Biol. Agric. Healthcare. 3(7): 163-169.

Widnyana IK and Javandira C, 2015. Activity Pseudomonas spp. and Bacillus sp. to stimulate Germination and Seedling Growth of tomato plants International Conference on Food, Agriculture and Natural resources (IC-FANRes, 2015) The University of Jember. August 31 - 2 September 2015. Agric. Agric. Sci. Procedia. 9: 419-4223.

Widnyana IK, Ngga M and Sapanca PLY, 2017. The Effect of Seed Soaking with Rhizobacteria *Pseudomonas alcaligenes* on the Growth of Swamp Cabbage (*Ipomoea reptans* Poir). OP Conf. Series: J. Physics: Conf. Series 95(2017): 12007.

12

# Effect of imidacloprid and bifenthrin on predation efficiency of *Coccinella septempunctata* (Coleoptera: Coccinellidae) under laboratory conditions

**Mubashar Hussain[1]*, Iqra Mustafa[1], Muhammad Faheem Malik[1], Sharon Zulifqar[1], Zaheer Abbas[2]**
[1]Department of Zoology, [2]Department of Statistics, University of Gujrat, Punjab, Pakistan

## Abstract

*Coccinella septempunctata* is an important natural enemy of several pest species attacking various crops in Pakistan. Mostly pests are managed by the use of pesticides in crops and ornamentals, thus, impact of pesticides needs to be evaluated on voracity of predatory beetle. The effect of Imidacloprid and Bifenthrin on predation of adult beetles predating on Russian wheat aphid, *Diuraphis noxia* (Mordvilko) was evaluated under laboratory conditions (25 ± 2 ºC and 65 ± 5 % RH) at Systematics and Pest Management Laboratory, Department of Zoology University of Gujrat. The significant differences in the mortality of *C. septempunctata* were observed amongst control and other treatments. The adult beetles showed highest mortality (91.66 % and 83.83%) when subjected to Bifenthrin (0.7%) after 48 and 24 h of exposure, respectively. Whereas the adult beetles demonstrated highest mortality (75 % and 72.66 %) when exposed to Imidacloprid (0.7%) after 48 and 24 h of exposure, respectively. The statistical analysis yielded non-significant differences were recorded at lower concentrations of both pesticides. *C. septempunctata* showed lower mean predation of 20.66, 18.66 and 16 when exposed to 0.01, 0.02 and 0.03 % Bifenthrin after 24 h, respectively. The voracity of *C. septempunctata* was significantly affected by Imidacloprid and Bifenthrin. The study emphasized that non target beneficial species are affected due to residual insecticidal spray effects reducing their population and increasing more dependency on hazardous chemical toxicants.

*Corresponding author email: dr.mubashar@uog.edu.pk

**Keywords:** Pesticides, Voracity, Non-target species, Coccinellids

## Introduction

Environmental hazards and costs are the key considerations while devising pest management strategies. Biological control agents, such as predators and parasitoids, are major component of integrated management of insect pests (Ahmad et al., 2011). Integrated pest management practices have been considered the best option for the management of insect pests. Pesticides as an effective and rapid control option is the most preferred strategy of pest management by the end users despite it poses serious threats to environment by developing resistance in pests, resurgence of various pests, pest outbreaks and a big menace to non-target beneficial insects (Solangi et al., 2007; Khaliq et al., 2007).

The integrated pest management programs are in the limelight from last three decades as the exclusive dependence on pesticides has been denting biodiversity and environment seriously. Coccinellid predators are the amongst major biological control agent that are naturally controlling pest populations in all natural terrestrial ecosystems and agro-ecosystems (Rain et al., 2016). In the current scenario of extensive use of pesticides in agro-ecosystems, the efficacy of predation and impact of pesticides on non-target species need to be evaluated time and again in order to develop sustainable IPM programs for agricultural and horticultural pests. This scenario explains the significance of evaluating existing and new pesticides for their effectiveness and environmental implications in the natural systems where biological control agents are acting as the friends of human by reducing the pests to their threshold levels. Coccinellidae family consists of about 5200 described species worldwide (Hawkeswood, 1987; Sarmad et al., 2015).

The perusal of literature indicated that about 75 species of predatory Coccinellids have been reported (belonging to 37 genera; 2 subgenera; 14 tribes and 05 subfamilies) from Pakistan (Rafi et al., 2005; Arif et al., 2011). Coccinellid beetles are amongst the effective predators of sucking insect pests (Dixon, 1998; Inayat et al., 2011).

Aphids, the destructive sucking pests of agricultural and horticultural crops, distributed worldwide with recorded species of about 4000 infesting over 250 crops (Ali and Rizvi, 2007). The ladybird beetles serve as natural control agents against insect pests of crops and in many cases are being utilized for the effective control of aphids and other insect pests all over the world. The perusal of literature indicated the harmful impact on the beetles predating on aphids but little work have been found on efficacy of predation, life cycle of predators and mortality of predators (Razaq et al., 2005; Cabral et al., 2011).

The study was undertaken to evaluate the impact of insecticidal spray on the mortality of adult beetles and efficacy of predation of *C. septempunctata* under laboratory conditions. The effect of Imidacloprid and Bifenthrin was evaluated on the predation efficiency of adults of *C. septempunctata* when sprayed on plant parts hosting aphid preys. The study emphasized on the impact non target species while managing pests through chemical control agents.

## Materials and Methods

The experiment was conducted in March-April, 2016 in the Laboratory of Systematics and Pest Management, Department of Zoology, University of Gujrat, Punjab, Pakistan. The research work was carried out to evaluate the impact of different concentrations of Imidacloprid and Bifenthrin on the mortality of *C. septempunctata* and its predatory potential of Russian Wheat Aphid, *Diuraphis noxia* (Mordvilko). The experiment was carried out under laboratory conditions (Table 1 & 2). Toxicity of Imidacloprid and Bifenthrin against adult ladybird beetles was evaluated by contact method using wheat leaves in the laboratory bioassay under laboratory conditions (25 ± 2 °C and 65 ± 5 % RH)

Adults of Coccinellid beetles, *C. septempunctata* were collected from fields adjacent to the Hafiz Hayat Campus, University of Gujrat, in 2016. These beetles were reared on wheat aphid under conditions suggested by earlier researchers to establish colonies in the laboratory (Naveed et al., 2007).

### Specimen collection and identification

The specimens of adult ladybird beetles and aphids were collected from wheat fields adjacent to the Hafiz Hayat Campus, University of Gujrat. The beetles were collected by using sweep net whereas aphids were collected by removing twigs from the plants. The beetles as well as aphids were collected from the same fields as to maintain the predator-prey preference. The specimens were brought in the laboratory soon after collection.

The experimental animals were kept in petri dishes (9 x 1.5 cm) covered with plastic lid having proper ventilation openings. The specimens were subjected to identification by following Taxonomic keys and using microscope (ZMS-9). The adult of *C. septempunctata* were exposed to dry residues of Imidacloprid and Bifenthrin the way suggested earlier (Cabral et al., 2011).

### Preparation of insecticidal treatments

The experimented was conducted in Completely Randomized Design (CRD) with three replications. The doses were prepared for the selected insecticides Imidacloprid and Bifenthrin by following formula:

$$\text{Volume of Insecticide(ml)} = \frac{\text{Total volume (L) x Percentage of insecticde required (\%)}}{\text{Formulation of insecticide}}$$

**Table 1: Treatments of Bifenthrin and Imidacloprid to Determine Mortality (%) of *C. septempunctata***

| Bifenthrin | | Imidacloprid | |
|---|---|---|---|
| Treatments | Concentrations (%) | Treatments | Concentrations (%) |
| $T_1$ | 0.0 (Control) | $T_1$ | 0.0 (Control) |
| $T_2$ | 0.1 | $T_2$ | 0.1 |
| $T_3$ | 0.2 | $T_3$ | 0.2 |
| $T_4$ | 0.3 | $T_4$ | 0.3 |
| $T_5$ | 0.4 | $T_5$ | 0.4 |
| $T_6$ | 0.5 | $T_6$ | 0.5 |
| $T_7$ | 0.6 | $T_7$ | 0.6 |
| $T_8$ | 0.7 | $T_8$ | 0.7 |

**Table 2: Treatments of Bifenthrin and Imidacloprid to Work out Voracity of *C. septempunctata***

| Bifenthrin | | Imidacloprid | |
|---|---|---|---|
| Treatments | Concentrations (%) | Treatments | Concentrations (%) |
| $T_1$ | Control (no insecticidal treatment) | $T_1$ | Control (no insecticidal treatment) |
| $T_2$ | 0.01 | $T_2$ | 0.01 |
| $T_3$ | 0.02 | $T_3$ | 0.02 |
| $T_4$ | 0.03 | $T_4$ | 0.03 |

**Exposure to insecticidal treatments**

The pre counted and treated aphids and beetles were placed in petri dishes along with wheat leaves as per procedure used by Desneux et al., (2007) with little modifications. Each replication contained 20 beetles and adequate number of pre-counted aphids. The treatment was applied on each group by topical exposure procedure to control individual dosage (and prohibited possible anti-feedant consequence of insecticides. For both pesticides, fresh solutions were prepared by diluting with distilled water at the doses recommended by the manufacturer for the control of aphids (Bifenthrin 250 ml/Acre; Imidacloprid 100- 250 ml / Acre). The solutions were prepared as following Fatima et al., (2016).

**Data collection and analysis**

The data of the mortality of beetles and aphids and predation was recorded after every 24 and 48 h of insecticidal exposures (Fatima et al., 2016). The data collected on the mortality of beetles was corrected by subjecting it to the Abbotts's Formula (Abbott, 1925):

**Corrected Mortality**

$$= \frac{\text{Mortality in Treatment [\%]} - \text{Mortality in Control [\%]})}{(100\% - \text{Mortality in control [\%]})}$$

Whereas the predation potential of *C. septempunctata* was determined by using voracity model (Soares et al., 2003).

The data collected on mortality and aphid voracity of adult coccinellid beetle, *C. septempunctata* was subjected to ANOVA by using SPPSS 21 and means were compared for significance (Zar, 1996).

## Results and Discussion

The series of concentrations of Imidacloprid and Bifenthrin were tested against the mortality of *C. septempunctata* in order to calculate the $LC_{50}$ and $LC_{90}$ by Probit Analysis using SPSS 21. On the basis of preliminary screening of pesticide concentrations, seven concentrations of each Imidacloprid and Bifenthrin were applied to find out mortality of beetles after 24 and 48 h in comparison with control (Table 3).

**Table 3: $LC_{50}$ and $LC_{90}$ Values for Imidacloprid and Bifenthrin at Different Times of Exposure**

| Insecticides | | Bifenthrin | Imidacloprid |
|---|---|---|---|
| 24 h | $LC_{50}$ | 0.053452 | 0.093287 |
| | $LC_{90}$ | 0.193536 | 0.33777 |
| 48 h | $LC_{50}$ | 0.029664 | 0.043858 |
| | $LC_{90}$ | 0.098883 | 0.146199 |
| Overall | $LC_{50}$ | 0.039593 | 0.062787 |
| | $LC_{90}$ | 0.145805 | 0.231222 |

The beetles were also subjected to insecticidal treatments to evaluate the impact of these pesticides and their concentrations on the non-target species (*C. septempunctata*) and predation efficiency.

**Mortality (%) of *C. septempunctata* after 24 and 48 h**

The data on mortality of adult *C. septempunctata* at 24 and 48 h after the application of treatments was given in (Figure 1). The data showed significant differences between various two duration were observed ($F_{(1, 89)}$ = 86.15, $P < 0.05$). The higher mortality was recoded at 48 h at all concentrations of both pesticides. There was no mortality in the control after 24 h, however, after 48 h non-significant mortality (2.66 & 4.16 % in Bifenthrin and Imidacloprid, respectively) after 48 h was observed which was non-significant with $T_2$ and $T_3$ (Figure 1). Overall mean mortality (%) of adult *C. septempunctata* at 24 (25.88 %) and 48 (45.55 %) h of insecticidal exposure indicated significant differences. The data on mortality exhibited significant differences between different concentrations ($F_{(14, 89)} = 33.89$, $P < 0.05$). Adult *C. septempunctata* showed highest mortality (91.66%) the when Bifenthrin (0.7%) after 48 h followed by beetle mortality (83.83%) after 24 h when exposed to same pesticide and concentration. Whereas lower mortality was recorded at lower concentrations of both pesticides and which was non-significant at $T_1$, $T_2$ and $T_3$ in Imidacloprid but significantly different from other treatments (Figure 1).The data indicated significant differences between two pesticides ($F_{(1, 89)} = 139.89$, $P < 0.05$).The highest values of mortality of 75 % were recorded followed by 72.66 % when beetles were exposed to Imidacloprid (0.7%) for 48 and 24 h, respectively.

The data indicated higher mortality (%) of adult *C. septempunctata* when the concentration was increased for both pesticides after 24 and 48 h. The findings of this study are in confirmation with earlier results presented by various researchers that Imidacloprid and Bifenthrin showed toxicity to both target and non-target species (Amer et al., 2010). Higher toxicity levels were observed in Bifenthrin as compared to Imidacloprid (Amin et al., 2014; Devee and Baruah, 2010). Predation rate decreased and mortality rate increased with the increase of insecticidal concentrations (Fernandes et al., 2016). Imidacloprid showed higher values of $LC_{50}$ and $LC_{90}$ as compared to Bifenthrin are in confirmation with the results presented in this study that depicted higher mortality in the beetles exposed to Bifenthrin. Higher mortality rates were observed after 24 and 48 h by exposure of Bifenthrin and Imidacloprid when mealy bugs were exposed to different concentrations (Fatima et al., 2016).

In addition to the direct effects presented in these results in the form of beetle mortality induced by pesticides, there may be some indirect effects of lethal and sub-lethal concentrations and repeated exposures on insect populations surviving leading to alterations in the longevity, reproduction, fecundity, development time, feeding behavior, mobility, and prey acceptance, etc.(Ruberson et al., 1998). Such sub-lethal effects more pronounced in the case of newly developed insecticide (Wright and Verkerk, 1995; Ochiai et al., 2007).

**Mean voracity by *C. septempunctata***

The data on the voracity of *C. septempunctata* adults was worked out for comparison between treatments by applying ANOVA ($P < 0.05$). The results presented in Fig. 2 showed significant results for treatments and duration ($F_{(2, 43)} = 23.57$, $P = 0.0000$, $F_{(1, 43)} = 10.40$, $P = 0.0024$). Predation rate decreased with increase in the concentration of insecticidal concentration highest predation was showed in control group (56 and 82.66 at 24 and 48 h, respectively) where there was no exposure of insecticides to the experimental unit. The voracity of *C. septempunctata* was higher when predators were added to control group where no insecticidal spray was carried out whereas lower voracity was observed when topical insecticidal spray was applied. The results showed positive increase in the voracity in control whereas a decreasing trend was noticed in all other treatment which may be attributed to non-feeding repellency and poisonous hazards caused by the pesticides. This may also have indirectly influenced their non-preference to their prey due to insecticidal effects on prey.

*C. septempunctata* showed significantly low predation by 20, 16, 18.66 after 24 h when exposed to Bifenthrin (0.01, 0.02 % & 03 %) respectively, whereas 36.33, 29.33, 21.66 after 48 h when exposed to Bifenthrin (0.01, 0.02 % & 03 %) respectively, as compared to control. This indicated that predation was lowest at higher concentrations of Imidacloprid and Bifenthrin (Figure 2). The aphid consumption of *C. septempunctata* decreased under insecticidal spray. The decrease in the aphid consumption with the increase in the concentrations of insecticidal from 0.01 to 0.03 % solutions of both pesticides may be attributed to physiological alterations caused by the pesticides in both predator and prey. This may be due to non-preference of Coccinellid beetles of the insecticidal sprayed aphids. The earlier studies endorsed the fact that the reduced consumption of prey resulted when insecticide treated prey were fed and changes in the consumption of aphids by Coccinellid beetles were recorded (Thornham et al., 2007; Qi et al., 2001). This may also be due to modified habitat and limited space for countering the stress factor due to insecticides. Topical exposure to different pesticides decreases significantly the rate of aphid consumption (Provost et al., 2005; Liu and Stansly, 2004). The research works have been carried out in various parts of the world on the ingestion insecticide treated prey by Coccinellid beetles that had demonstrated negative effects of various pesticides at different concentrations on non-target species directly by causing mortality or indirectly by reducing longevity, fecundity, oviposition period and behavior (Angeli et al., 2005; Wang et al., 2005; Thornham et al., 2008; Ujjan et al., 2014). The studies also demonstrated that the efficacy of predation decreases with the use of insecticides to control pests. The result showed that there is serious decline in the aphid consumption in the insecticide treated groups which may be due to knockdown effect, locomotory and behavioral recorded by Yuxian et al., 2012. The work carried out by them reported sublethal effects of imidacloprid affect adversely development and reproductive capacity and may lead to lower population of the coccinellid predator. Imidacloprid residue lowers predator voracity.

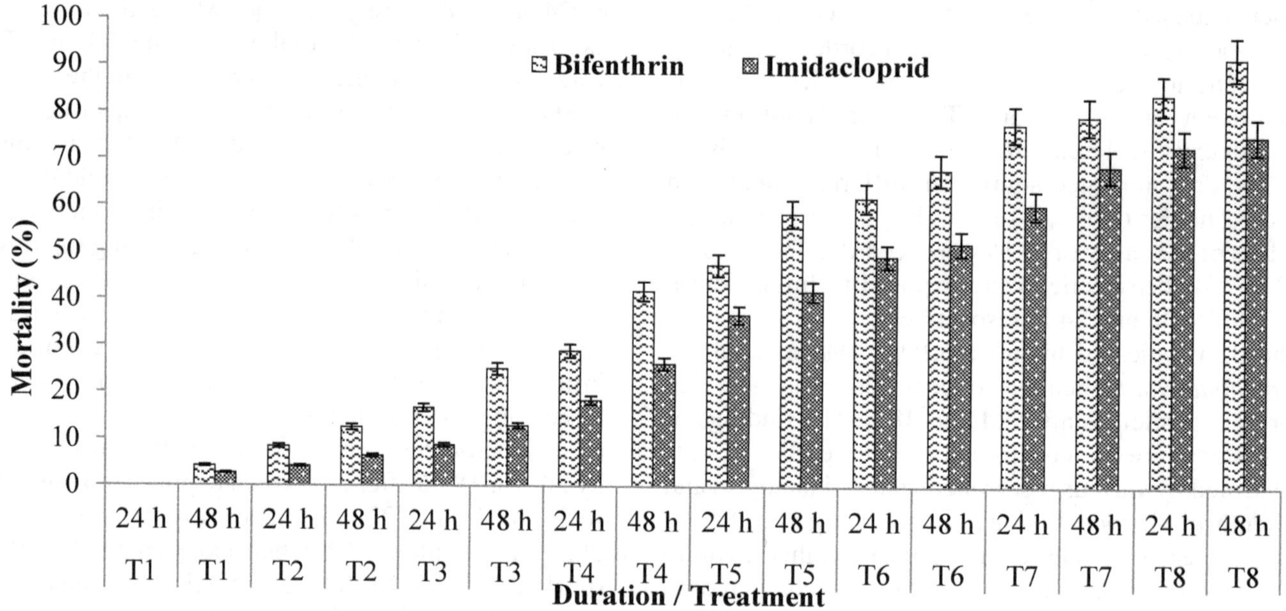


**Figure 1: Mortality (%) of Adult *C. septempunctata* after 24 and 48 h of Exposure to Different Concentrations of Imidacloprid and Bifenthrin**

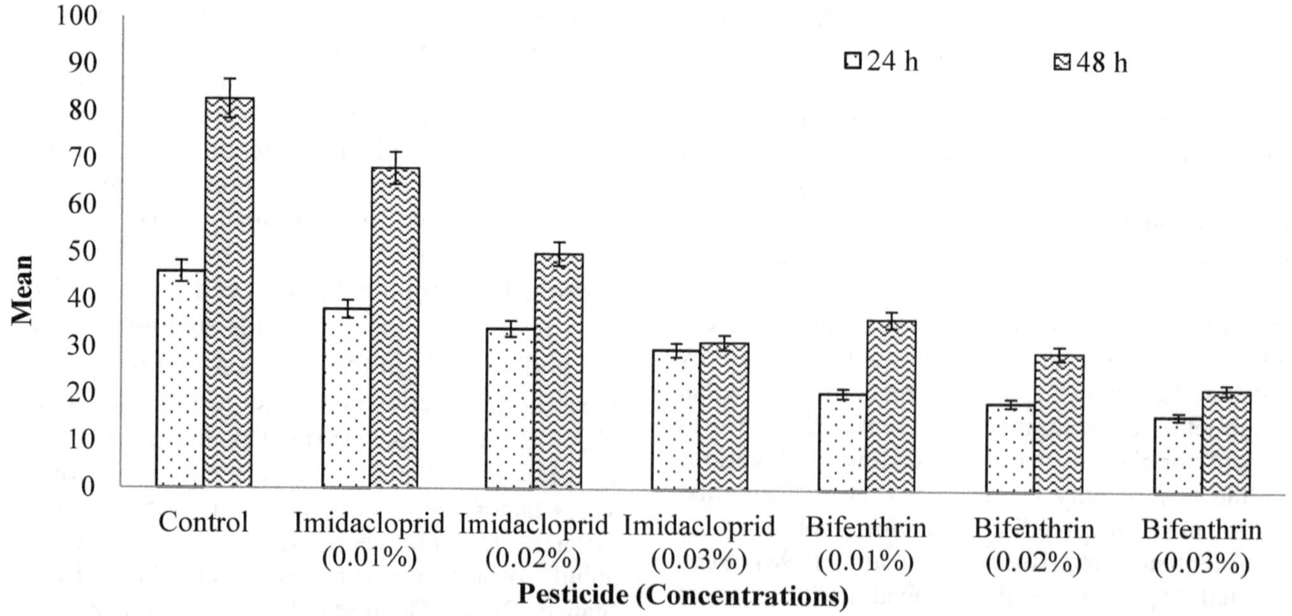


**Figure 2: Mean Voracity of *C. septempunctata* adults when treated with Bifenthrin and Imidacloprid or distilled water (control) using *Diuraphis noxia* as prey**

## Conclusion

*C. septempunctata* showed negative response to the insecticidal treatments for both beetle mortality and decreased aphid consumption. The exposure of coccinellid beetles to Imidacloprid and Bifenthrin, may affect physiology and behavior of beetles as a result of ingestion of treated prey. Our study focused on the two aspects of insecticidal effects on non-target organisms at higher concentrations of pesticides in accordance with the $LC_{50}$ and $LC_{90}$ ($LC_{50}$ for Bifenthrin 0.039593and $LC_{50}$ 0.062787for Imidacloprid whereas $LC_{90}$ for Bifenthrin 0.145805 and $LC_{90}$ 0.231222 for Imidacloprid). The study suggested that there is need to protect natural enemies from residual effect of chemical sprays for the protection, management and conservation of biodiversity. The sub-lethal doses may also affect the biology of Coccinellid beetles significantly, thus, biological control agents need to be reintroduced and augmented in the areas where integrated pest management programs are dominated by chemical control of insect pests. The toxic effects of insecticidal exposures to ladybird beetles have agricultural and ecological importance due to their predatory significance in crop protection (Yu et al., 2014).

## References

Abbott WS, 1925. A method of computing the effectiveness of an insecticide. J. Econ. Entomol. 18: 265-267.

Ahmad M, Rafiq M, Arif MI and Sayyed AH, 2011. Toxicity of Some Commonly Used Insecticides against *Coccinella undecimpunctata* (Coleoptera: Coccinellidae). Pak. J. Zool. 43(6): 1161-1165.

Ali A and Rizvi PQ, 2007. Development and Predatory Performance of *Coccinella septempunctata* L. (Coleoptera: Coccinellidae) on Different Aphid Species. J. Biol. Sci. 7(8): 1478-1483.

Amer M, Aslam M, Razaq M and Ali SS, 2010. Effect of Conventional and Neonicotinoid Insecticides against Aphids on Canola, *Brassica napus* L. at Multan and Dera Ghazi Khan. Pak. J. Zool. 42(4): 377-381.

Amin MA, Hameed A, Rizwan M and Akmal M, 2014. Effect of Different Insecticides against Insect Pests and Predators Complex on *Brassica napus* L., Under Field Conditions. Int. J. Sci. Res. Environ. Sci. 2(9): 340-345.

Angeli G, Baldessari M, Maines R and Duso C, 2005. Side-effects of pesticides on the predatory bug *Orius laevigatus* (Heteroptera: Anthocoridae) in the laboratory. Biocont. Sci Tech. 15(7): 745–754.

Arif MJ, Gogi MD, Abid AM, Imran M, Shahid MR, Husain S and Arshad M, 2011. Predatory potential of some native coccinellid predators against *Phenacoccus solenopsis*, Tinsely (Pseudococcidae: Hemiptera). Pak. Entomol. 33(2): 97-103.

Cabral S, Soares A and Garcia P, 2011. Voracity of *Coccinella undecimpunctata*: effects of insecticides when foraging in a prey/plant system. J. Pest. Sci. 84: 373–379.

Desneux N, Decourtye A and Delpuech JM, 2007. The sub-lethal effects of pesticides on beneficial arthropods. Annu. Rev. Entomol. 52: 52–81.

Devee A and Baruah AA, 2010. Relative toxicity of Imidacloprid and Bifenthrin against *Lipaphis erysimi* (Kalt.). Pesticide Res. J. 22(1): 86-87.

Dixon AFG, 1998. Aphid Ecology, 2nd edn. Chapman and Hall, London, pp. 300.

Fatima S, Hussain M, Malik MF, Noureen N and Abbas Z, 2016. Field efficacy of some insecticides against hibiscus mealybug, *Maconellicoccus hirsutus* (Hemiptera: Pseudococcidae). J. Entomol. Zool. Stud. 4(1): 240-244.

Fernandes MES, Alves FM, Pereira RC, Aquino LA, Fernandes FL and Zanuncio JC, 2016. Lethal and sub-lethal effects of seven insecticides on three beneficial insects in laboratory assays and field trials. Chemosphere. 156: 45-55.

Hawkeswood T, 1987. Beetles of Australia. Augus and Robertson, Sydney, Australia.

Inayat TP, Rana SA, Rana N, Ruby T, Sadiqui MJI and Abbas MN, 2011. Predation rate in selected coccinellid (Coleoptera) predators on some major aphidid and cicadellid (Hemiptera) Pests. Int. J. Agric. Biol. 13: 427–430.

Khaliq A, Attique MNR and Sayyed AH, 2007. Evidences of resistance to pyrethroids and organophosphate in *Plutella xylostella* from Pakistan. Bull. Entomol. Res. 97: 191-200.

Liu TX and Stansly PA, 2004. Lethal and sub-lethal effects of two insect growth regulators on adult *Delpgatus catalinae* (Coleoptera: Coccinellidae), a predator of whiteflies (Homoptera: Aleyrodidae). Biol. Control. 30: 298–305.

Naveed M, Salam A and Saleem MA, 2007. Contribution of cultivated crops, vegetables, weeds and ornamental plants in harboring of *Bemisia tabaci* (Homoptera: Aleyrodidae) and associated parasitoids (Hymenoptera: Aphelinidae) in cotton agroecosystem in Pakistan. J. Pestic. Sci. 80: 191–197.

Ochiai N, Mizuno M, Mimori N, Miyake T, Dekeyser M, Canlas LJ and Takeda M, 2007. Toxicity of Bifenazate and its Principal Active Metabolite, Diazene, to *Tetranychus urticae* and *Panonychus citri* and Their Relative Toxicity to the Predaceous Mites, *Phytoseiulus persimilis* and *Neoseiulus californicus*. Exper. App. Acarol. 43: 181-197.

Provost C, Coderre D, Lucas E, Chouinard G and Bostanian NJ, 2005. Impact of intraguild predation and lambda-cyalothrin on predation efficacy of three acarophagous predators. Pest Manag. Sci. 61: 532–538.

Qi B, Gordon G and Gimme W, 2001. Effects of neem-fed prey on the predacious insects *Harmonia conformis* (Boisduval) (Coleoptera: Coccinellidae) and *Mallada signatus* (Schneider) (Neuroptera: Chrysopidae). Biol. Control. 22: 185–190.

Rafi MA, Irshad M and Inayatullah M, 2005. Predatory ladybird beetles of Pakistan. PARC/NWFP Agric. Univ. Roohani Art Press, Islamabad, Pakistan, pp. 105.

Rain FF, Abu Faiz MD. Aslam, Sringki S, Sultana N, Akhter N and Howlader AJ, 2016. Coccinellid predators of aphid and their phylogenetic analysis using coi gene sequences. Int. J. Appl. Sci. Biotechnol. 4(3): 408-416.

Razaq M, Suhail A, Aslam M, Arif J, Saleem, AM, and Khan MHA, 2005. Evaluation of neonicotinoids and conventional insecticides against cotton jassid, *Amrasca devastans* (Dist.) and cotton whitefly, *Bemisia tabaci* (Genn.) on cotton. Pak Entomol.. 27(1):75-78.

Ruberson JR, Nemoto H, Hirose Y, 1998. Pesticides and Conservation of Natural Enemies in Pest Management. In: Barbosa P. (ed.) Conservation Biological Control. Academic Press, San Diego, CA. pp. 207-220.

Sarmad SA, Afzal M, Raza AM, Khalil, MS, Khalid H, Aqueel MA and Mansoor MM, 2015. Feeding efficacy of *Coccinella septempunctata* and *Propylea quatuordecimpunctata* against *Macrosiphum rosae*. Sci. Agri. 12(2):105-108.

Soares A, Coderre D and Schanderl H, 2003. Effect of temperature and intraspecific allometry on predation by two phenotypes of *Harmoniaaxyridis pallas* (Coleoptera: Coccinellidae). Environ Entomol. 32(5): 939–944.

Solangi BK, Lanjar AG and Lohar MK, 2007. Comparative toxicity of some insecticides on 4th instar grub of *Coccinella septempunctata* under laboratory conditions. Sarhad J. Agric. 23(4): 1091-1096.

Thornham DG, Blackwell A, Evans KA, Wakefield M and Walters KFA, 2008. Locomotory behaviour of the seven-spotted ladybird, *Coccinella septempunctata*, in response to five commonly used insecticides. Ann. Appl. Biol. 152:.349–359.

Thornham DG, Stamp C, Walters KFA, Mathers JJ, Wakefield M, Blackwell A and Evans KA, 2007. Feeding responses of adult seven-spotted ladybirds, *Coccinella septempunctata* (Coleoptera; Coccinellidae) to insecticide contaminated prey in laboratory arenas. Biocontrol Sci. Techn. 17(10): 983–994.

Ujjan ZA, Khanzada MA and Shahzad S, 2014. Insecticide and papaya leaf extract toxicity to mustard aphid, Lipaphis erysimi (kaltenbach). J Agri Food App.Sci. 2(2): 45-48.

Wang L, Huang J, You M, Guan X, Liu B, 2005. Effects of toxins from two strains of *Verticillium lecanii* (Hyphomycetes) on bioattributes of a predatory ladybeetle *Delphastus catalinae* (Col., Coccinellidae). J Appl. Entomol. 129 (1):32–38. doi:10.1111/j.1439-0418.2005.00929.32-38

Wright DJ and Verkerk RHJ, 1995. Integration of Chemical and Biological Control Systems for Arthropods: Evaluation in a Multitrophic Context. Pesticide Sci. 44: 207-218.

Yu C, Fu M, Lin R, Zhang Y, Yongquan L, Jiang H and Brock TCM, 2014. Toxic effects of hexaflumuron on the development of *Coccinella septempunctata*. Environ. Sci. Poll.Res. Int. 21(2): 1418–1424.

Yuxian, H., Juanwei, Z., Yu, Z., Desneux, N., Kongming, W, 2012. Lethal effect of imidacloprid on the Coccinellid predator Serangium japonicum and sublethal effects on predator voracity and on functional response to the whitefly *Bemisia tabaci*. Ecotoxicity, 21(5): 1291-1300. doi: 10.1007/s10646-012-0883-6

Zar JH, 1996. Biostatistical analysis. 3. London, UK: Prentice. pp. 662.

13

# Growth and Egg Production Performance of Dekalb Layers (*Gallus gallus domesticus*) supplemented with Cassava Leaf Meal (*Manihot esculenta* Crantz)

**Eloiza May Galon*[1,4], Dante Lorenzo[1,2] and Florencia Claveria[1,3]**

[1*]AgriVet Sciences Institute, De La Salle Araneta University, Upper Ciudad Real, City of San Jose Del Monte, Bulacan, 3023 Philippines

[2]Department of Animal Science, College of Agriculture, Central Luzon State University, Science City of Muñoz, Nueva Ecija, 3119 Philippines

[3]Biology Department, College of Science, De La Salle University, 2401 Taft Avenue, 0922 Manila, Philippines

[4]National Research Center for Protozoan Diseases, Obihiro University of Agriculture and Veterinary Medicine, Obihiro, Hokkaido, 080-8555 Japan

*Corresponding author email: emsgalon@yahoo.com

## Abstract

A feeding trial was done to determine the best level of supplementation of cassava leaf meal (CLM) on the growth, productivity and egg quality of Dekalb layers at Salikneta Farm, San Jose del Monte, Bulacan, Philippines. The treatments were control (no CLM), T1 (3% CLM supplement), T2 (5% CLM supplement) and T3 (7% CLM supplement). The study was laid in a single factor Completely Randomized Design, with three replicates, consisting of ten birds per replicate. Results show highest feed consumption of birds at T3 ($p<0.05$) compared to control, T1, and T2. Weight gain was not affected by CLM supplementation. However, the FCR, percentage egg production, and cumulative number of eggs per bird in T1, and albumen height in T2 and T3 were higher ($p<0.05$) compared to the control group. Likewise, the supplementation of CLM up to 7% has no adverse effect on the health of the birds. Results revealed that 3% supplementation of CLM can improve FCR, percent egg production, and the cumulative number of eggs per bird. Moreover, increasing the level of CLM to 5% and 7% supplement may further improve birds' performance in terms of egg weight, yolk color, and albumen height.

**Keywords:** Cassava leaf meal, Dekalb layers, Egg quality, Productivity, Salikneta Farm

## Introduction

The increasing demand for eggs can be attributed to the egg as an affordable protein-rich food. However, one of the major constraints that hinders the country's egg producers to address this huge demand efficiently is the high cost for animal feeds. Considering that animal feeds account for 70-75% of the cost of production, providing high quality feeds is essential in determining the quality of animal produce, specifically for poultry (Aderinola et al., 2013). Thus, it is important that cheap and easily obtainable resources containing essential nutrients be tapped in order to improve the performance and productivity of layers.

Cassava (*Manihot esculenta* Crantz) leaves are the by-product of cassava root production and is either thrown away or is utilized as a fertilizer or compost. It is estimated that 10 tons of dry cassava foliage is produced per one hectare of cassava production area (Morghan and Choct, 2016). The leaves possess a relatively high crude protein and essential amino acids required for nutrition (Iheukwumere et al., 2008). Although cassava leaves also contain compounds that can be toxic to human and other animals like the glucosinolate linamarin, it can still be utilized by detoxification process through drying or turning the leaves into a meal (Dolan et al., 2010).

Cassava leaf meal has been demonstrated to be an efficient feed source for poultry. While several studies had been conducted on the effects of cassava leaf meal (CLM) on broilers, only a few studies in the recent years had reported effects of CLM on layers' performance, productivity and egg quality (Oludare, 2006; Zanu et al., 2012).

The present study was conducted to assess the value of cassava leaf meal as a supplement to the layers' diet. Cassava leaf meal is a feed that is affordable, readily available and efficient, and can be utilized as well to improve the growth performance, productivity and quality of eggs produced. The use of cassava leaf meal can be adopted and can be integrated with cassava root production, thus minimizing waste in farm operations. Furthermore, the study aimed to generate new knowledge by exploring the potential of cassava leaves as feeds for layers.

## Materials and Methods

*Cassava Leaf Meal Preparation:* Mature cassava leaves were collected from the cassava plantation area at Salikneta Farm. After collecting, the leaves were shredded using a shredding machine and were subjected to sun drying for 3-4 days. The dried leaves were crushed to produce a fine cassava leaf meal. The leaf meal was stored in perforated sacks in a dry place.

*Feeding Management and Production Monitoring*: The layers were acquired at 18 weeks of age and were used when they were 29 weeks old. The commercial feed Robina™ layer crumble, which has a composition of 17% crude protein, 17% crude fat, 5% crude fiber, 12% moisture, 3.8-4.2% calcium and 0.7% phosphorus, was used. For the nine-week feeding trial, a total of 756 kg of commercial feed was used. The feed was mixed with cassava leaf meal. Feeding of the layers was done twice a day: in the early morning and afternoon. Half of the formulated feed was given in the morning, and the other half in the afternoon. During early morning, cleaning of the feed trough was done to collect the feed refusal. Clean and fresh water was provided *ad libitum* using the installed drip drinkers. Production of each bird was monitored and recorded. Eggs were collected in the afternoon post-feeding. The collected eggs were placed in trays and individual egg weights were recorded using the digital weighing scale and classified using the commercial egg scale.

*Experimental Set-up:* One hundred and twenty Single Comb White Leghorn Dekalb layers were randomly chosen and allocated to four different treatments and laid in a single factor Completely Randomized Design experiment. Each treatment consisted of 30 birds and three replicates, with 10 birds each replicate. The following treatments were used: Control -Commercial Layer Feeds only (100g/day); T1 - Commercial Layer Feeds (100g/day) + 3% Cassava Leaf Meal (3g); T2 - Commercial Layer Feeds (100g/day) + 5% Cassava Leaf Meal (5g) and T3 - Commercial Layer Feeds (100g/day)+ 7% Cassava Leaf Meal (7g).

*Data Collection:* The following parameters for growth and productivity were obtained: (1) Feed consumption (g) = feed provided –feed refusal; (2) Weight gain (g) = final weight – initial weight; (3) FCR= total amount of feed consumed ÷ (# of eggs x mean egg weight); (4) % Egg production = (# of eggs ÷ # of housed layers) x 100; (5) Cumulative number of eggs per bird = total # of eggs/treatment ÷ 30; (6) Mean egg weight = sum of egg weights ÷ # of eggs weighed; and (7) Mortality rate = (# of dead birds ÷ # of housed birds) x 100. For the assessment of egg quality, random samples of eggs were chosen from each treatment. Egg samples were cracked on a Petri dish. Using the DSM™ Yolk Color Fan, yolk samples were assessed and recorded. After evaluating the intensity of the yolk color, albumen height was measured. The height of the thick albumen and shell thickness were measured using a Vernier wheel type calliper. The conduct of the study is summarized in Figure 1.

**Figure 1:** Flowchart of methodology used in the study

*Statistical Analysis:* The data recorded for each parameter were statistically analyzed using Analysis of Variance (ANOVA) and differences among the means obtained were evaluated using Duncan's Multiple Range Test (DMRT).

## Results and Discussion

Birds fed with CLM consumed more feed and had higher egg production compared to the control (Table 1). A direct relationship is conveyed between the amount of CLM supplementation and the feed consumed. This is because the higher the amount of CLM mixed in the diet, more feed is made available to the layers. Contrary to the results of the feed consumption, T2 and Control recorded the highest gain in weight (Table 1). Layers supplemented with CLM recorded a higher egg production and cumulative number of eggs per bird compared to the control. T1 consistently recorded the highest values for the two parameters throughout the nine weeks of feeding trial. Meanwhile, T2 consistently recorded the heaviest egg weight. There is a similar trend for the eggs produced by the birds fed with CLM, which steadily increased reaching the peak on Week 7, after which the egg weights started to decrease.

Current observations do not align with the established direct correlation between daily feed intake and body weight (Pérez-Bonilla et al., 2012). Bird mortality was negligible.

**Table1:** Comparison of the productivity and egg quality characteristics of layers fed with various amounts of cassava leaf meal (CLM).

| Parameters | Control | T1 (3%) | T2 (5%) | T3 (7%) |
|---|---|---|---|---|
| Feed Consumption (g) | $99 \pm 0.13^{b}$ | $101 \pm 1.85^{b}$ | $100 \pm 1.83^{b}$ | $105 \pm 0.63^{a}$ |
| Weight Gain (g) | $203 \pm 53$ | $170 \pm 43.59$ | $207 \pm 33.29$ | $195 \pm 52.35$ |
| FCR | $2.27 \pm 0.25^{a}$ | $1.96 \pm 0.29^{b}$ | $2.03 \pm 0.075^{ab}$ | $2.07 \pm 0.023^{ab}$ |
| Percent Production (%) | $80.44 \pm 8.47^{b}$ | $92.33 \pm 1.80^{a}$ | $86.55 \pm 6.10^{ab}$ | $89.63 \pm 1.85^{ab}$ |
| Cumulative # of eggs/bird | $51 \pm 5.57^{b}$ | $58 \pm 1.15^{a}$ | $54 \pm 3.79^{ab}$ | $57 \pm 1.53^{ab}$ |
| Egg weight (g) | $55.04 \pm 1.02^{b}$ | $55.92 \pm 0.53^{ab}$ | $57.39 \pm 0.98^{a}$ | $56.81 \pm 0.94^{a}$ |
| Yolk color | $9 \pm 0.58^{b}$ | $9 \pm 0.58^{ab}$ | $10 \pm 0.58^{a}$ | $10 \pm 0^{a}$ |
| Albumen height (cm) | $0.68 \pm 0.0076^{b}$ | $0.69 \pm 0.013^{b}$ | $0.72 \pm 0.019^{a}$ | $0.74 \pm 0.01^{a}$ |
| Shell thickness (cm) | $0.0364 \pm 0.00076^{a}$ | $0.0391 \pm 0.0025^{a}$ | $0.0389 \pm 0.001^{a}$ | $0.0392 \pm 0.00138^{a}$ |
| Mortality (%) | $3.33 \pm 5.78$ | 0 | $3.33 \pm 5.78$ | 0 |

Analysis on yolk color ratings of eggs from the control and experimental groups revealed no appreciable difference (Figure 2). An earlier study reported that cassava leaves possess carotenoids that increase yolk pigmentation (Montagnac et al., 2009). Egg albumen height increased with increasing levels of CLM supplementation. The differences in the albumen height may be explained by the presence of high crude protein content brought about with increasing levels of CLM supplementation as clearly noted in T3. The thicker shells of eggs given CLM supplement can be attributed to the greater availability of calcium, which is the primary component of egg shell (An et al., 2016). It was also emphasized that calcium content is 100 times more concentrated in cassava leaves than in roots (Montagnac et al., 2009).

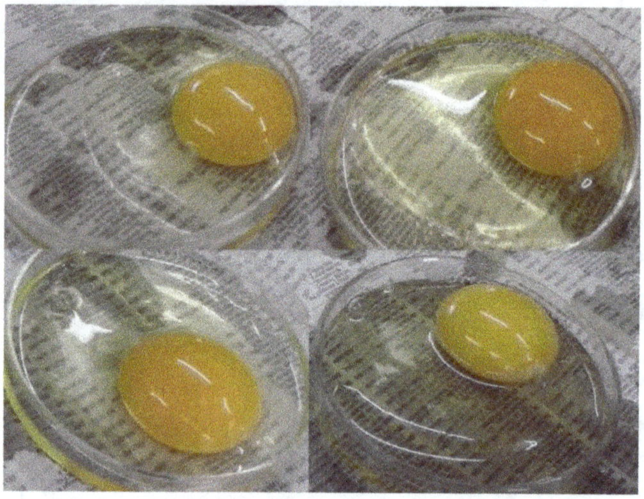

**Figure 2:** Egg samples for yolk color evaluation

In summary, the supplementation of CLM to the feed ration of Dekalb layers was better in terms of feed conversion ratio, percent production, number of eggs per bird, egg weight, yolk color and albumen height, and shell thickness, compared to the control group. The study also found no adverse effects on the livability of the layers up to 7% supplementation of CLM, as cassava leaves are often discouraged as a supplement due to high content of linamarin and lotaustralin (Chauynarong et al., 2009). Present data revealed that food of layers supplemented with 3-7% CLM produced better and acceptable results relative to the control, with the most desirable results for egg productivity and FCR in T1 (3% CLM) and enhanced egg quality characteristics in T3 (7% CLM). With the present findings, the incorporation of CLM as a feed supplement alongside the conduct of a follow-up study on the use of higher proportion of supplementation is recommended.

## Acknowledgment

This research was supported by Salikneta Farm, De La Salle Araneta University, which provided the facilities and materials required. The researcher would also like to thank the poultry farm staff and students for the technical assistance during the course of the experiment. Acknowledgment is also given to Dr. Pablo Rafael, Dr. Pedrito Nitural, and to other research advisers for the advice and help in the completion of the study.

## References

Aderinola OA, Rafiu TA, Akinwumi AO, Alabi TA and Adeagboo OA, 2013. Utilization of *Moringa oleifera* as feed supplement in broiler diet. Int. J Food Agri. Vet. Sci. 3: 94-102.

An SH, Kim DW and An BK, 2016. Effects of Dietary Calcium Levels on Productive Performance, Eggshell Quality and Overall Calcium Status in Aged Laying Hens. Asian Australas. J. Anim. Sci. 29:1477-1482.

Chauynarong N, Elangovan AV and Iji PA, 2009. The potential of cassava products in diets for poultry. Worlds Poult. Sci. J. 65: 23–36.

Dolan LC, Matulka RA and Burdock GA, 2010. Naturally Occurring Food Toxins. Toxins 2: 2289-2332.

Iheukwumere FC, Ndubisi EC and Etusiim PE, 2008. Performance, nutrient utilization and organ characteristics of broilers fed cassava leaf meal (*Manihot esculenta* Crantz). Pakistan J. Nutr. 7: 13-16.

Morgan NK and Choct M, 2016. Cassava: Nutrient composition and nutritive value in poultry diets. Anim. Nutr. 2: 253-261.

Montagnac JA, Davis CR and Tanumihardjo SA, 2009. Nutritional Value of Cassava for Use as Staple Food and Recent Advances for Improvement. Compr. Rev. Food Sci. Food Saf. 8: 181–194.

Oludare FA, 2006. Protein replacement value of cassava (*Manihot esculenta* Crantz) leaf protein concentrate (CLPC): effects on egg quality, biochemical and haematological indices in laying birds. J. Food, Agri. and Env. 4:54-59.

Pérez-Bonilla A, Novoa S, García J, Mohiti-Asli M, Frikha M and Mateos GG, 2012. Effects of energy concentration of the diet on productive performance and egg quality of brown egg-laying hens differing in initial body weight. Poult. Sci. 91: 3156-3166.

Zanu HK, Kagya-agyemang JK and Avukpor CM, 2012. Effects of enzyme supplementation on the performance of laying hens fed diets containing different levels of cassava leaf meal (*Manihot esculenta* Crantz). Online J. Anim. Feed Res. 3: 9-14.

14

# Co-inoculation of auxin producing PGPR and rhizobia enhanced growth of *Vigna mungo* (L) under cadmium stress

**Rabiya Ikram, Basharat Ali***
Department of Microbiology and Molecular Genetics, University of the Punjab, Lahore, Pakistan

*Corresponding author email: basharat.ali.mmg@pu.edu.pk

## Abstract

The main objective of this study was to evaluate the effect of cadmium resistant bacteria to enhance the growth of black gram (*Vigna mungo*) under cadmium ($CdCl_2$) stress. For this purpose, plant growth promoting rhizobacteria (PGPR) and strains of rhizobia were isolated from the rhizosphere of different plants. 16S rRNA gene sequencing of bacteria confirmed the presence of the strains of *Bacillus cereus, B. thuringiensis, Pseudomonas fluorescens* and *Rhizobium*. L-tryptophan dependent auxin production was determined colorimetrically by growing strains in Nutrient broth. *B. Cereus* HG, *B. thuringiensis* JAF *and Rhizobium sp.* IBA showed highest levels of auxin production that ranges from 140 to 146 µg $ml^{-1}$. For pot trials, soil was amended with 50, 100 and 200 mg $kg^{-1}$. In single bacterial inoculations, significant shoot length response was up to 65% and 39% with *P. fluorescens* WN-1 and *B. cereus* CF at 200 mg $kg^{-1}$, over respective $CdCl_2$ treated control. Similarly, for fresh biomass, 2.2 and 1.3 fold increases were shown by *B. cereus* CF at 100 and 200 mg $kg^{-1}$ $CdCl_2$ stress, respectively. For mixed culture of PGPR, combination MC-3 (CF-WN-1, A-2) significantly promoted shoot length (1 fold) at 50 mg $kg^{-1}$ $CdCl_2$. In case of mixed culture of PGPR and rhizobia, up to 1 fold increases were observed for shoot length, root growth and fresh biomass with MCR-1 (IBA, HG, BS-1) at 100 mg $kg^{-1}$ $CdCl_2$. Finally, it can be concluded that strains isolated from the rhizosphere of different pants showed promising results for plant growth promotion in cadmium amended soils.

**Keywords:** Bacterial auxin, Cadmium stress, PGPR, Growth promotion, Phytoremediation, Vigna mungo

## Introduction

Among all the heavy metals, cadmium (Cd) is wide spread and most toxic pollutant present in the top soil. It is present with Zinc (Zn) and the increased demand of Zn has also increased the amount of Cd in the environment. The use of Cd has also increased in industries as it has many physical and chemical properties that make it useful (Hussain and Islam, 2010). Cadmium releases into environment through mining, smelting, burning of fuel, sewage sludge and application of phosphate fertilizers. In nature volcanic eruption, forest fires and soil erosion are the main sources of cadmium release (Figueira et al., 2005). Cadmium particles travel as dust and fall with rain, there they enter into soil, crops and water supplies. It is very toxic and it is also considered as carcinogenic. It causes certain diseases such as lung cancer, osteomalacia, renal damage by affecting tubular cells, bone demineralization by renal dysfunction (Godt et al., 2006). It also causes damage to liver, reproductive system, emphysema, hypertension, testicular atrophy and cardiovascular system even at low level of exposures (Bazrafshan et al., 2006). Once it enters in the body it tends to remain there for a longer period of time. It enters in human bodies from diet as many crops like lentils, potato, tobacco, leafy vegetables, sunflowers and root vegetables take up large amount of cadmium (Qing et al., 2007). Cadmium poses risk because it absorbs very easily in the body and its half-life is 10-30 years. Its amount in adults of 45 years of age is about 30-40 mg per total body weight. Its declared lethal dose is 2 mg $kg^{-1}$ body weight that is quit less than other toxic metals. In plants it cause chlorosis, stunted growth of roots, disturbed homeostasis and affect nutrient uptake (Jankiewicz et al., 2000).

In order to relief soil from metal contamination, plant growth promoting rhizobacteria (PGPR) are being used. The plant growth promoting traits of these bacteria are nitrogen fixation, siderophores and phytohormone production (Ladwal et al., 2012; Raheem et al., 2015). Plant resists to heavy metal contamination in the soil by expressing shock absorbent proteins. But when the concentration of such metals increases beyond the limit, plants start to become deficient in certain nutrients. Low iron absorption that leads to under development of chloroplast and hence plants becomes chlorotic exhibiting yellow coloration of leaves. PGPR able to produce siderophores resolves this problem for plants by providing them a large amount of iron to absorb by their roots. (Jing et al., 2007). In addition, it has also been observed that the PGPR also help the seedlings to germinate in the stressed environment. Once the seedling has established, the rhizobacteria help them to grow by providing them with solubilised phosphates and phytohormones (Rajkumar et al., 2010; Raheem et al., 2017).

Bacteria use quorum sensing mechanism for plant growth promotion. They sense the molecular signal, amplify it and attract other microbial species thus forming a biofilm over the roots of plants. Bacteria in return get food and nutrients from the plant root exudates as well as shelter for living (Khan, 2005). Inoculated bacterial strains first share the metal load with plants and after emergence of seedling, these PGPR starts to produce phytohormones like auxin that helps the plant roots to establish for better nutrient uptake (Madhaiyan et al., 2007; Ali, 2015).

Heavy metals interfere with the uptake of essential nutrients by roots of plants. Metal resistant bacteria are able to convert the heavy metals from one oxidation state to another thus, lessen the impact of metals on roots and allow them to absorb nutrients. In addition, they also produce phytohormone auxin that increases the surface area of root by enhancing the root hair growth that helps the plants to absorb nutrients far away in the soil. (Belimov et al., *2005*). The main aim of this study was the isolation of cadmium resistant bacteria from the rhizosphere of plants and their application to enhance the growth of black gram i.e. *Vigna mungo* (L.) Hepper in cadmium amended soil. For co-inoculation experiments, rhizobia were also isolated from the root system of *Vicia faba* L. to evaluate their combined effect on the growth of leguminous plant.

## Materials and Methods

### Isolation and characterization of bacterial strains

For the isolation of plant growth promoting rhizobacteria (PGPR), soil samples were collected from the rhizosphere of *Rosa indica, Trifolium alexandrinum, Jasminum officinale, Solanum tuberosum* and *Brassica campestris*. One gram of soil was dissolved in 99 ml adutoclaved distilled water to prepare $10^{-2}$ dilution and so on. From each dilution, 50 µl sample was spread on Nutrient agar (N-agar) plates and incubated at 30°C for 24 h. Afterwards, diverse bacterial colonies were selected and purified by quadrant streaking. For rhizobia isolation, roots of *Vicia faba* were collected and sterilized by dipping in 1% solution of bleach for 2-3 min. Afterwards, nodules were washed with distilled autoclaved water three times. The nodules were crushed in 1 ml autoclaved distilled water and loopful inoculum was streaked on Yeast Mannitol Agar (YMA). Finally, different colonies with varied appearance were purified after several rounds of streaking. Gram and spore staining, methyl red Vogus Proskauer test (MR-VP), starch hydrolysis, citrate utilization, nitrogenase activity and Congo red tests were performed.

### 16S rRNA gene sequencing

Final taxonomic status of a few bacterial strains was determined by 16S rRNA gene sequencing. Genomic DNA was extracted by using Tissue Genomic DNA Extraction Kit (Favorprep™, Favorgen). About 1.5-Kb DNA fragment containing 16S rRNA gene was amplified using forward 27f and reverse 1522r primers as described earlier (Akhtar and Ali, 2011). After PCR amplification, product was purified using FavorPrep™

Gel Purification Kit. Purified fragments were sequenced from First Base Laboratories (Singapore).

**Colorimetric analysis of bacterial auxin production**

Auxin production in liquid medium was quantified in the presence of L-tryptophan. N-broth was amended with 200 µg $ml^{-1}$ of L-tryptophan. Then, flasks were inoculated with 100 µl of bacterial cell suspension in duplicate. Cultures were incubated around 30°C for 72 h on shaker at 120 rpm. Then, 1 ml of bacterial culture was centrifuged (at 5000 rpm) and mixed with 2 ml of Salkowski reagent. Test tubes were kept in dark for 25 min and optical density of the samples was recorded at 535 nm with spectrophotometer. Standard curve from different concentrations of authentic auxin was constructed for the estimation of bacterial auxin production.

**Analysis of bacterial auxin by HPLC**

Auxin was also detected qualitatively by using High Performance Liquid Chromatography (HPLC). For this purpose, bacterial culture supernatants from the above experiment were also used for HPLC analysis. Ten ml of bacterial supernatant was adjusted to pH 3 and mixed with equal volume of ethyl acetate. Afterward, upper aqueous layer was separated in a flask and allowed to dry by placing it in water bath at 50°C. Then, 5 ml of HPLC grade methanol was added to dissolve the auxin content. Auxin was analyzed qualitatively after injecting 20 µl of each sample in HPLC (Model 203; Sykam, Eresing, Germany) equipped with hypersil-keyston ODS column (5 µm; 4.6 × 250 nm). The mobile phase was methanol/water (90:10) at a flow rate of 1 ml $min^{-1}$. Elutes were detected at 220 nm and IAA was detected by integrating the areas under the peaks of authentic IAA.

**Bacterial cadmium resistance**

Bacterial $CdCl_2$ resistance in solid media was determined qualitatively by assessing the abundance of growth appeared on the surface of media. To determine resistance, N-agar was supplemented with different concentrations of metal that included 0.25, 0.5, 1, 1.25 and 1.5 mg $ml^{-1}$. Bacterial strains were inoculated in duplicate and incubated at 37°C for 24 h. Bacterial cadmium resistance was determined by evaluating the abundance of growth.

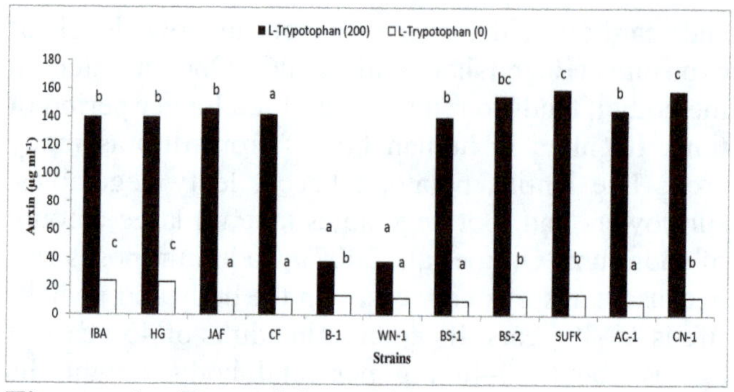


**Figure 1. Bacterial auxin production in the presence and absence of L-tryptophan.**

Different letters on bars indicate significant difference between treatments using Duncan's multiple range test ($P \leq 0.05$).

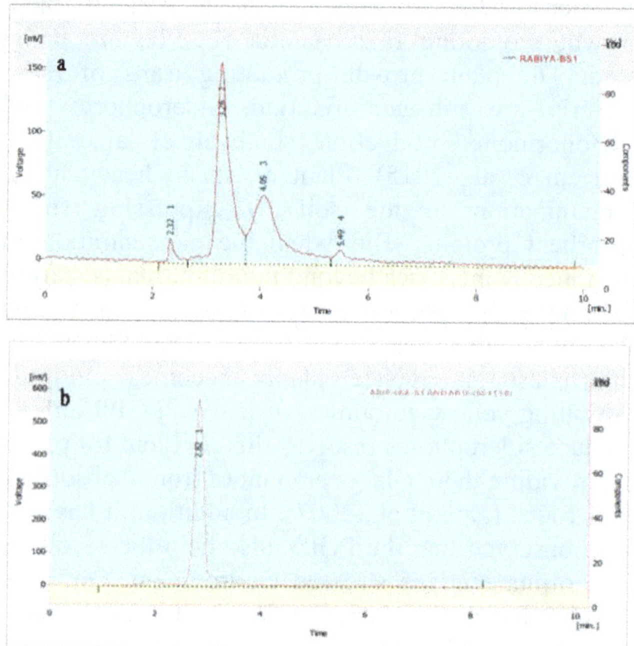


**Figure 2. HPLC analysis of bacterial auxin production. (a)-extracts from P. fluorescens WN-1, (b)- standard IAA**

**Nitrogenase activity of bacterial strains**

Bacterial nitrogenase activity was determined in nitrogen free mineral salt medium. Strains were inoculated on the media plate and placed in incubator at 37°C for 48 h to allow the appearance of growth or hollow zone of growth. Bacterial Congo red absorption was determined in order to indicate the presence of rhizobia by supplementing the yeast mannitol agar with Congo red indicator. *Rhizobium* species took up the colour of Congo red in the presence of $CaCO_3$. Strains were inoculated on the plates and incubated at 28 °C for 48 h. After incubation in dark, the plates were incubated in light for 24 h to allow the strains to absorb the Congo red colour. Appearance of pink colouration in colonies upon incubation in light showed the presence of *Rhizobium.*

**Pot trails**

For pot trials, inoculum was prepared by dissolving the bacterial growth from N-agar to 25 ml of autoclaved distilled water. Optical density of the cultures were adjusted $10^7$ CFU $ml^{-1}$. Sterilized seeds of *Vigna mungo* were inoculated with single bacterial cultures such as HG, WN-1, JAF, B-1, CF and SUFK. Mixed cultures (MC) combinations included MC-1 (WN-1, JAF, A-2), MC-2 (HG, JAF, CF), MC-3 (CF-WN-1, A-2) and MC-4 (CF, HG, JAF). For rhizobia single cultures included Nj-2, Swm-1, Sdm-1, IBA and Wn-2. Mixed cultures of rhizobia with other PGPR strains were also used that included MCR-1 (IBA, HG, BS-1), MCR-2 (IBA, WN-1, AC-1), MCR-3 (IBA, JAF, CN-1) and MCR-4 (IBA, BF-1, SUFK). For bacterization, seeds were treated with bacterial suspensions (single or mixed) for 25 min. For control treatment, seeds were dipped in distilled autoclaved water for the same duration. Soil was prepared by mixing equal ratio of sand and soil (1:1). After mixing, soil was autoclaved to eliminate the growth of previously present microbes. After autoclaving, soil was filled in pots (6×6 cm) and amended with three different concentrations of $CdCl_2$ to a final amount 50, 100 and 200 mg $kg^{-1}$ of soil. Positive controls were inoculated only with bacterial suspensions. Six pots were used for each strain, metal treatments or control. Five seeds were inoculated in each pot at the depth of 1 cm. Pots were moistened with sterilized water immediately after sowing and placed in plant growth chamber at 12 h photoperiod. After 7 days, plants were harvested to record their shoot length, root length and fresh weight. For dry weight, plants were dried at 80°C in oven for 24 h.

**Statistical analysis**

Data for auxin production and plant growth parameters was subjected to analysis of variance (ANOVA) and means were separated by using Dnncan's multiple range test ($P \leq 0.05$).

## Results

**Biochemical characterization**

After purification of colonies, 11 strains were selected for further study. Majority of the strains (HG, JAF, B-1, CF) were found to be gram positive, while some were gram negative (WN-1, IBA). Among gram positive strains, majority showed starch hydrolysis and citrate utilization as well as spore formation. Gram negative strains WN-1, AC-1, BS-1 and CN-1 showed positive Congo red test and nitrogenase activity. For MR-VP test HG, SUFK and JAF were found to be MR positive and VP negative. Bacterial strains grown under different concentration of $CdCl_2$ showed varied pattern of resistance. Strains HG, IBA, SUFK and JAF showed high resistance at 0.5 mg $ml^{-1}$. Whereas, WN-1 showed sensitivity against $CdCl_2$ as metal concentration increased from 0.7 to 5.62 mg $ml^{-1}$.

**Identification of bacterial strains**

16S rRNA gene sequences were compared with online available sequences in GenBank through BLAST. Majority of the strains showed 99% resemblance with their respective genes. Strain HG isolated from *R. indica* showed resemblance with *Bacillus cereus*. WN-1 associated with *T. alexandrinum* showed homology with *Pseudomonas fluorescence*. Similarly, JAF isolated from *J. officinale* showed homology with *Bacillus thuringiensis*. On the other hand, B-1 and CF isolated from *S. tuberosum* and *B. campestris* were similar to *B. cereus*. Strain IBA isolated from *V. faba* showed homology with *Rhizobium* species. The sequences from strains HG, WN-1, JAF, B-1, CF and IBA have been submitted in the GenBank under accession numbers KM206792 to KM206797 (Table 1).

**Bacterial auxin production**

Bacterial strains were grown in the presence and absence of L-tryptophan. For majority of the strains, auxin production was very low in the absence of L-tryptophan. However, in the presence of substrate higher levels of auxin were recorded. For example *B. Cereus* HG, *B. thuringiensis* JAF *and Rhizobium sp.* IBA showed highest levels of auxin production that ranges from 140 to 146 µg $ml^{-1}$(Fig. 1). About 4.9 fold increase in auxin production was recorded by *Rhizobium sp.* IBA as compared to un-supplemented medium. *B. cereus* CF and *B. cereus* B-1 also showed 11 and 5 fold increases, respectively. Similarly, 14 fold increases were recorded by *B. thuringiensis* JAF, over respective un-amended control. Bacterial auxin production was further confirmed by analysing the bacterial crude extracts with HPLC. Figure 2 showed the chromatogram of auxin production by *P. fluorescens* WN-1 in comparison with standard.

**Pot trials with monocultures**

Plants grown under cadmium stress showed a very poor growth as compared to bacterized or water treated treatments. Overall, plant growth enhancement by inoculation with monocultures of PGPR was up to 78% as compared to their respective water treated control (Table 2). For shoot length, single bacterial inoculations in cadmium non-amended soils showed statistically comparable results to water treated controls. However, when plants were subjected to 50, 100 and 200 mg $kg^{-1}$ of $CdCl_2$ stress, significant reduction in shoot length (up to 41%) was observed over water treated control. On the other hand, when plants were grown in $CdCl_2$ amended soils under bacterial inoculations, significant reduction in metal toxicity was recorded in comparison with respective control. For instance, *B.* cereus HG showed up to 60% increase in shoot length over 100 mg $kg^{-1}$ metal treated control. However, maximum shoot length response was up to 65% and 39% with *P. fluorescens* WN-1 and *B. cereus* CF at 200 mg $kg^{-1}$. Similarly, *B. cereus* B-1 and *B. thuringiensis* JAF showed highest root growth response that was up to 1.2 and 1.8 fold at 50 mg $kg^{-1}$, respectively, over metal treated control. On the other side, at 100 mg $kg^{-1}$ of stress SUFK showed 1.7 fold increases over respective metal treated control. Similarly, for fresh biomass *B. cereus* CF showed highest response in cadmium non-amended soil as compare to water treated control. However, when plants were exposed to cadmium stress, best recorded plant growth promotion was up to 78% by strain SUFK at 50 mg $kg^{-1}$. Moreover, 2.2 and 1.3 fold increases were shown by *B. cereus* CF at 100 and 200 mg $kg^{-1}$ cadmium stress respectively. For dry biomass, highly significant increase (4 fold) was shown by *B. cereus* HG at 50 mg $kg^{-1}$.

**Growth response with mixed bacterial culture inoculations**

Pot trials were also conducted by using mixed cultures of PGPR and rhizobia. In cadmium non-amended soils, plants showed significant growth under bacterial inoculations as compared to water treated seeds (Table 3). However, under cadmium stress, plants recorded retarded growth response, over control. Nevertheless, mixed culture combination MC-3 significantly performed well at 50 mg $kg^{-1}$ by promoting the shoot length up to 1 fold over respective metal treated control. Similarly, MC-2 and MC--4 enhanced shoot length up to 1.2 fold and 60 % at 100 and 200 mg $kg^{-1}$ of cadmium stress, respectively. As far as root length promotion is concerned, mixed culture induced growth was up to 38% and 26% with combinations MC-3 and MC-1 under 50 and 100 mg $kg^{-1}$ of $CdCl_2$ stress, respectively. The increase in fresh weight was up to 1.3, 2.2 and 1.1 fold with MC-3, MC-4 and MC-2 under 50, 100 and 200 mg $kg^{-1}$. For dry weight, up to 53% increase was observed with MC-3 at 50 mg $kg^{-1}$ $CdCl_2$ (Table 3).

Mixed cultures of rhizobia (MCR) with other PGPR also recorded positive results for the growth of *V. mungo* as compare to metal treated controls (Table 4). For shoot length at 100 mg $kg^{-1}$ $CdCl_2$ up to 1 fold increase was observed with MCR-1, over respective control. Similarly, for root length, 1 fold increase was also recorded by MCR-1 combination at 100 mg $kg^{-1}$. For fresh biomass, significant increase of 1.3 fold was recorded at 50 mg $kg^{-1}$ with MCR-1 combination over metal treated control. In case of dry weight, MCR-1 also performed well at 100 mg $kg^{-1}$ by increasing the weight up to 76% as compare to metal treated control.

**Table 1. 16S rRNA gene sequencing of PGPR and Rhizobium isolated from the rhizosphere of plants**

| S. No. | Source/ Plant | Strain | Identified as | Accessions |
|---|---|---|---|---|
| 1 | *Rosa indica* L. | HG | *Bacillus cereus* HG | KM206792 |
| 2 | *Trifolium alexandrinum* L. | WN-1 | *Pseudomonas fluorescens* WN-1 | KM206793 |
| 3 | *Jasminum officinale* L. | JAF | *B. thuringiensis* JAF | KM206794 |
| 4 | *Solanum tuberosum* L. | B-1 | *B. cereus* B-1 | KM206795 |
| 5 | *Brassica Campestris* L. | CF | *B. cereus* CF | KM206796 |
| 6 | *Vicia faba* L. | IBA | *Rhizobium* sp. IBA | KM206797 |

**Table 2. Effect of single cultures of PGPR strains on the growth of *V. mungo* in cadmium amended soils**

| S. No. | Strains/ Treatmetns | Shoot length (cm) | Root length (cm) | Fresh weight (g)* | Dry weight (g)* |
|---|---|---|---|---|---|
| 1 | Control | 19.92 de | 11.93 i | 2.15 hi | 0.233 bc |
| 2 | 50 mg $kg^{-1}$ | 11.61 ab | 5.33 ab | 0.83 a-e | 0.131 b |
| 3 | 100 mg $kg^{-1}$ | 11.10 a | 3.94 a | 0.63 c | 0.12 a |
| 4 | 200 mg $kg^{-1}$ | 11.60 ab | 4.10 a | 0.80 a-e | 0.40 cd |
| 5 | HG | 19.50 de | 9.27 e-i | 3.45 j | 0.39 a-d |
| 6 | HG+50 mg $kg^{-1}$ | 16. b-e | 7.85 b-f | 1.28 b-g | 0.67 d |
| 7 | HG+100 mg $kg^{-1}$ | 17.80 cde | 8.70 e-h | 1.46 c-h | 0.21 abc |
| 8 | HG+200 mg $kg^{-1}$ | 16.55 cde | 3.31 a | 1.80 f-i | 0.45 abcd |
| 9 | WN-1 | 16.10 b-e | 8.0 c-g | 1.55 d-h | 0.58 cd |
| 10 | WN-1+50 mg $kg^{-1}$ | 15.65 a-d | 10.05 f-i | 0.84 a-e | 0.20 abc |
| 11 | WN-1+100 mg $kg^{-1}$ | 16.17 b-e | 9.16 e-i | 0.75 a-e | 0.15 ab |
| 12 | WN-1+200 mg $kg^{-1}$ | 19.22 de | 5.53 abc | 1.48 a-h | 0.33 ad |
| 13 | JAF | 16.69 cde | 8.31 d-h | 0.481 ab | 0.21 abc |
| 14 | JAF+50 mg $kg^{-1}$ | 18.65 cde | 11.61 i | 0.31 a | 0.24 abc |
| 15 | JAF+100 mg $kg^{-1}$ | 15.56 a-d | 9.43 e-i | 0.31 a | 0.14 ab |
| 16 | JAF+200 mg $kg^{-1}$ | 13.80 abc | 3.18 a | 1.04 a-f | 0.35 a-d |
| 17 | B-1 | 16.49 cde | 10.46 f-i | 1.41 c-h | 0.33 a-d |
| 18 | B-1+ 50 mg $kg^{-1}$ | 15.56 a-d | 11.83 i | 1.49 d-h | 0.42 a-d |
| 19 | B-1+ 100 mg $kg^{-1}$ | 16.89 cde | 10.76 ghi | 1.57 e-h | 0.18 abc |
| 20 | B-1+200 mg $kg^{-1}$ | 20.70 e | 5.71 a-d | 1.60 e-h | 0.33 a-d |
| 21 | CF | 15.92 b-e | 9.53 e-i | 2.53 i | 0.34 a-d |
| 22 | CF+ 50mg $kg^{-1}$ | 16.65 cde | 6.95 b-e | 1.37 c-h | 0.15 ab |
| 23 | CF+ 100 mg $kg^{-1}$ | 17.81 cde | 10.55 f-i | 2.05 g-i | 0.25 abc |
| 24 | CF+ 200 mg $kg^{-1}$ | 16.13 b-e | 3.16 a | 1.85 f-i | 0.52 bcd |
| 25 | SUFK | 16.26 b-e | 8.58 e-h | 0.71 abcd | 0.13 ab |
| 26 | SUFK + 50 mg $kg^{-1}$ | 18.03 cde | 9.40 e-i | 0.85 a-e | 0.26 abc |
| 27 | SUFK + 100 mg $kg^{-1}$ | 16.05 b-e | 10.86 hi | 0.78 a-e | 0.20 abc |
| 28 | SUFK + 200 mg $kg^{-1}$ | 17.44 cde | 4.11 a | 0.14 c-h | 0.32 abcd |

Mean of 30 plants. Different letter within same column indicate significant difference between treatments using Duncan's multiple range test ($P \leq 0.05$).
* Fresh and dry weights represented by three replicates of 5 plants

**Table 3. Effect of mixed cultures of PGPR on growth of *V. mungo* in cadmium amended soils**

| S. No. | Strains/ Treatments | Shoot length (cm) | Root length (cm) | Fresh weight (g)* | Dry weight (g)* |
|---|---|---|---|---|---|
| 1 | Control | 19.92 cdef | 11.93 e | 2.15 ab | 0.233 a-e |
| 2 | 50 mg $kg^{-1}$ | 11.61 ab | 5.33 ab | 0.83 a | 0.131 ab |
| 3 | 100 mg $kg^{-1}$ | 11.10 a | 3.94 a | 0.63 a | 0.12 ab |
| 4 | 200 mg $kg^{-1}$ | 11.60 ab | 4.10 a | 0.80 a | 0.4 cde |
| 5 | MC-1** | 15.03 abc | 7.97 d | 0.98 a | 0.36 b-e |
| 6 | MC-1+50 mg $kg^{-1}$ | 14.86 abc | 4.86 ab | 1.75 ab | 0.17 a-d |
| 7 | MC-1+100 mg $kg^{-1}$ | 14.29 abc | 4.99 ab | 1.11 ab | 0.09 a |
| 8 | MC-1+200 mg $kg^{-1}$ | 18.59 cdef | 3.91 a | 1.48 ab | 0.14 ab |
| 9 | MC-2** | 17.79 b-e | 9.90 de | 2.51 b | 0.44 e |
| 10 | MC-2 +50 mg $kg^{-1}$ | 18.16 c-f | 5.90 abc | 1.41 ab | 0.12 ab |
| 11 | MC-2 +100 mg $kg^{-1}$ | 24.20 f | 4.72 a | 1.5 ab | 0.11 ab |
| 12 | MC-2 +200 mg $kg^{-1}$ | 18.54 c-f | 4.14 a | 1.7 ab | 0.24 a-e |
| 13 | MC-3** | 15.76 abc | 9.21 d | 1.22 ab | 0.42 de |
| 14 | MC-3+50 mg $kg^{-1}$ | 23.23 ef | 7.39 bcd | 1.95 c ab | 0.2 a-d |
| 15 | MC-3+100 mg $kg^{-1}$ | 19.95 c-f | 3.94 a | 1.45 ab | 0.15 abc |

| 16 | MC-3+200 mg kg$^{-1}$ | 17.01 a-e | 3.51 a | 1.45 ab | 0.14 ab |
|---|---|---|---|---|---|
| 17 | MC-4** | 15.57 abc | 7.95 cd | 2.51 b | 0.44 e |
| 18 | MC-4+50 mg kg$^{-1}$ | 16.77 a-d | 3.53 a | 1.35 ab | 0.12 ab |
| 19 | MC-4+100 mg kg$^{-1}$ | 23.06 def | 3.95 a | 2.05 ab | 0.18 a-d |
| 20 | MC-4+200 mg kg$^{-1}$ | 18.67 c-f | 3.1 a | 1.4 ab | 0.121 ab |

Mean of 30 plants. Different letter within same column indicate significant difference between treatments using Duncan's multiple range test ($P \leq 0.05$).
* Fresh and dry weights represented by three replicates of 5 plants
** Mixed culture combinations: MC-1 (WN-1, JAF, A-2), MC-2 (HG, JAF, CF), MC-3 (CF, WN-1, A-2), MC-4 (CF, HG, JAF)

**Table 4. Effect of mixed culture of PGPR and rhizobia on the growth of *V. mungo* in cadmium amended soils**

| S. No. | Strains | Shoot length (cm) | Root length (cm) | Fresh weight (g)* | Dry weight (g)* |
|---|---|---|---|---|---|
| 1 | Control | 19.92 cd | 11.93 f | 2.15 d | 0.233 ab |
| 2 | 50 mg kg$^{-1}$ | 11.61 b | 5.33 bcd | 0.83 abc | 0.131 ab |
| 3 | 100 mg kg$^{-1}$ | 11.10 b | 3.94 abc | 0.63 ab | 0.116 ab |
| 4 | 200 mg kg$^{-1}$ | 11.60 b | 4.10 abc | 0.80 abc | 0.40 ab |
| 5 | MCR- 1** | 23.40 d | 11.30 f | 1.57 bcd | 0.36 ab |
| 6 | MCR-1+50 mg kg$^{-1}$ | 19.88 cd | 5.34 bcd | 1.95 cd | 0.13 ab |
| 7 | MCR-1 +100 mg kg$^{-1}$ | 22.11 d | 7.94 de | 1.08 abcd | 0.14 ab |
| 8 | MCR-1 +200 mg kg$^{-1}$ | 6.4 a | 2.16 a | 0.09 a | 0.24 ab |
| 9 | MCR -2** | 22.20 d | 10.84 f | 0.78 abc | 1.21 c |
| 10 | MCR -2 +50 mg kg$^{-1}$ | 22.31 d | 6.31 cd | 1.08 abcd | 0.23 ab |
| 11 | MCR-2 +100 mg kg$^{-1}$ | 19.70 cd | 5.74 bcd | 1.12 abcd | 0.11 ab |
| 12 | MCR-2 +200 mg kg$^{-1}$ | 15.91 b | 4.08 abc | 0.53 ab | 0.05 a |
| 13 | MCR -3** | 23.68 d | 10.02 ef | 1.59 bcd | 0.25 ab |
| 14 | MCR-3 + 50 mg kg$^{-1}$ | 22.05 d | 5.22 bcd | 1.36 abcd | 0.16 ab |
| 15 | MCR-3 + 100 mg kg$^{-1}$ | 19.16 cd | 5.56 bcd | 0.9 abcd | 0.1 ab |
| 16 | MCR-3 + 200 mg kg$^{-1}$ | 13.66 b | 2.85 ab | 1 abcd | 0.11 ab |
| 17 | MCR -4** | 22.0 d | 10.74 f | 1.3a-d | 0.82 bc |
| 18 | MCR -4 + 50 mg kg$^{-1}$ | 22.88 d | 7.1 d | 1.28 abcd | 0.2 ab |
| 19 | MCR -4 + 100 mg kg$^{-1}$ | 22.00 d | 5.71 bcd | 0.8 abc | 0.12 ab |
| 20 | MCR-4 + 200 mg kg$^{-1}$ | 14.45 b | 2.36 a | 1 abcd | 0.15 ab |

Mean of 30 plants. Different letter within same column indicate significant difference between treatments using Duncan's multiple range test ($P \leq 0.05$).
* Fresh and dry weights represented by three replicates of 5 plants
** Mixed culture combinations: MCR-1 (IBA, HG, BS-1), MCR-2 (IBA, WN-1, AC-1), MCR-3 (IBA, JAF, CN-1), MCR-4 (IBA, BF-1, SUFK)

## Discussion

This study was conducted to demonstrate the cadmium stress alleviation of *Vigna mungo* after inoculation with rhizobacteria. The metal especially cadmium contamination of soil is hazardous for animals as well as humans and plants. In the present study, cadmium resistant strains that have the ability to produce auxin were isolated. Finally, selected strains were used to determine their potential of plant growth promotion under cadmium stress. Highest auxin production by rhizobacteria was ranged from 140-160 μg ml$^{-1}$. Auxin is a phytohormone that is involved in increasing root growth of plants. Increase in adventitious roots causes an increased level of nutrient absorption from the soil and thus enhance the plant growth parameters like shoot growth and leaf development (Iqbal and Hasnain, 2013). Bacteria have a number of mechanisms to resist the presence of metal either by converting it into less toxic compound or by up taking it in combined form with minerals. Bacteria associated with the rhizosphere have the ability to resist the cadmium and other metal stress. They do so either by immobilizing, pumping, blocking or binding the metal to certain specific parts of the cells (Balestrazzi et al., 2009).

After screening, strains were used to observe their growth promoting ability in the presence of metal stress. Experiments demonstrated that the plants recorded suppression in growth when sown solely in metal amended soils. However, plants showed statistically good growth response under metal stress after bacterial inoculations. But, when these seeds were sown along with the bacterial co-inoculation, plants showed better growth. It was recorded that increase in plant root and shoot lengths were up to 1.8 fold and 65% under metal stress conditions. Cadmium metal inhabits the root and shoots growth by imposing damages to the plants normal physiological actions (Rathour et al., 2013). Cadmium can cause oxidative damage in *Vigna mungo* that can be mitigated by eliciting certain antioxidant defence mechanisms (Molina et al., 2008). Recorded increment in fresh and dry biomass of plants with single culture inoculation was 2.2 fold. Ali et al. (2013) has reported that co-inoculation of seeds with auxin producing PGPR increase the biomass of plants by promoting the nutrient availability. Depending upon the applied concentration of metal stress, PGPR immobilize the metal ions in the rhizosphere, thus inhabiting their uptake by plants. In addition these bacteria produce auxin hormone which cause the root enlargement and hence promoting the mineral uptake by the roots (Belimov et al., 2004).

When the bacterial strains were used as mixed culture, plant growth promotion was recorded that was up to 1 fold in terms of plant shoot and root length under cadmium stress. Similarly, fresh weight enhancement was up to 2.2 fold and dry biomass increase was up to 76% in cadmium amended soil. Bacteria associated with the plant roots enhance plant growth by various mechanisms like nitrogen fixation, nutrients conversion and phosphorous availability. PGPR also resist cadmium and can enhance plant growth in the presence of metal stress (Belimov et al., 2005). Co-inoculation with *Bacillus* and *Rhizobium sp.* can enhance the plant growth by enhancing root nodulation or biological nitrogen fixation. In addition, *Bacillus sp.* cause change in composition of rhizosphere leading to enhanced yield of plants in terms of height and weight (Venkatesan et al., 2011; Raheem et al., 2017). Sajid et al. (2017) also reported the plant growth enhancement by using single or mixed bacterial inoculations of *Bacillus* strains under stress conditions. Use of mixed culture of different PGPR either belonging to different species or from same species enhance the performance of these bacterial strains. One property of a strain complements the other property of the strain. As *Rhizobium* fix the nitrogen in the roots of plants, and the other bacteria with the auxin producing ability will enhance their work and plant will be benefited as compared to single strain inoculation (Yu et al., 2012).

## Conclusion

In conclusion, strains isolated in this study showed promising results for plant growth promotion under axenic conditions in cadmium amended soils. Majority of the strains showed resistance to cadmium stress. Highest auxin production was ranged from 140- 160 μg $ml^{-1}$ by *B. cereus* HG and *B. thuringiensis* JAF. In pot trials, *B. cereus* HG, *B. cereus* CG and *P. fluorescens* WN-1 showed significant growth response in the presence of cadmium stress. In mixed culture of PGPR, combinations of MC-2 and MC-3 performed to enhance vegetative growth parameters. In co-inoculation experiments of PGPR and rhizobia, mixed culture combination MCR-1 was the most promising as compared to other combinations. Overall, strains showed good potential to facilitate crop growth under cadmium contaminated environment.

## References

Ali B, 2015. Bacterial auxin signaling: Comparative study of growth induction in *Arabidopsis thaliana* and *Triticum aestivum*. Turk. J. Bot. 39: 1-9.

Ali T, Mahmood S, Khan MY, Aslam A, Hussain MB, Asghar HN and Akhtar JA, 2013. Phytoremediation of cadmium contaminated soil by auxin assisted bacterial co-inoculation. Asian J. Agri. Biol. 1: 79-84.

Akhtar S and Ali B, 2011. Evaluation of rhizobacteria as non-rhizobial inoculants for mung beans. Aust. J. Crop Sci. 5: 1723-1729.

Balestrazii A, Bonadei M, Quarttrini E and Carbonera D, 2009. Occurrence of multiple metal-resistance in bacterial isolates associated with transgenic white poplars (*Populous alba* L.). Ann. Microbiol. 59: 17-23

Bazrafshan E, Mahvi AH, Nasseri S, Mesdaghinia AR, Vaezi F and Nazmara S, 2006. Removal of cadmium from industrial effluents by electrocoagulation process using iron electrodes, Iran. J. Environ. Health Sci. Eng. 3: 261-266.

Belimov AA, Hontzeas N, Safronova VI, Demchinsakaya SV, Piluzza G, Bullitta S and Glick BR, 2005. Cadmium-tolerant plant growth-promoting bacteria associated with the roots of Indian mustard (*Brassica juncea L. Czern.*). Soil Biol. Biochem. 37: 241-250.

Belimov AA, Kunakova AM, Safronova VI, Stepanok VV, Yudkin LY, Alekseev VY and Kozhemyakov AP, 2004. Employment of rhizobacteria for the inoculation of barley plants cultivated in soil contamination with lead and cadmium. Microbiol. 73: 99-106.

Figueira PAME, Lima GIA and Rereira SIA, 2005. Cadmium tolerance plasticity in *Rhizobium leguminosarum* bv. viciae: glutathione as a detoxifying agent. Can. J. Microbiol. 51: 7-14.

Godt J, Scheidig F, Grosse-Siestrup C, Esche V, Brandenburg P, Reich A and Groneberg DA, 2006. The toxicity of cadmium and resulting hazards for human health. J. Occup. Med. Toxicol. 1: 22. doi: 10.1186/1745-6673-1-22.

Hussain Z and Islam M, 2010. Leaching of heavy metals from contaminated soils using inductively coupled plasma optical emission spectrometer (ICP-OES) and atomic adsorption spectrometer (ASS). J. Sci. Res. 2: 47-53.

Iqbal A and Hasnain S, 2013. Auxin producing bacilli: Potential candidates in growth stimulation of *Vigna radiata* (L.). Int. Proc. Soc. Behav. Sci. 1: 145-151.

Jankiewicz B, Ptaszynski B and Wieczorek M, 2000. Spectrophotometric determination of cadmium (II) in soil of allotment gardens in lodz. Pol. J. Environ. Stud. 9: 83-86.

Jing YD, He ZL and Yang XE, 2007. Role of soil rhizobacteria in phytoremediation of heavy metal contaminated soil. J. Zhejiang Univ. Sci. *B*. 8: 192-207.

Khan GA, 2005. Role of soil microbes in the rhizosphere of plants growing in trace metal contaminated soils in phytoremediation. J. Trace Elem. Med. Biol. 18: 355-364.

Ladwal A, Bhatia D and Malik KD, 2012. Effect of co-inoculation of *Mesorhizobium cicer* with PGPR on *Cicer arietinum.* Aust. J. Basic Appl. Sci. 6: 183-187.

Madhaiyan M, Poonguzhali S and Sa T, 2007. Metal tolerating methylotrophic bacteria reduces nickel and cadmium toxicity and promotes plant growth of tomato (*Lycopersicon esculentum* L.). Chemosphere. 69: 220-228.

Molina SA, Nievas C, Chaca PVM, Garibotto F, Gonza U, Marsa MS, Luna C, Gimenez SM and Zirulnik F, 2008. Cadmium-induced oxidative damage and antioxidative defense mechanisms in *Vigna mungo* (L.). Plant Growth Reg. 56(3): 285-295.

Rahim A, Shaposhnikov A, Belimov AA, Dodd IC and Ali B, 2017. Auxin production by rhizobacteria was associated with improved yield of wheat (*Triticum aestivum* L.) under drought stress. Arch. Agron. Soil Sci. doi: 10.1080/03650340.2017.1362105 (in press).

Raheem A and Ali B, 2015. Halotolerant rhizobacteria: Beneficial plant metabolites and growth enhancement of *Triticum aestivum* L. in salt amended soils. Arch. Agron. Soil Sci. 61: 1691-1705.

Rajkumar M, Ae N, Prasad MNV and Freitas H, 2010. Potential of siderophore-producing bacteria for improving heavy metal phytoextraction. Trends Biotechnol. 28: 142-149.

Rathour P, Ramteka WP, Raja W and John AS, 2013. Isolation and characterization of nickel and cadmium tolerant plant growth promoting rhizobacteria from rhizosphere of *Withania somnifera.* J. Biodivers. Environ. Sci. 6: 253-261.

Sajid S, Raheem A and Ali B, 2017. Phylogenetic diversity of drought tolerant *Bacillus* spp. and their growth stimulation of *Zea mays* L. under different water regimes. Res. J. Biotechnol. 12: 38-46.

Qing HU, Dou MN, Qi HY, Zie XM, Zhuang GQ and Yang M, 2007. Detection, isolation and identification of cadmium resistant bacteria based on PCR-DGGE, J. Environ. Sci. 19: 1114-1119.

Venkatesan S, Kirithika M, Rajapriya R, Ganesan R and Muthuchelian K, 2011. Improvement of economic phytoremediation with heavy metal tolerant rhizosphere bacteria. Int. J. Environ. Sci. 1: 1864-1873.

Yu X, Liu X, Zhu H-T, Liu GH and Mao C, 2012. Co-inoculation with phosphate-solubilizing and nitrogen-fixing bacteria on solubilization of rock phosphate and their effect on growth promotion and nutrient uptake by walnut. Eur. J. Soil Microbiol. 50: 112-117.

# 15

# Effect of insecticides on *Coccinella septempunctata* (Coleoptera; Coccinellidae)

**Aqsa Sattar[1], Iqra Azam[1*], Muhammad Kaleem Sarwar[1], Afifa Amjad[1], Muhammad Faheem Malik[2]**

[1] Department of Zoology, University of Gujrat, Sub-Campus Narowal, Narowal, Pakistan

[2] Department of Zoology, University of Gujrat, Hafiz Hayat Campus, Gujrat, Pakistan

*Corresponding author email: iqraazam27@gmail.com

## Abstract

*Coccinella septempunctata* is the generalist, representative predator that is potentially used as biological control agent in many crop pest management programs. A comprehensive and critical review of published and formerly unpublished studies was accomplished to evaluate the effects of insecticides such as imidacloprid, hexaflumuron, dimethoate, deltamethrin, pyriproxyfen, spinosad, lambda-cyhalothrin, bifenthrin, acetamiprid, fenazaquin, quinalphos and neemix on *C. septempunctata.* Data generated clearly indicates severe effects of used insecticides on its neurophysiological and behavioral responses indicating loss of co-ordination, paralysis as well as reduce fecundity, longevity and growth. Further studies are required to focus mainly on the measure to evaluate their real effects under field conditions

**Keywords:** *C. septempunctata*, Biological control agent, Predatory efficacy, Imidacloprid, Pyriproxyfen

## Introduction

The name of coccinellids was firstly originated in Britain where they were commonly called 'Our Lady's bird' or the Lady beetle. They belong to family Coccinellidae of order Coleoptera (Samaha, 2010). There are approximately 6000 known species worldwide and out of which 300 species are found in Indo-Pakistan subcontinent (Akhavan et al., 2013; Rahatullah et al., 2010). Because of their predacious nature, about 90% of approximately 4,200 coccinellids species are known as beneficial. Family Coccinellidae represents six sub-families i.e. Coccinellinae, Chilocorine, Scymninae, Coccidulinae, Sticholotidinae and Epilachninae. Five sub-families (Coccinellinae, Chilocorine, Scymninae, Coccidulinae and Sticholotidinae) are predacious in nature but only one sub-family Epilachninae is phytophagus in nature (Singh and Bras, 2004; Ullah et al., 2012).

Coccinellids have distinctive striking appearance. They can be small upto 1 mm and not longer than 4-9 mm (Koren et al., 2012). Their body may be rounded, elongated, oval and flattened with different colors such as black, orange, red or straw color. They are successfully used in past in controlling many crop pests like aphids, leafhoppers, scale insects mites and other soft bodied insects (Slipinski, 2007). They are distributed all over the world including agro-ecosystems, plains, mountains, forests and grasslands (Urooj, 2016).

### *C. septempunctata*

With family Coccinellidae, the most efficient natural enemy of crop pests is *C. septempunctata* that is potentially used as biological control agent (Hodek and Honek, 2009). Both larvae and adults of *C. septempunctata* are active predators and are highly polyphagus (Yu et al., 2014). Forster (1995) observed the data for insecticidal analysis on *C. septempunctata* and stated that seven-spotted ladybird beetle was the fifth best studied natural predator in insecticides registration framework in Germany.

**Habitat**

Coccinella septempunctata is a generalist ladybird, present in a wide range of plants and aphid infested crops for instance potato, corn leaf, green bug and pea. Aphids are plant feeders and are a main food source of *C. septempunctata. C. septempunctata* feeds on large number of aphids. Aphids damage the plants in several ways such as; they curl and deform the leaves, and stunt their shoots as well as they affect plants as viruses (Raboudi et al., 2002; kaits et al., 2007).

**Reproduction and Development**

The female of *C. septempunctata* laid their eggs on the surface of leaves that are elongated and oval shaped. The eggs take about four days to hatch and as the eggs hatch; they give rise to four sequential stages of instar larvae before they develop into pupa. The fourth instar larvae are more insatiable than the other instar larvae (Solangi et al., 2007). At high temperature, it develops into light orange pupae. Contradictory, at low temperature it develops into dark-brown or blackish pupae. The adults of *C. septempunctata* are red in color and possess seven black spots on their body (Majerus and Kearns, 1989).

**Integrated Pest Management (IPM)**

Smith and Van den Bosch was the first to use the term "Integrated Pest Management". IPM is a broad methodology to control pests by using other living organisms known as natural predators (Ehler, 2006). Integrated Pest Management can decrease the dependence upon insecticides that are used to control many insect problems. Moreover, coccinellids have been successfully organized in various integrated management programs and biological control programs against many insect pests on different crops (Ahmed et al., 2016).

**Insecticides:** Insecticides are toxic chemicals which are used for preventing, destroying, controlling or modifying any pest. Insecticides not only affect the nervous system of insect pests but also harm their exoskeleton (NPIC). Several insecticides are introduced in the market for instance organophosphates (OP) were introduced from 1960s to late 1980s (Akhtar et al., 2009). Insecticides vary in their activity i.e. they are neurotoxins, respiration inhibitors, growth inhibitors and regulators and stomach poisons. The most commonly used insecticides are imidacloprid, hexaflumuron, dimethoate, deltamethrin, pyriproxyfen, spinosad, lambda-cyhalothrin, bifenthrin, acetamiprid, fenazaquin, quinalphos and pimicarb. They are applied in different doses: $LC_{50}$ (Median lethal concentration), $LD_{50}$ (Lethal median dose) or $EC_{50}$ (Median effective concentration) to arthropods (Sánchez-Bayo, 2012).

**Effect of Imidacloprid**

Imidacloprid is a systematic neonicotinoid, used to control different insects e.g. aphids, whiteflies, termites, fleas, soil insects and some beetles. It affects neurophysiological responses in terms of lack of co-ordination, shocks and cause paralysis in insects (Papachristos and Milonas, 2008). Imidacloprid, a broad-spectrum insecticide, has gained registration in almost 120 different countries (Drobne et al., 2008). Imidacloprid has greater ability to control pest insects when applied in the form of foliar sprays, microcosm or by treating glass plates (Tan et al., 2012). According to Bozsik (2006), imidacloprid appears to be harmless for *C. septempunctata.*

**a) Effect on Egg Production**

In control microcosm test, on exposure to highest concentration of imidacloprid, the numbers of egg production were reduced for the females. From the control microcosm, the hatching percentage of eggs from females was on normal 87.1%. But on exposure to highest application of imidacloprid, the hatching percentage of egg was decreased to 75.2% for females. The egg hatching rate decreased with increasing the imidacloprid application (Yu et al., 2014).

**b) Effects on 2nd instar Larvae**

In long-term microcosm toxicity test, single application of imidacloprid causing 50% affects. Mortality in the control group was observed and only 62.5% of control individuals had persisted. In short-term toxicity test, second instar larvae were not affected at all and mortality in the control group was not examined (Yu et al., 2014). According to Yu et al., (2014) higher rate of imidacloprid delayed the development of second and third instar larvae.

**c) Effects on 4th instar Larvae**

After current application of sub-lethal dose of imidacloprid ($LD_{10}$) on the fourth instar larvae considerably affected the weight of adult ladybird beetle. On exposure to imidacloprid, the feeding behavior of fourth instar larvae of seven spotted ladybird beetle was reduced. As a result, mortality was greater, development and weight was reduced, pupae and adult size was smaller (Yu et al., 2014).

**d) Effect on Survival Rate**
In long-term laboratory microcosm test, Yu et al., (2014) observed that, survival rate of *C. septempunctata* larvae was started to decrease from second day after single imidacloprid application.

**e) Effect on Feeding Behaviour**
According to a hypothesis, imidacloprid may affect the capability of F1 insects by decreasing the activity of their parents. After ladybird beetles being exposed to sub-lethal concentrations of imidacloprid then they were not capable to feed appropriately (Ramirez-Romero et al., 2008; Han et al., 2010). After the application of imidacloprid, *C. septempunctata* showed longer developmental period (from fourth instar larvae to adult), decreased aphid consumption rates as well as adult weight (Skouras et al., 2017).

**Effect of Hexaflumuron**
Hexaflumuron is a benzoylphenyl urea-type insecticide and an insect growth regulator (IGR) that affects the insect's endocrine system, inhibit growth, disrupts molting and cuticle formation as well as affects the development of seven spotted ladybird beetle (Dhadialla et al., 1998). Hexaflumuron is commonly used insect growth regulator in China with 120 products and have been registered as first active ingredient as reduced-risk insecticide (Yu et al., 2013).

**a) Effect on Egg Production:** The females produced 1,042 eggs and the number of eggs reduced with increasing the concentration of the hexaflumuron mainly at the two highest application levels (Yu et al., 2013).

**b) Effect on Egg Hatching:** The percentage of egg hatching was on normal 88.3%. But on exposure to highest concentration of hexaflumuron, the percentage of egg hatching was reduced to 76.9%. By increasing the concentration of hexaflumuron at various levels the mean egg hatching percentages were 88.0, 88.7, 86.1 and 83.57 (Yu et al., 2013).

**c) Effects on Larval stages**
The long-term toxicity test yield the $LR_{50}$ (304 g a.i. $ha^{-1}$) for the 2nd instar larvae of ladybird beetle on exposure to single treatment of hexaflumuron and the application rate causing 50% deaths among larvae of *C. septempunctata*. It is effective in immature stages and restrict with chitin synthesis at the time of shedding (Desneux et al., 2007). Highest application of hexaflumuron was related with larger delays in developmental period of *C. septempunctata* (from second instar to next generation). In the controls, the total developmental period of *C. septempunctata* was 28.43 days. On exposure to highest treatment level of hexaflumuron, the time was about of 1.15 longer than in the control. At the highest treatment level, the development times of second (2.60), third (2.90), fourth instar larvae (5.77), the pupae (6.80) and the total generation (32.75) were all considerably longer than the controls (Yu et al., 2013).

**d) Effect on Pupal Stage**
Yu et al., (2013) examined that concentration up to 1.52 g a.i. $ha^{-1}$ of both control and hexaflumuron resulted in 100% pupation. The mean pupation percentage that received in the test systems was 3.04 g a.i. $ha^{-1}$ and reduced to 88.9% (considerably not different from the control group) and at the maximum application rate, the mean pupation percentage was 73.80% (considerably different from the control group.

**e) Effect on Adult Emergence**
In the controls and two highest application rates of hexaflumuron, the emergence of ladybird beetle was 100%. On the other hand, in the control test systems and three highest application rates of hexaflumuron, the mean emergence percentages of *C. septempunctata* were 96.7, 95.8, and 94.4%, respectively, although not considerably different from the controls (Yu et al., 2013).

**g) Effect on Survival Rate**
In test period (0-21 days) performed by Yu et al., (2013), the survival rate of larvae of seven spotted ladybird beetle was affected and decreased from 2nd day after single hexaflumuron application. On 6-day, rate of survival was reduced to 83.3 and 76.7% respectively. At the end of 21day, the survival in the blank control was 96.7%. But at the concentration of 6.08 g a.i. $ha^{-1}$, survival was 53.3%. He concluded that,

effect of hexaflumuron on the survival rate of *C. septempunctata* was larger.

**Effect of Dimethoate**

Dimethoate is an organophosphorous and used for the control of substantial number of insects such as whiteflies, aphids, mites and plant hoppers. Due to more efficacy and rapid environmental degradation, dimethoate is used in urban areas and agriculture across the world (Mirajkar et al., 2005). Singh (2001) used dimethoate (organophosphate) under laboratory conditions and observed comprehensive locomotory responses in seven spotted ladybird beetles. Dimethoate suppressed the locomotion of *C. septempunctata* and interrupts the normal behavior pattern of seven spotted ladybird beetle.

**a) Effect on Locomotory Behaviour**

Thornham et al. (2007) reported that male *C. septempunctata* travel more distance than they did in control group, during individual walking events conducted on substrates Al treated with dimethoate. It was observed by Singh (2001) that dimethoate treated AI increased the locomotion of *C. septempunctata*.

**b) Time Spent on Walking and Resting**

In an experiment it was found that, when upper half of the plant was treated with dimethoate then *C. septempunctata* spent greater period on lower half of the plant. *C. septempunctata* spent greater period walking than on the controls ($F\ 4,119 = 9.52$, $P < 0.01$) and spent much time resting on dimethoate treated plants. When upper lower half of the plant was treated with dimethoate then seven spotted ladybird beetles spent much time resting on lower half of the plant than the upper and successful treatments (Singh, 2001).Under laboratory conditions, Singh (2001) noted considerable modifications among treatments of dimethoate regarding to time spent on different parts of the plants ($F16,\ 342 = 2.647$, $P < 0.0$). When dimethoate applied to the upper half of the plant (apex), then seven spotted ladybird beetles spent less time on the upper half of the plant (LSD $P < 0.05$). On the other hand, when dimethoate applied to the lower half of the plant then, coccinellids spent greater period on upper half of the plant.

**c) Time Spent on Stem**

Coccinellids spent more time on the stem on dimethoate treated plants and much less time on the control plants (LSD $P < 0.05$). Coccinellids spent considerably much on leaf margins than those on the apex and lower treated plants (Singh, 2001).

**d) Predatory Efficacy**

Thornham et al., (2007) observed that, predatory efficacy of larvae and adults of *C. septempunctata* were decreased considerably on exposure to dimethoate residues and treated prey.

**Effect of Deltamethrin**

Deltamethrin is a pyrethroid that kills insects on exposure and through stomach action. However, deltamethrin is a broad-spectrum insecticide and promptly affect the nervous system of insects (Tomlin, 2006; Rehman et al., 2014).

**a) Effect on Walking Behavior**

In an experiment, for five days the walking behavior of *C. septempunctata* was observed between deltamethrin treated (36 and 82%) plot and untreated plot (14-58%). Observations of seven spotted ladybird beetle resting behavior considered for between 12 and 60% in the treated plot; 38 and 85% in the untreated plot. It was reported that, *C. septempunctata* are present at the tip of plants because aphids feed on the apex of the plant (Majerus and Kearns, 1989). On exposure to deltamethrin residues, reorganization of seven spotted ladybird beetle from the lower end of the canopy toward the ground was observed. This relocation increased walking behavior and reduced food accessibility. First three days after spray, fewer *C. septempunctata* were examined on the upper surface of plant than the lower surface. This may show that deltamethrin causes short-term irritant effect. On 1$^{st}$, 2$^{nd}$ and 3$^{rd}$ day, the difference in the behavior pattern of ladybird beetles indicate that ladybird beetles walked considerably more often in the plots treated with deltamethrin than those on the untreated. The walking behaviour of adult Coccinella septempunctata becomes stronger and resting was reduced on exposure to deltamethrin (Wiles and Jepson, 1994).

**b) Effect on Grooming Behavior**

The seven spotted ladybird beetles were observed to groom considerably more regularly in the treated plot on day 1 and on day 1 to 4 and feed less frequently than those in the plot not treated with deltamethrin (Wiles and Jepson, 1994).

The adult behavior of *C. septempunctata* was affected when contact to deltamethrin. On exposure to deltamethrin, grooming of *C. septempunctata* become stronger however, they stimulated toward the lower part of the shelter (Wilis and Jepson, 1994). According to Bozsik (2006), deltamethrin seems to be moderately harmful to seven spotted ladybird beetles and causes greater mortality from 1st to all larval instars of ladybird beetles.

**Effect of lambda-Cyhalothrin**

Lambda-cyhalothrin is a synthetic pyrethroid and similar to natural insecticide pyrethrin (Muhammad, 2009). Lambda-cyhalothrin affects the nervous system and cause paralysis or death (Tomlin, 1997).

**a) Effect on Locomotory Behaviour**

It was observed that lambda-cyhalothrin AI treatment reduced the locomotion in seven spotted ladybird beetles. In response to lambda-cyhalothrin PP but not the AI treatment, large locomotory changes were detected (Thornham et al., 2007).

**b) Mortality Rate**

Lambda-cyhalothrin seems to be moderately harmful to *C. septempunctata* adult because of acute surface contact effects (Bozsik, 2006). In an experiment, lambda-cyhalothrin was moderately harmful according to their $LT_{50}$ values. Lambda-cyhalothrin act as axonal poison and is sodium channel modulator (Bozsik, 2006).

The efficacy of lambda-cyhalothrin on coccinellids population may demonstrate that, the effect of insecticide pressure and their susceptibility to certain insecticides are similar across the world (Sterk et al., 1999).

**Effect of Pyriproxyfen**

Pyriproxyfen is an insect growth regulator that affects enormous number of insects such as fleas, cockroaches, ticks, ants, carpet beetles and mosquitoes. Pyriproxyfen is influenced by juvenile hormone so that it mainly affects young insects and eggs and restricts their growth (Hallman et al., 2015).The toxicity of pyriproxyfen against third instar and adult of *C. septempunctata* was estimated at three and seven days post treatment interval. At 3-day post treatment interval, pyriproxyfen was moderately safe against 3rd instar and adults of *C. septempunctata* as the $LC_{50}$ value of pyriproxyfen was lower than their field recommended dose. Moreover, at 7-day treatment interval, pyriproxyfen was moderately safe against 3rd instar and adults of *C. septempunctata* as the $LC_{50}$ value of pyriproxyfen was lower than their field recommended dose. The result is that, pyriproxyfen would be used where ladybird beetle was used for the controlling of whiteflies and jassids (Asrar et al., 2014).

It was observed that pyriproxyfen had no side-effects on adult *C. septempunctata*. According to Olszak et al., (1994), in Poland pyriproxyfen had minimal impact on the adult of seven spotted ladybird beetle. According to Bozsik (2006), pyriproxyfen found to be safe for adults of *C. septempunctata*. Asrar et al. (2014) suggested that pyriproxyfen has less lethal effect on *C. septempunctata*.

**Effect of Spinosad**

Spinosad is first active ingredient as well as broad-spectrum insecticide, belongs to chemical group naturalyte. Due to high efficacy, spinosad has low use generally. Spinosad residue on plant vegetation found to be harmless to foraging honeybees and has less effect on beneficial arthropods such as *C. septempunctata*. Spinosad expressed as solid or liquid (DWO).Under laboratory conditions, toxicity of different insecticides applied under various water levels against *C. septempunctata* was observed after 96 hours of three treatments. Spinosad (30.00%) found to be less toxic against seven spotted ladybird beetles after 96 hours of application. Moreover, it was found that low application of water 2ml/lit caused higher mortality of predators (72.86%). On the other hand, toxicity of different insecticides under various water levels against *C. septempunctata* was also evaluated after one week of their treatment. After one week of their treatment, all insecticides including spinosad (38%) increase their toxicity to predators and causing greater mortality as compared to 96 hours of their treatment. Moreover, it was further found that insecticides with low application of water caused higher mortality of *C. septempunctata* (84.29%) and insecticides applied with high application of water (2ml/5liter) caused less mortality of seven spotted ladybird beetle (Solangi et al., 2007).Spinosad possess impartial effect on the Coccinellidae population. In an experiment, during the dose of first week spinosad highlighted moderately toxic effect and in the second week presented neutral effect (Singh et al., 2004).

**Effect of Bifenthrin**

Bifenthrin is an insecticide, belongs to pyrethroid family of chemicals and similar to natural insecticide pyrethrin. Bifenthrin affects the nervous system and

kills arthropods through stomach action (ingestion). Bifenthrin is more toxic to insects than mammals because mammals have larger body size, higher body temperature and low sensitivity of the ion channel sites (Bradberry, 2005).
Skouras et al.(2017) evaluated the $LD_{50}$ values for all larval instars (first, second, third and fourth) from the Kalamata population and Argos respectively. The effect of bifenthrin on seven spotted ladybird beetle larvae was different on exposure but bifenthrin was proved to be more toxic and caused high mortality to all larval instars of seven spotted ladybird beetle. The susceptibility of first instar larvae to bifenthrin was greater due to larval weight. Moreover, larval weight increases from first to fourth instar larvae and resistant factor between Kalamata and Argos population for fourth instar larvae was 4.49 and 2.19.
Under laboratory conditions, toxicity of bifenthrin applied under various water levels against *C. septempunctata* after 96 hours of three treatments was observed. Bifenthrin (56%) was found to be toxic against seven spotted ladybird beetles after 96 hours of application. On the other hand, toxicity of different insecticides under various water levels against *C. septempunctata* was also evaluated after one week of their treatment. After one week of their treatment, all insecticides including bifenthrin (76%) increase their toxicity to predators and causing greater mortality as compared to 96 hours of their treatment. Moreover, it was further found that insecticides with low application of water caused higher mortality of *C. septempunctata* (84.29%) and insecticides applied with high application of water (2ml/5liter) caused less mortality of seven spotted ladybird beetle (Solangi et al., 2007).

**Effect of Acetamiprid**

Acetamiprid also known as ethanimidamide is an insecticide belongs to neonicotinoid. It interrupts the function of nervous system of insects thereby, causing paralysis (USEPA).
It was observed that acetamiprid was less toxic to seven spotted ladybird beetle larvae. But according to Youn et al. (2003) acetamiprid caused 100% mortality at 40 mg a.i. /L. Further he noticed that fourth instar was more susceptible to acetamiprid than the second, third and fourth instar respectively. In a laboratory bioassay, performed by Asrar et al. (2014) residual toxicity of different insecticides including acetamiprid against *C. septempunctata* at 3 and 7 days post interval was determined. At 3-day post treatment interval, acetamiprid (neonicotinoid) was toxic against 3$^{rd}$ instar larvae and adults of seven spotted ladybird beetles as the $LC_{50}$ values of acetamiprid were lower than their field recommended doses. Moreover, acetamiprid was toxic against third instar and adults of *C. septempunctata* at 7-day post treatment interval.

**Effect of Fenazaquin and Quinalphos**

Fenazaquin is an insecticide and miticide belongs to chemical class Quinazoline. It is used to control mites and insects mainly whiteflies and quinalphos is also used to control pests (U.S. EPA 2007).
Ahmad and Ahmad (2009) reported the comparative toxicity of two insecticides to adults of *C. septempunctata*. Quinalphos and fenazaquin was applied with different concentrations. For each spray one ml of each insecticide concentration was applied. The percentages were recorded after 24, 48 and 72 h as well as toxicity of quinalphos and fenazaquin was also recorded.
The two insecticides (quinalphos and fenazaquin) were proved to be highly toxic to adults of *C. septempunctata*. The fenazaquin with $LC_{50}$ (0/0123055%) value was found to be highly toxic against adults of *C. septempunctata* in 72h and lower and upper fiducial limits were found to be 0.00740146 and 0.0204589, correspondingly at 95% confidence. Contradictory, the quinalphos with $LC_{50}$ (0.0453297%) value was found to be moderately toxic against adults of *C. septempunctata* in 72h and lower and upper fiducial limits were found to be 0.0301508 and 0.0681241 and, correspondingly at 95% confidence. According to Thomas and Phadke (1991), quinalphos was proved to be less toxic from first instar to adults of the Coccinellids than chlropyrifos but Shukla et al., (1994) proposed that it was found to be toxic to seven spotted ladybird beetles.

**Effect of Neemix**

Neemix, a neem insecticide, is used to control pea aphid (Stark and Rangus, 1994). The azadirachtin is the principle active ingredient in neem and found to be oxidized limonoid (Mordue and Blackwell, 1993). Azadirachtin affects chemoreceptors of insects in several ways such as, prevent feeding and oviposition as well as affects overall fitness. 1000 ppm azadirachtin and 250 ppm azadirachtin and 30% neem oil was reported by Kaethner (1991) and on exposure to dried residues on bean plants those two neem insecticides caused no damage to the eggs, second instar or adults of seven spotted ladybird beetle. On the other hand, neem was considered to be toxic, causing mortality and morphogenetic deficiencies to seven spotted ladybird beetles on direct sprays.

Kaethner (1991) identified that neem insecticides may be harmless to seven spotted ladybird beetles on the basis of criteria developed by the International Organization of Biological Control (IOBC) thereby, causing less mortality (30%) in laboratory studies (Hassan, 1992). Kaethner (1991) reported that the addition of neem oil to 1000 ppm azadirachtin and 250 ppm azadirachtin did cause greater morphogenetic abnormalities but did not cause greater mortality. At $LC_{50}$ (1,120 ppm), Neemix was considered to be more toxic to fourth instars.

At the time of larval stage of untreated insects, two hormones are produced such as ecdysone and juvenile hormone and these are carefully balanced during all larval stages. But at the time of final instar, the level of juvenile hormone decreases, causing ecdysone level to surge (Rosomer and Stoffolano, 1994). Azadirachtin, an active ingredient, affects the normal concentration of ecdysone and juvenile hormone. At $LC_{50}$, toxicity studies of adults show that Neemix is not toxic to *C. septempunctata* (Kaethner, 1991). Individual-level bioassays of one life stage are very beneficial for gaining information on the total effect of insecticides (Cairns, 1983) but according to Stark and Wennergren (1995) when difference in the susceptibility of life stages was present then total risk of a insecticides could not be assessed by observing one life stage. Further results suggest that it is necessary to observe complete life stage (from first instar to adult) of a species to evaluate the entire risk of insecticides (Banken and Stark, 1997).

**Table - 1: Characterization of relative toxicity of insecticides on *C. septempunctata***

| Insecticides Common Name | Chemical Group | Toxicity Class | References |
|---|---|---|---|
| Imidacloprid | Neonicotinoids | Toxic | (Yu et al., 2014) |
| Hexaflumuron | Benzoylureas | Toxic | (Yu et al., 2013) |
| Dimethoate | Organophosphate | Safe | (Singh et al., 2001) |
| Deltamethrin | Pyrethroids | Toxic | (Wiles and Jespon, 1994) |
| Lambda-cyhalothrin | Pyrethroids | Toxic | (Bozsik, 2006) |
| Pyriproxyfen | Pyriproxyfen | Safe | (Asrar et al., 2014) |
| Spinosad | Naturalyte | Toxic | (Solangi et al., 2007) |
| Bifenthrin | Pyrethroids | Toxic | (Solangi et al., 2007) |
| Acetamiprid | Neonicotinoids | Toxic | (Skouras et al., 2017) |
| Fenazaquin | Quinazoline | Highly Toxic | (Ahmad and Ahmad, 2009) |
| Quinalphos | - | Highly Toxic | (Ahmad and Ahmad, 2009) |
| Neemix | - | Safe | (Kaethner, 1991) |

## Conclusion

Insecticides and their natural enemies i.e. *C. septempunctata* are used for controlling pests mainly aphids meanwhile; these insecticides have severe effects on the biological and physiological processes of *C. septempunctata.* The effects of insecticides on natural enemies' especially *C. septempunctata* should be carefully taken into considerations when they are incorporated into IPM programs. Many scientists have performed appreciable experimental work to evaluate the toxic effects of different insecticides on *C. septempunctata.* Further research is also required to gain more knowledge to prevent *C. septempunctata* from the harmful effects of insecticides

## References

Ahmed KS, Majeed AM, Haidary AA and Haider N, 2016. Integrated pest management tactics and predatory coccinellids: A review. J. Entomol. Zool. Stud. 4(1): 591-600.

Akhavan E, Jafari R, Jafai R and Afrogheh S, 2013. Biodiversity and distribution of predaceous ladybird (Coleoptera: Coccinelliae). Int. Res. J. Basic appl. Sci. 5: 705-709.

Ahmad MM and Ahmad SB, 2009. Comparative toxicity of two pesticides to adults of *Coccinella septempunctata* Linn. (Coleoptera: Coccinellidae). Insect Environment. 14 (4): 187-189.

Akhtar MdW, Sengupta D and Chowdhury A, 2009. Impact of pesticides use in agriculture: their benefits and hazards. Interdisc. Toxicol. 2(1): 1–12.

Asrar M, Ghafoor A, Hassan M and Zia K, 2014. Laboratory bioassay for determination of residual toxicity of different insecticides against two predators, *Chrysoperla carnea* (Chrysopidae: Neuroptera) and *Coccinella septempunctata* (Coccinellidae: Coleoptera). Pak. Entomol. 36 (1):61-67.

Banken JAO and Stark JD, 1997. Stage and age influence on the susceptibility of *Coccinella septempunctata* (Coleoptera: Coccinellidae) after direct exposure to Neemix, a neem Insecticide. J. Econ. Entomol. 90 (5): 1102-1105.

Bozsik A, 2006. Susceptibility of adult *Coccinella septempunctata* (Coleoptera: Coccinellidae) to insecticides with different modes of action. J. Pest Manag. Sci. 62:651–654.

Bradberry SM and Cage SA, Proudfoot AT and Vale JA, 2005. Poisoning due to Pyrethroids. Toxicol. Rev. 24 (2): 93-106.

Cairns JJr, 1983. Are single species toxicity tests alone adequate for estimating environmental hazard? Hydrobiologia. 100: 47-57.

Drobne D , Blazic M , Van Gestel CA , Leser V , Zidar P , Jemec A and Trebse P, 2008. Toxicity of imidacloprid to the terrestrial isopod, *Porcellio scaber* (Isopoda, Crustacea). Chemosphere. 71 (7): 1326–1334.

Desneux N, Decourtye A and Delpuech JM, 2007. The sublethal effects of pesticides on beneficial arthropods. Annu. Rev. Entomol. 52: 81–106.

Ehler LE, 2006. Integrated pest management (IPM): definition, historical development and implementation, and the other IPM. J. Pest Manag. Sci. 62(9):787-789.

Forster R, 1995. Auswirkungen von Pflanzenschutzmitteln auf Nutzorganismen – Kennzeichnung im Rahmen des Zulassungsverfahrens. Nachrichtenbl Dtsch Pflanzenschutzdienste. 47:233–236.

Han P, Niu CY, Lei CL, Cui JJ and Desneux N, 2010. Quantification of toxins in a Cry1Ac + CpTI cotton cultivar and its potential effects on the honey bee Apis mellifera L. J. Ecotoxicol. 19:1452–1459.

Hallman A, Bond C, Buhl K and Stone D, 2015. *Pyriproxyfen General Fact Sheet*; National Pesticide Information Center, Oregon State University Extension Services. http://npic.orst.edu/factsheets/pyriprogen.html

Hassan SA, 1992. Guidelines for testing the effects of pesticides on beneficial organisms. Bull. IOBC/WPRS. 15 (3): 186.

Hodek I and Honek A, 2009. Scale insects, mealybugs, whiteflies and psyllids (Hemiptera, Sternorrhyncha) as prey of ladybirds. Biol. Contr. 51: 232-243. https://doi.org/10.1016/j.biocontrol.2009.05.018

Kaethner M, 1991. Keine Nebenwirkungen von Niempriiparaten auf die aphidophagen Pradatoren *Chtllsoper/a carnea* (Steph.) und *Coccinella septumpuncatata* L. Anz. Schiiedlingskd. Pflanzenschutz Umweltschutz. 64: 97-99.

Katis NI, Tsitsipis JA, Stevens M and Powell G, 2007. Transmission of plant viruses. In H. van Emden & R. Harrington (Eds.), Aphids as crop pests. Wallingford: CAB International, pp. 353–377.

Koren TD, Hlavat, Rojko I and Zadravec M, 2012. First checklist of ladybird beetle of Croatia along with new faunistical records. Acta Entomol. Serb. 17: 101-122.

Majerus M and Kearns PK, 1989. Ladybirds. Naturalists' Handbooks Slough, vol. 10. Richmond Publishing Co. Ltd, UK.

Muhammad F, Akhtar M, Rehman ZU, Javed I and Anwar MI, 2009. Role of veterinarians in providing residue free animal food. Pak. Vet. J. 29:42–46.

Mirajkar N and Pope CN, 2005. Dimethoate. Encyclopedia of Toxicology. pp. 47-4.

Mordue (Luntz) AJ and Blackwell A, 1993. Azadirachtin: an update. J. Insect Physiol. 39: 903-924.

National Pesticides Information (Center http://npic.orst.edu/ingred/ptype/insecticide.html npic@ace.orst.edu.).

Obrycki JJ and Kring TJ, 1998. Predaceous Coccinellidae in biological control. Annu. Rev. Entomol. 43: 295- 321.

Olszak RW, Pawlik B and Zajac RZ, 1994. The influence of some insect growth regulators on mortality and fecundity of the aphidophagous coccinellids Adalia bipunctata L. and *Coccinella septempunctata* L. (Coleoptera: Coccinellidae). J Appl. Entomol. 177:58–6.

Papachristos D and Milonas P, 2008. Adverse effects of soil applied insecticides on the predatory coccinellid *Hippodamia undecimnotata* (Coleoptera: Coccinellidae). Biol. Contr. 47: 77–81.

Pawlik WB and Zajac RZ, 1994. The influence of some insect growth regulators on mortality and fecundity of the aphidophagous coccinellids *Adalia bipunctata* L. and *Coccinella septempunctata* L. (Col., Coccinellidae). J. Appl. Entomol. 117: 58–63.

Raboudi F, Benmoussa A, Makni H and Marrakchi M, 2002. Serological detection of plant viruses in their aphid vectors and host plants in Tunisia. OEPP/EPPO Bulletin. 32: 495–498.

Rehman H, Aziz AlT, Saggu S, Abbas ZK, Mohan A and Ansari AA, 2014. Systematic review on pyrethroid toxicity with special reference to deltamethrin. J. Entomol. Zool. Stud. 2(6): 60-70.

Ramirez-Romero R, Desneux N, Decourtye A, Chaffiol A and Pham-Delègue MH, 2008. Does Cry1Ab protein affect learning performance of the honey bee Apis mellifera L. (Hymenoptera, Apidae)? J. Ecotoxicol. Environ. Saf. 70: 327–333.

Rahatullah, Ahmad H and Inayatullah M, 2010. Species diversity of coccinellidae of Dir valley. M. Phil. Thesis. Department of Zoology, Hazara University, Mansehra, Pakistan

Rosomer WS and Stoffolano JG, 1994. The science of entomology, 3rd ed. Wm. C. Brown, Dubuque, IA.

Samaha JM, 2010. Marian roots of the name. Our Lady's Bug. Dayton, Ohio: International Marian Research Institute. University of Dayton, USA.

Sánchez-Bayo F, 2012. Insecticides Mode of Action in Relation to Their Toxicity to Non-Target Organisms. J. Environ. Analytic Toxicol. 4: 1-11.

Shukla RM, Kumar A, Chandra M and Saini ML, 1994. J. Plant Prot. Bull. 46(1): 48-49.

Singh SR, Walters KFA, Port GR and Northing P, 2004. Consumption rates and predatory activity of adult and fourth instar larvae of the seven-spot ladybird, *Coccinella Septempunctata* L. following contact with dimethoate residue and contaminated prey in laboratory arenas. Biol. Contr. 30: 127-133.

Singh SR, Walters KFA and Port GR, 2001. Behaviour of the adult seven spot ladybird, *Coccinella septempunctata* (Coleoptera: Coccinellidae), in response to dimethoate residue on bean plants in the laboratory. Bull. Entomol. Res. 91: 221–226.

Singh SR, 2001. The behaviour of the ladybird, *Coccinella septempunctata*, in response to sub-lethal doses of insecticide, Ph.D. Thesis, University of Newcastle-upon-Tyne, UK.

Singh J and Bras KS, 2004. Mass production and biological control potential of coccinellids in India. Indian insect predators in biological control. Edn Daya Publishing House, Delhi, India. pp. 204-260.

Skouras PJ, Stathas GJ, Voudouris CCh, Darras AI, Tsitsipis JA and Margaritopoulos JT, 2017. Effect of synthetic insecticides on the larvae of *Coccinella septempunctata* from Greek populations. Phytoparasitica. 45(2): 165-173.

Slipinski A, 2007. Australian ladybird beetles (Coleoptera: Coccinellidae), their biology and classification. Austr. J. Biol. Res. Study. Coll. Illus. pp. 288.

Solangi BK, Lanjar AG and Lohar MK, 2007. Comparative toxicity of some insecticides on 4$^{th}$ instar grub of *Coccinella septempunctata* L. Under Laboratory conditions. Sarhad J. Agric. 23(4): 1091-1096.

Sterk G, Hassan SA, Baillod M, Bakker F, Bigler F, Blumel S et al., 1999. Results of the seventh joint pesticide testing programme carried out by the IOBC/WPRS Working Group 'Pesticides and Beneficial Organisms'. Biol. Contr. 44:99–117.

Stark JD and Wennergren U, 1995. Can population effects of pesticides be predicted from demographic toxicological studies? J. Econ. Entomol. 88: 1089-1096.

Stark JD and Rangus TM, 1994. Lethal and sublethal effects of the neem insecticide formulation, Margosan-O, on the pea aphid. J. Pest. Sci. 41: 155–160.

Spinosad Technical Bulletin, Dow AgroSciences LLC, Form No. Y45-000-001 (01/01) CBK.

Tan Y, Biondi A, Desneux N and Gao XW, 2012. Assessment of physiological sublethal effects of imidacloprid on the mirid bug *Apolygus lucorum* (MeyerDür). Ecotoxicology. 21: 1989–1997.

Thomas J and Phadke KG, 1991. Comparative Toxicity of Two Pesticides to Adults of *Coccinella septempunctata* Linn. (Coleoptera: Coccinellidae). Indian J. Entomol. 53(3): 405-411.

Thornham DG, Wakefield ME, Blackwell A, Evans KA and Walters KFA, 2007. Sexual dimorphism in the distribution and biometrics of the palpal sensilla of *Coccinella septempunctata*, and a description of a new sensillum. Acta Entomol. Sin. 50: 667–674.

Thornham DG, Blackwell A, Evans KA, Wakefield M and Walters KFA, 2007. Locomotory behaviour of the seven-spotted ladybird, *Coccinella septempunctata*, in response to five commonly used insecticides. J. Ann. Appl. Biol. 152: 349-359.

Tomlin CDS, 1997. The Pesticide Manual: A World Compendium, 11th ed.; British Crop Protection Council: Farnham, Surrey, UK, pp 300-302.

Tomlin CDS, 2006. The Pesticide Manual: A World Compendium, ($14^{th}$ ed); British Crop Protection Council: Farnham, UK, pp 286-287.

Ullah R, Haq F, Ahmad H, Inayatullah M, Saeed K and Khan S, 2012. Morphological characteristics of ladybird beetles collected from District Dir Lower, Pakistan. African J. Biotechnol. 11(37): 9149-9155.

Urooj, Z and A. Ali. 2016. Ladybird beetles (Coleoptera: Coccinellidae) fauna of district Swabi, Nowshera and Mardan. Int. J. Agric. Environ. Res. 2(1): 86-92.

(U.S.EPA) United States Environmental Protection Agency, 2004. Office of Prevention, Pesticides and Toxic Substances. Ecological Risk Assessment for Proposed Use of Acetamiprid on Cucurbit, Stone Fruit and Tree Nut Crops. U.S. EPA, Europe.

(U.S. EPA) United States Environmental Protection Agency, 2007. Fenazaquin, Pesticide Fact Sheet. U.S. EPA, Europe.

Wilis JA and Jepson PC, 1994. Sub-lethal effects of deltamethrin residues on the within-crop behavior and distribution of *C. septempunctata*. J. Entomol. Exp. Appl. 72: 33-45.

Youn YN, Seo MJ, Shin JG, Jang C and Yu YM, 2003. Toxicity of greenhouse pesticides to multicolored Asian lady beetles, *Harmonia axyridis* (Coleoptera: Coccinellidae). Biol. Control. 28: 164.

Yu C, Lin R, Fu M, Zhou Y, Zong F, Jiang H, Lv N, Piao X, Zhang J, Liu Y and Brock TCM, 2014. Impact of imidacloprid on life-cycle development of *Coccinella septempunctata* in laboratory microcosms. J. Ecotoxicol. Environ. Saf. 110: 168-173.

Yu C, Fu M, Lin R, Zhang Y, Yongquan L, Jiang H and Brock TCM, 2013. Toxic effects of hexaflumuron on the development of *C. septempunctata*. Environ. Sci. Pollut. Res. 21: 1418–1424.

# 16

# Functional response of the predatory mite *Cydnoseius negevi* (Swirski & Amitai) (Acari: Phytoseiidae) to the *Oligonychus afrasiaticus* (Mcgregor) and *Tetranychus urticae* Koch (Acari: Tetranychidae)

**Fahad J. Alatawi[1], Syed Zain ul Abidin[1], Jawwad H. Mirza[1], Muhammad Kamran[1]**
[1]Department of Plant Protection, College of Food and Agriculture Sciences, King Saud University, KSA

**Abstract**
The study of functional response characteristics of natural enemies is one of the most important approaches to assess their efficiency in regulating the pest population as they are key aspects in the dynamics of predator-prey interactions. In the present study, the functional response type and parameters, and maximum feeding capacity of the predatory mite *Cydnoseius negevi* (Swirski & Amitai) (Acari: Phytoseiidae) deutonymph and adult female were assessed at different densities of movable stages of the date palm mite (DPM) *Oligonychus afrasiaticus* (McGregor) and two-spotted spider mite (TSSM) *Tetranychus urticae* Koch (Acari: Tetranychidae). The experiments were carried out at 30 ± 2°C and relatively low humidity of 35 ± 5 % RH under controlled conditions, for the first time. The logistic regression model was used to determine the type of functional response exhibited by *C. negevi*. Both the stages of predator showed Type II functional response when fed the DPM and TSSM individuals at different stages. Roger's random-predator equation was used to define the handling time (*Th*) and attack rate (*a*) coefficients of type II functional response by fitting the equation into the non-linear least square regression model. The longest *Th* was shown by deutonymph and adult females of *C. negevi* when they fed on the DPM and TSSM females, and TSSM females, respectively. Furthermore, no significant differences were recorded for most of the *a* values of the deutonymph and adult female of *C. negevi* among different stages of DPM and TSSM. The maximum feeding capacity of *C. negevi* adult female was significantly higher for DPM female when compared with that of TSSM. According to previous and current studies, *C. negevi* could be a potential predator at a wide range of humidity.

*Corresponding author email: falatawi@ksu.edu.sa

**Keywords:** Biological control, Prey, Density, Feeding capacity, Handling time, Attack rate

## Introduction

The date palm mite (DPM) *Oligonychus afrasiaticus* (McGregor) and two-spotted spider mite (TSSM) *Tetranychus urticae* Koch (Acari: Tetranychidae) are severe pests of the date palm, *Phoenix dactylifera* and various agricultural crops in Saudi Arabia (SA), respectively (Alatawi, 2011; Alatawi and Kamran, 2018). They severely affect both the quality and quantity of crop production by impairing their commercial value in national and international markets (Khoualdia et al., 1997; Palevsky et al., 2005). The arid and hot climate of SA, which supports the outbreak of spider mites, is one of the main factors that hinders the application of bio-control programs using phytoseiid predators against DPM and TSSM pests. Furthermore, most of these predators have been used successfully against spider mites only at relatively humid climatic condition in the world (Hoy, 2011). Therefore, searching for local phytoseiid species adapted to arid and hot climate, which is prevalent in most of SA, are highly necessary (Negm et al., 2012). The predatory mite *Cydnoseius negevi* (Swirski & Amitai) (Acari: Phytoseiidae) is distributed over a wide range of hot and arid areas in SA and other Middle East countries. Its abundance and association with the DPM are indications of an adaptation that could positively enhance the efficiency of predators against mite pests (Alatawi et al., 2017).

The biological aspects, life table parameters, and predation rate of *C. negevi* have been studied against several pests and different diet sources, such as plant pollen (El-Banhawy et al., 1999; Negm et al., 2014; Alatawi et al., 2018). In particular, *C. negevi* successfully reproduced and developed on date palm pollen (Alatawi et al., 2018), and on the DPM under low relative humidity (35% RH). Furthermore, its performance was better than that of *Neoseiulus barkeri* Hughes (Acari: Phytoseiidae) (Negm et al., 2014).

Studying the functional response characteristics is one of the most important approaches to assess the efficiency of biocontrol agents in regulating the pest population as they are key aspects of predator-prey interactions (Murdoch and Oaten, 1975; Berryman, 1992). Furthermore, it enhances the practical predictive potential of a predator used as a candidate for biocontrol programs (Sepúlveda and Carrillo, 2008).

Usually, the moderate to high relative humidity is necessary for phytoseiid mites to establish their population (Helle and Sabelis, 1985). Thus, most of their functional response studies have been conducted at relatively high humidity ranging from 60% to 80% RH. However, studies on the fluctuations in predator and prey populations under natural conditions are necessary (Vangansbeke et al., 2013; Audenaert et al., 2014).

The functional response of *C. negevi* has been reported only against the TSSM nymphs at relatively high humidity 70 ± 5% RH (Rasmy et al., 2014). Despite a strong association with the DPM in the date palm orchards, to the best of our knowledge, the functional response of the predatory mite *C. negevi* against DPM has not been reported.

The present study aimed to define the functional response type and parameters, and feeding capacity of the predatory mite *C. negevi* against moveable life stages of DPM and TSSM at low humidity.

## Material and Methods

### Predator and prey sources

The species of mite used in the present study—*C. negevi*, TSSM, and DPM—were acquired from the date palm orchards at Riyadh, SA. Initial populations were collected from Bermuda grass *Cynodon dactylon* (Poaceae), black nightshade *Solanum nigrum* L. (Solanaceae) (both growing under the date palm trees), and highly infested date fruits, respectively.

### Maintenance of colonies

The colony of *C. negevi* was maintained in the laboratory and were fed the pollen of date palm and different stages of the two prey species TSSM and DPM at 27 ± 2°C in small arenas as described by Marafeli et al. (2011). The TSSM colony was maintained on eggplants *Solanum melongena* L. (Solanaceae) in a green house. The DPM individuals were collected from infested date palm fruits in field.

### Assessment of functional response, predation rate, and maximum feeding capacity of *C. negevi*

The experiments were performed in a climate-controlled chamber (BINDER®, Germany) at 30 ± 2°C, 35 ± 5 % RH, and 14:10 h light/dark period. Functional response and maximum feeding capacity were assessed using newly emerged *C. negevi* deutonymphs and adult females feeding on the larvae, nymphs, and females of the pest mites DPM and TSSM in small rearing units. The experimental unit consisted of a black acrylic lid of 2 cm diameter positioned on a Petri dish of 15 cm diameter. The different stages of DPM and TSSM were released on the abaxial surface of date palm and common bean leaf discs, placed on the black acrylic lids, respectively. In addition, the leaf discs were surrounded with damp cotton wool. Furthermore, the Petri dishes were constantly filled with distilled water to stop mites from leaving the arena and to keep the leaf disc fresh following Marafeli et al. (2011).

Until the maximum feeding capacity of each stage of the predator *C. negevi* (deutonymphs and adult females) was reached, different densities of each prey stage (larvae, nymphs, and adult females) of DPM (5, 10, 25, 35, 45, 60, and 70) and TSSM (5, 10, 25, 35, 45, and 60) were offered separately. That is, each prey stage of DPM and TSSM was offered individually to each stage of the predator. The control treatment (without predator) for each prey stage and density of DPM and TSSM was maintained to determine natural mortality. To standardize the functional response of *C. negevi* adult female and deutonymph, both the stages were starved individually for 8 h in separate rearing arenas, and then transferred to different experimental units containing different densities and stages of prey species. The predatory mite was taken out of the arena after 24 h and the live prey were counted.

**Statistical analysis of data**

A logistic regression of the proportion of prey consumed (*Na/N0*) as a function of initial prey density presented (*N0*) was used to determine the type and curve shapes of the functional response exhibited by *C. negevi* deutonymph and adult females to different stages of DPM and TSSM. After correcting the natural mortality of prey according to Xia et al. (2003), the data were analyzed by polynomial function following Juliano (2001), to calculate *Na/N0* and *N0*.

$$\frac{N_a}{N_0} = \frac{\exp\ (p_0 + p_1 N_0 + \ p_2\ N_0^2 + p_3 N_0^3)}{1 + \exp\ (p_0 + p_1 N_0 + \ p_2\ N_0^2 + p_3 N_0^3)}$$

Whereas, *P1*, *P2*, *P3* and *P0* stand for linear, quadratic, cubic coefficients and intercept, respectively. These coefficients were estimated by the maximum likelihood with PROC logistic regression model using the SigmaPlot® software version 12.2. The linear and quadratic coefficients were used to determine the functional response types (II and III). If *P1* value is less than zero, it indicates that the prey proportion fed declines monotonically with its initial prey density provided, and this trend reflects type II functional response. If *P1* > 0 and *P2* < 0, it indicates that the prey proportion fed is proportional to the density of prey, and this trend displays type III functional response (Juliano, 2001).

After defining functional response type, its parameters; handling time (*Th*) and attack rate (*a*) coefficients, of type II functional response were calculated by Roger's random-predator equation (Roger, 1972) after fitting the equation parameters into nonlinear least square regression (NLIN Procedure, SAS Institute, 2012).

$$Na = No(1 - \exp[a\ (Th\ \times\ Na\ -\ T)]$$

Where, *Na* = number of preys fed, *N0* = number of offered preys, *Th* = handling time, *a* = attack rate, and *T* = searching time (24 h). After determining the *a* and *Th* from the original data (*mt*), the differences among the values of each *a* and *Th* were verified for significance by estimating the variance by the Jackknife technique (Meyer et al. 1986). The Jackknife pseudo-value (*mj*) was estimated for the samples (*n*) using the equations:

$$m_{ja} = n\ \times\ m_{ta} -\ (n\ -\ 1)m_{ia}$$
$$m_{jT_h} =\ n\ \times\ m_{tT_h} -\ (n-1)m_{iT_h}$$

The mean values of (*n* – 1) Jackknife pseudo-values of *a* and *Th* for each prey stage were subjected to one-way analysis of variance (ANOVA), followed by the least significant difference (LSD) comparison test (P < 0.05) (SPSS, Version 21.0). The maximum feeding capacity of each prey stage (larva, nymph, and adult female) of DPM and TSSM consumed by each *C. negevi* deutonymphs and adult female was analyzed by one-way ANOVA followed LSD comparison test (P < 0.05) (SPSS, version 21.0).

## Results

The logistic regression analysis showed a significantly negative linear coefficient (P1 < 0) for both deutonymph and adult female of *C. negevi* (Table 1). The prey proportion fed by the predator declined as the prey density increased for all life stages of DPM and TSSM (Fig. 1, A and B). These results indicated that both deutonymph and adult female of *C. negevi* exhibited a type II functional response (Fig. 2, A and B). However, the graphical analysis of number of preys consumed versus offered prey density showed that the shape of functional response curves for the deutonymph and adult female of *C. negevi* to TSSM adult female were in between type II and type III functional response (Fig. 2, B).

The functional response parameters, handling time (*Th*) and attack rate (*a*) coefficients, for the deutonymph and adult female of *C. negevi* feeding on different stages of DPM and TSSM are demonstrated numerically as shown in Table 2.

The *Th* of *C. negevi* deutonymph was not significantly different among the immature individuals of both DPM and TSSM (0.62–1.34), except for the adult females of DPM and TSSM, which showed longer *Th* of 1.98 and 2.66, respectively. A similar trend was observed for the *C. negevi* female (0.33–0.82), except on the TSSM female where the *Th* was longer (2.54). Non-significant differences were recorded in the attack rate (*a*) values of *C. negevi* deutonymph and adult female among most treatments of DPM and TSSM (Table 2, A and B).

The maximum feeding capacity of the deutonymph of *C. negevi*. for each prey stage (larva, nymph, and adult female) of DPM and TSSM was 21,17, and 13; and 19, 19, and 8; respectively, at a prey density of 45. Whereas,the adult female of *C. negevi* was 34, 30, and 25; and 24, 18, and 8; respectively, at a prey density of 60 and 45 (Table 3). The maximum feeding capacity of *C. negevi* adult female was significantly different within different stages of each prey species and between each stage of DPM and TSSM (Table 3, A and B). However, the maximum feeding capacity of each *C. negevi* deutonymph and adult female was higher for DPM female when compared with that of TSSM female (Table 3, A and B).

**Table 1: Maximum likelihood estimates from logistic regression of proportion of both prey species *O. afrasiaticus* (DPM) and *T. urticae* (TSSM) consumed as a function of initial prey densities by deutonymph and female of *Cydnoseius negevi*.**

| Predator stage | Prey | Prey stage | Model summary | | | | Estimated parameter | | | |
|---|---|---|---|---|---|---|---|---|---|---|
| | | | $R^2$ | F-value | d.f | P | *P0** | *P1** | *P2** | *P3** |
| Deutonymph | DPM | Larva | 0.82 | 30.66 | 3,20 | <0.001 | 5.99 | -0.04 | 0.01 | -0.01 |
| | | Nymph | 0.83 | 33.55 | 3,20 | <0.001 | 5.39 | -0.17 | 0.02 | -0.02 |
| | | Female | 0.70 | 15.50 | 3,20 | <0.001 | 5.21 | -0.06 | 0.01 | -0.02 |
| | TSSM | Larva | 0.84 | 34.81 | 3,20 | <0.001 | 5.06 | -0.03 | 0.02 | -0.03 |
| | | Nymph | 0.86 | 42.05 | 3,20 | <0.001 | 4.42 | -0.14 | 0.02 | -0.01 |
| | | Female | 0.62 | 10.83 | 3,20 | <0.001 | 2.53 | -0.23 | 0.02 | -0.02 |
| Female | DPM | Larva | 0.90 | 73.85 | 3,24 | <0.001 | 5.99 | -0.13 | 0.03 | -0.01 |
| | | Nymph | 0.85 | 45.14 | 3,24 | <0.001 | 6.35 | -0.12 | 0.02 | -0.02 |
| | | Female | 0.72 | 20.39 | 3,24 | <0.001 | 5.55 | -0.05 | 0.02 | -0.02 |
| | TSSM | Larva | 0.95 | 57.85 | 3,20 | <0.001 | 5.10 | -0.01 | 0.02 | -0.03 |
| | | Nymph | 0.92 | 38.06 | 3,20 | <0.001 | 4.74 | -0.13 | 0.02 | -0.03 |
| | | Female | 0.59 | 7.95 | 3,16 | <0.001 | 3.67 | -0.02 | 0.01 | -0.01 |

Maximum likelihood estimated coefficients: *P0* = intercept, *P1* = linear, *P2* = quadratic and *P3* = cubic.

**Table 2: Mean estimates values ± SE of functional response parameters, attack rate (*a*) and handling time (*Th*) for *C. negevi*; A) deutonymph and B) adult female feeding on each stage (larva, nymph, and female) of *O. afrasiaticus* (DPM) and *T. urticae* (TSSM).**

**A) *C. negevi* deutonymph**

| Prey stage | Prey species | (*a*) | (*Th*) | P | $R^2$ |
|---|---|---|---|---|---|
| **Larva** | DPM | 0.040 ± 0.023 a | 0.623 ± 0.416 a | 0.001 | 0.82 |
| | TSSM | 0.083 ± 0.047 b | 1.121 ± 0.232 a | 0.001 | 0.71 |
| **Nymph** | DPM | 0.030 ± 0.012 a | 0.761 ± 0.349 a | 0.001 | 0.92 |
| | TSSM | 0.041 ± 0.014a | 1.336 ± 0.254 a | 0.001 | 0.68 |
| **Female** | DPM | 0.109 ± 0.092 b | 1.982 ± 0.336 ab | 0.001 | 0.50 |
| | TSSM | 0.013 ± 0.008 a | 2.658 ± 1.294 b | 0.003 | 0.55 |

**B) *C. negevi* adult female**

| Prey stage | Prey species | (*a*) | (*Th*) | P | $R^2$ |
|---|---|---|---|---|---|
| **Larva** | DPM | 0.078 ± 0.024 ab | 0.339 ± 0.108 a | <.0001 | 0.94 |
| | TSSM | 0.080 ± 0.031 b | 0.717 ± 0.158 a | <.0001 | 0.86 |
| **Nymph** | DPM | 0.042 ± 0.013 a | 0.395 ± 0.175 a | <.0001 | 0.94 |
| | TSSM | 0.072 ± 0.043 ab | 1.315 ± 0.289 a | 0.001 | 0.62 |
| **Female** | DPM | 0.054 ± 0.025 ab | 0.821 ± 0.225 a | <.0001 | 0.81 |
| | TSSM | 0.049 ± 0.020 ab | 2.539 ± 0.368 b | 0.001 | 0.56 |

Means followed by different small letters within the column(s) are significantly different (ANOVA followed by LSD: P<0.05).

**Table 3: The average number (*Na*) ± S.E consumed and maximum mean capacity of *C. negevi* A) deutonymph and B) adult female for *O. afrasiaticus* (DPM) and *T. urticae* (TSSM) consumed at different stages and density levels (*No*) during 24 hours.**

**A) *C. negevi* deutonymph**

| | | Prey stages | | |
|---|---|---|---|---|
| **Prey Density** | **Prey Species** | **Larvae** | **Nymph** | **Adult (Female )** |
| **5** | DPM | 4.25 ± 0.25 a A | 4.00 ± 0.41 a A | 3.75 ± 0.25 bc A |
| | TSSM | 4.25 ± 0.48 a A | 3.25 ± 0.75 a B | 1.25 ± 0.45 a B |
| **10** | DPM | 9.50 ± 0.29 b A | 6.50 ± 0.65 ab B | 8.13 ± 0.41 e B |
| | TSSM | 8.25 ± 0.48 b A | 5.75 ± 0.48 ab B | 2.25 ± 1.45 ab C |
| **25** | DPM | 10.71 ± 0.54 b A | 8.25 ± 0.85 bc A | 6.88 ± 0.86 de B |
| | TSSM | 10.50 ± 0.65 b A | 7.75 ± 0.48 bc B | 3.50 ± 2.58 bc C |
| **35** | DPM | 11.25 ± 0.95 b A | 10.75 ± 0.48 c A | 11.43 ± 0.71 fg A |
| | TSSM | 16.75 ± 0.48 c A | 11.00 ± 0.91 c B | 4.25 ± 2.73 bc C |
| **45** | DPM | 20.89 ± 1.12 d A | 16.25 ± 1.93 d AB | 12.73 ± 0.46 g B |
| | TSSM | 18.75 ± 1.65 cd A | 16.50 ± 1.85 d B | 7.50 ± 2.06 e B |
| **60** | DPM | 21.25 ± 0.86 d A | 17.00 ± 1.29 d B | 10.36 ± 0.71 f C |
| | TSSM | 16.00 ± 1.83 c A | 19.00 ± 1.78 d B | 5.00 ± 3.70 cd B |

**B) *C. negevi* adult female**

| | | Prey stages | | |
|---|---|---|---|---|
| **Prey Density** | **Prey species** | **Larvae** | **Nymph** | **Adult (Female )** |
| **5** | DPM | 4.69 ± 0.31 a A | 4.75 ± 0.25 a A | 4.75 ± 0.25 a A |
| | TSSM | 4.75 ± 0.25 a A | 4.00 ± 0.41 a AB | 3.50 ± 0.29 a B |
| **10** | DPM | 9.16 ± 0.53 ab A | 8.75 ± 0.48 ab A | 7.49 ± 0.70 ab A |
| | TSSM | 8.50 ± 0.50 ab A | 6.25 ± 0.25 ab B | 4.50 ± 0.28 a C |
| **25** | DPM | 15.22 ± 2.39 c A | 12.50 ± 1.55 cd A | 11.41 ± 1.80 bc A |
| | TSSM | 13.00 ± 1.47 bc A | 10.50 ± 1.32 bc B | 5.75 ± 0.25 a B |
| **35** | DPM | 21.17 ± 5.12 d A | 16.09 ± 1.25 de A | 16.13 ± 1.81 d A |
| | TSSM | 20.75 ± 2.10 d A | 13.50 ± 0.65 cd B | 8.00 ± 0.82 ab C |
| **45** | DPM | 30.00 ± 3.42 e A | 20.81 ± 2.04 f A | 15.80 ± 3.57 cd B |
| | TSSM | 23.75 ± 1.49 d A | 17.50 ± 0.96 ef B | 8.25 ± 1.65 ab C |
| **60** | DPM | 33.67 ± 1.90 e A | 29.72 ± 3.53 g A | 24.54 ± 0.95 e B |
| | TSSM | 23.00 ± 1.08 d A | 12.75 ± 1.65 cd B | - |

| 70 | DPM | 28.95 ± 1.94 e A | 27.15 ± 2.00 g B | 18.69 ± 2.13 d C |
|---|---|---|---|---|

Means followed by different small letters within the column(s) are significantly different (ANOVA followed by LSD: P<0.05).

Means followed by different capital letters within the row (s) are significantly different (ANOVA followed by LSD: P<0.05).

**A) *C. negevi* deutonymph**

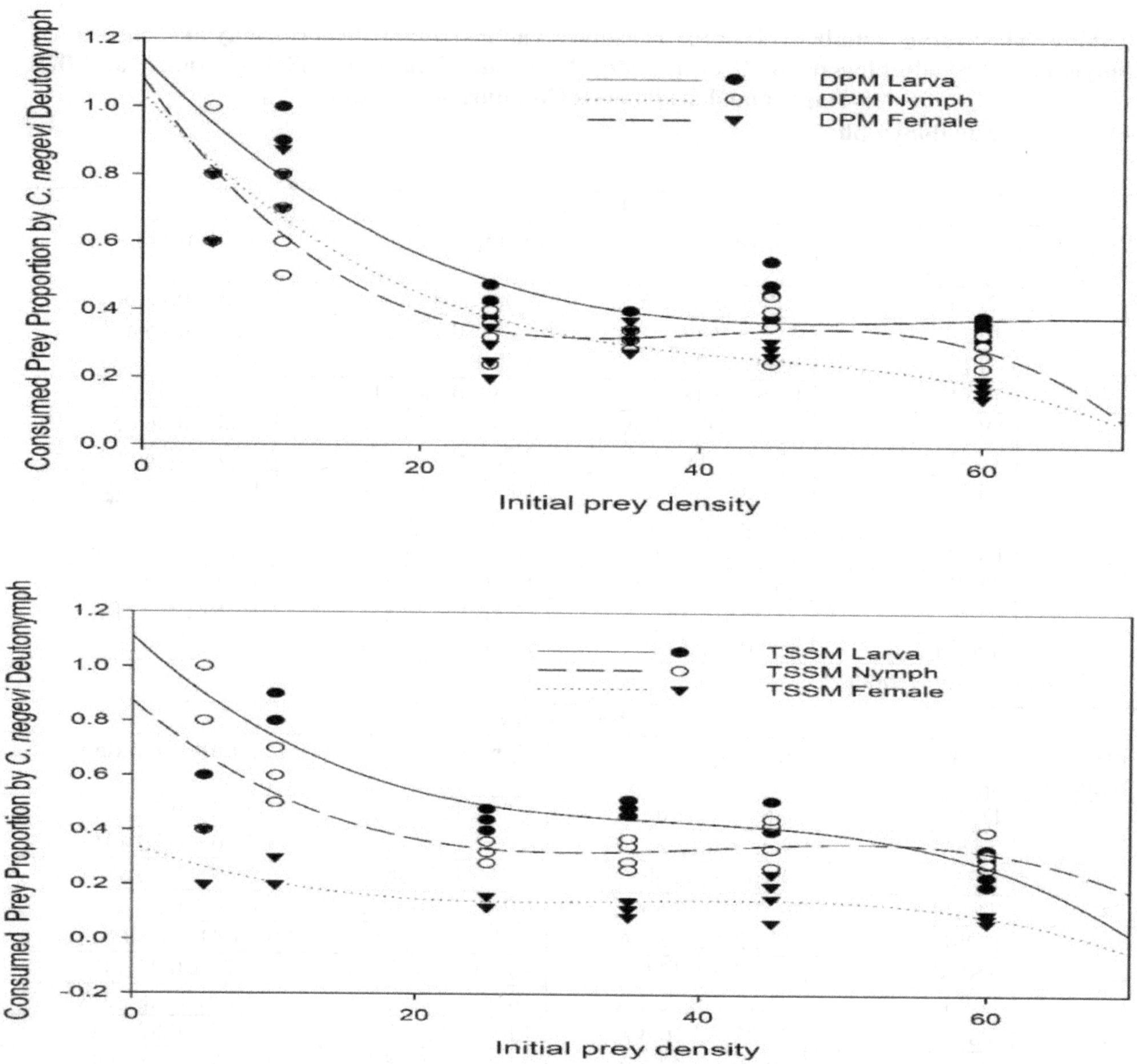

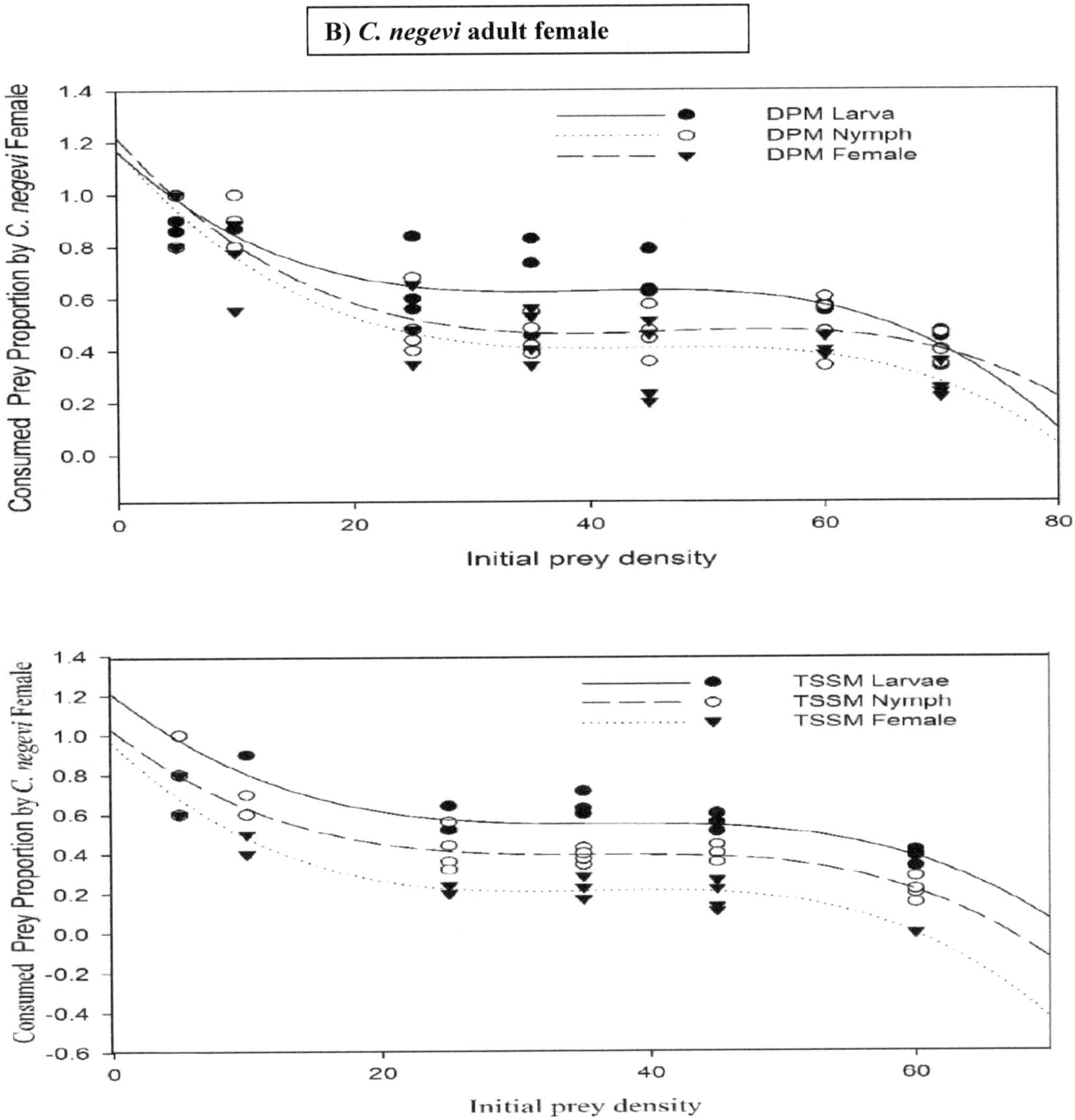


**Figure 1: Mean proportion of each prey stage (larva, nymph, and female) of *O. afrasiaticus* (DPM) and *T. urticae* (TSSM) consumed by *C. negevi* A) deutonymph and B) adult female at different prey densities.**

**A) *C. negevi* deutonymph**

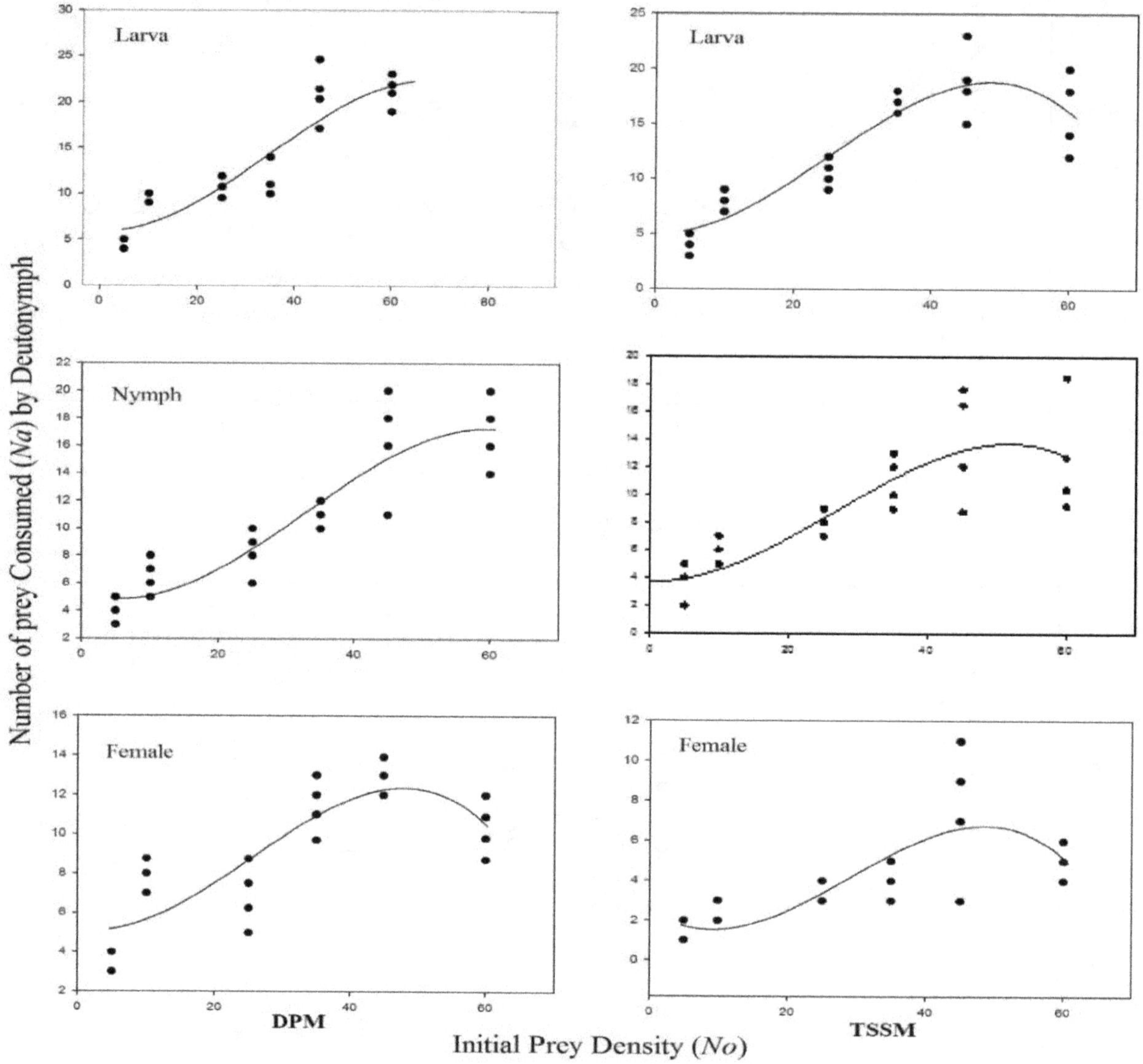

**B) *C. negevi* adult female**

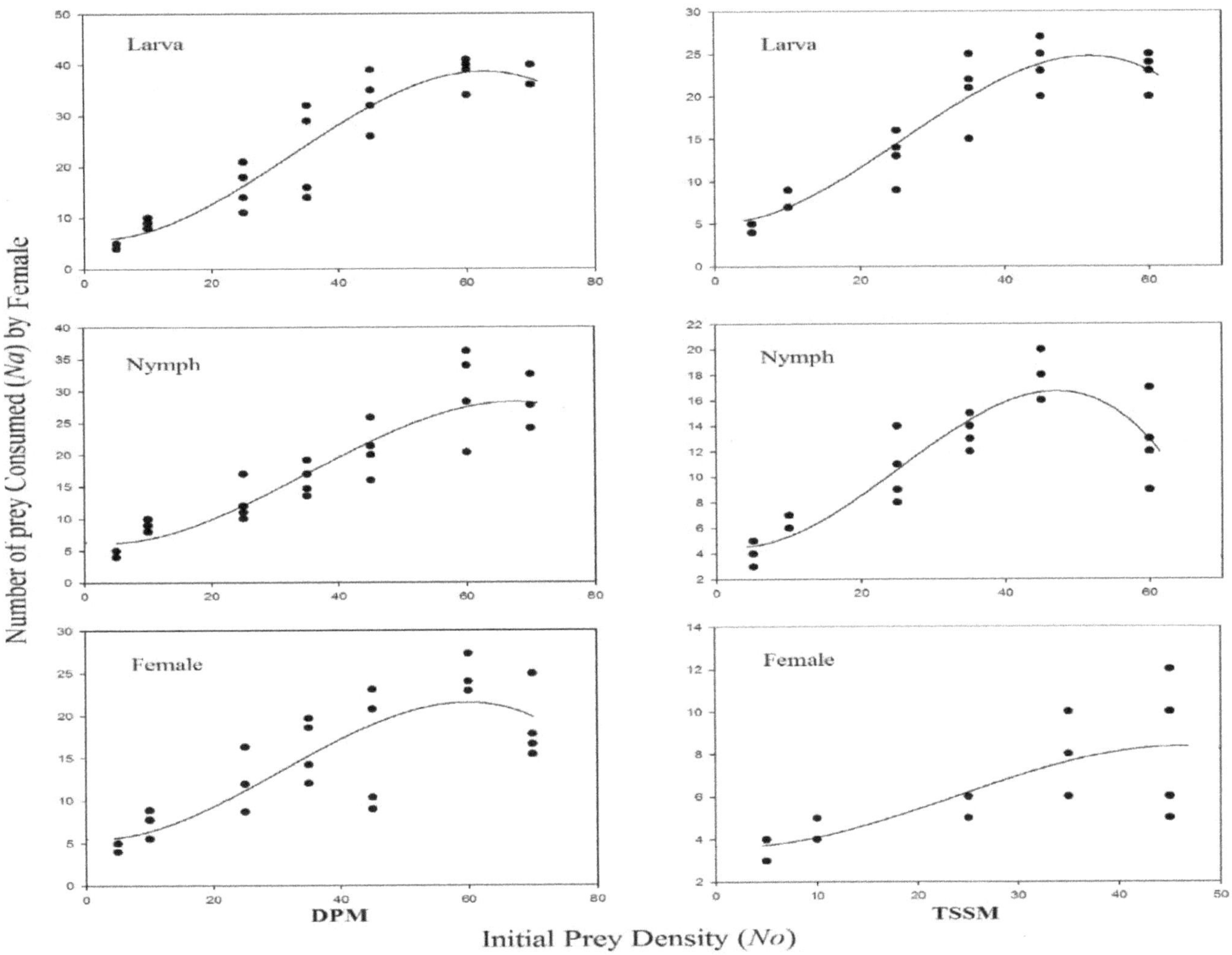


**Figure 2: Functional responses curves of *C. negevi* A) deutonymph and B) adult female to each stage (larva, nymph, and female) of *O. afrasiaticus* (DPM) and *T. urticae* (TSSM) prey species (Points represent the observed numbers of prey 'consumed at each initial prey density, and lines were predicted by the random predator equation).**

## Discussion

To the best of our information, the present study is the first to evaluate the functional response of the phytoseiid predator *C. negevi* to all mobile life stages of DPM and TSSM at a relatively low humidity 35 ± 5% RH.

However, few functional response studies on phytoseiid mites have been conducted at relatively low humidity i.e., *Metaseiulus flumenis* (Chant) against immature stages of a major pest of date palm fruit, the banks grass mite *O. pratensis* (Banks), at 50 ± 10% RH (Ganjisaffar and Perring, 2015*);* *Neoseiulus californicus* (McGregor)against the eggs of TSSM at extreme humidity levels (i.e., 30, 50, 70, and 90% RH) (Döker et al., 2016); *Neoseiulus picanus* (Ragusa) against the eggs of TSSM at 50% RH (Tello-Mercado et al., 2017); and *N. fallacis* (Garman), *Phytoseiulus persimilis* Athias-Henriot, and *N. californicus* against the eggs of TSSM in a greenhouse at 40% RH during day and 70% RH during night (Audenaert et al., 2014). In the present study, the polynomial logistic regression confirmed that type II functional response specified and designated all the data tested (Table 1). A high proportion of each life stage (larva, nymph, and adult female) of DPM and TSSM was consumed at their low densities (inverse density-dependent) (Fig. 1). Type II functional response has been observed in most of the phytoseiid predators (Sabelis, 1986). At relatively low and extreme humidity, the functional response type II was recorded for all the phytoseiid predatory mites against the eggs of TSSM and immature stages of *O. pratensis* in all studies mentioned above.

Furthermore, the functional response type II was also found for each life stage of *C. negevi*—protonymph, deutonymph, and adult female—that fed on the TSSM nymphs, but at high relative humidity 70 ± 5% RH (Rasmy et al., 2014). In the present study, the functional response of *C. negevi* was not affected by different levels of humidity. A similar observation was made by Döker et al. (2016) in *N. californicus* against the eggs of TSSM at extremely high humidity.

However, in the present study, the graphical relationship between the number of preys consumed and prey density of both deutonymphs and females of *C. negevi* against the females of TSSM exhibited an inclination toward type III response (sigmoid curve). That is, the shape of type II decelerating curve was in between types II and III as shown in Fig. 2, A and B. Synonymous results have also been reported by other studies (Trexler et al., 1988; Omkar and Pervez, 2004). Generally, type III functional response exhibits a sigmoidal increase only at the initial level, but not throughout the range (positive-density dependence). Later, it shows negative-density dependence, as shown in Type II curves, due to predator satiation (Pervez and Omkar, 2005). Though very rarely, type II response can be determined with a positive *P1* when *P2* or *P3* values are negative and sufficiently large (Juliano, 2001).

Such likelihood usually causes difficulties in discriminating the functional response types because the curve lies between types II and III (Trexler et al., 1988; Casas and Hulliger, 1994; Juliano, 2001). To avoid unclear inferences, functional responses type II and III should be distinguished at low prey densities offered and high number of replications (Pervez and Omkar, 2005; Dick et al., 2014). This would boost the reliability of curve-fitting of each tested functional type.

However, mathematically, the logistic model can easily differentiates the type II and III functional responses and increases the credibility of correctness of the curves (Shah and Khan, 2013). Furthermore, it should be used before analyzing the data through a particular Holling equation.

The coefficient of attack rate (*a*) and handling time (*Th*) coefficients are the parameters used to determine the magnitude of functional response (Pervez and Omkar, 2005). The *Th* parameter usually provides more information about the efficiency of a predator against its prey (Holling, 1959; Veeravel and Baskaran, 1997).

In the present study, the *a* and *Th* values did not vary considerably within and between the stages of DPM and TSSM when exposed to both deutonymph and adult female of *C. negevi* (Table 2, A and B). However, the *Th* of each life stage of *C. negevi* was shorter for all the stages of DPM when compared with that of TSSM. Furthermore, there were variations in the *Th* within different stages of DPM when compared with that of TSSM, with the shortest *Th* on the larvae of both the prey species (Table 2, A and B).

Significantly longer *Th* was required for each predator stage (deutonymph and adult female of *C. negevi*) against TSSM females (Table 2, A and B). The predators require more *Th* when the prey is large (Holling, 1961). This also explains the very low prey predation rate and maximum capacity for the TSSM females when compared with those of the DPM females (Table 3, A and B). According to Sabelis and Helle (1985), the phytoseiid mites rarely feed on spider mite adults.

The *Th* of both deutonymph and adult female of *C. negevi* against the nymphs of TSSM observed in the present study (1.34 and 1.32, respectively) was similar to those reported by Rasmy et al. (2014) (1.79 and 1.07, respectively) at 35% and 70% RH, respectively. Furthermore, at extreme humidity, *N. californicus* against the eggs of TSSM showed no significant difference in its *Th* (Döker et al., 2016). This indicates that the predatory mite *C. negevi* can adapt to a wide range of humidity.

In the present study, the *a* coefficient did not vary significantly among the treatments (Table 2, A and B), probably because of similar and small-sized experimental arenas (Shah and Khan, 2013). Such non-significant differences in the *a* coefficient have been reported by different studies (Montserrat et al., 2000; Pervez and Omkar, 2005).

The maximum predation rate of the *C. negevi* deutonymph was significantly higher for the DPM female than the TSSM female. Furthermore, the *C. negevi* adult consumed significantly more number of DPM individuals at different stages (Table 3, A and B).

The maximum number of preys consumed by the *C. negevi* deutonymph and adult female based on the functional response curves made by Rasmy et al.

(2014) indicated that the average number of TSSM nymphs consumed were about 12 and 19, respectively, at maximum density level (50 preys). In the present study, it was 19 and 18, respectively. This suggests that the maximum feeding capacity can be better understood based on *Th* and is not affected by the RH. The differences in the maximum feeding capacity of *C. negevi* for DPM and TSSM females indicates its association with the DPM in date palm agro-ecosystem (Negm et al., 2014; Alatawi et al., 2017). The TSSM females were more active and larger in size than the DPM females, which was elucidated by the longer *Th* of *C. negevi* on the TSSM than on the DPM (Table 3B). It was also reported that more time is required by the predators to attack and kill larger preys (Holling, 1961) and the attack rate could reduce with increase in size of prey due to improved defense response or ability of prey to escape (Pastorok, 1981; Sabelis, 1992). Furthermore, the slower mobility of *Eutetranychus orientalis* (Klein) as compared to TSSM was considered as one of the main reasons for the increase in predation rate of *Euseius gossipi* (El-Badry) on *E. orientalis* (ElBadry et al., 1968).

The present study shows that the *C. negevi* female fed more efficiently on immature individuals of DPM and TSSM. Moreover, maximum feeding capacity and shorter *Th* were observed on different stages of DPM when compared with those of the TSSM. This indicates the association of *C. negevi* with the DPM. Furthermore, *C. negevi* predated efficiently at a wide range of RH, and could be a potential biological control agent of the DPM

## Acknowledgment

The authors would like to extend sincere appreciation to the Deanship of Scientific Research and Researchers Support and Services Unit (RSSU) at King Saud University for their technical support.

## References

Alatawi FJ, 2011. Phytophagous and predaceous mites associated with vegetable crops from Riyadh, Saudi Arabia. Saudi J. Biol. Sci. 18: 239–246.

Alatawi FJ, Basahih J and Kamran M, 2017. The superfamily Phytoseioidea (Acari: Mesostigmata) from Saudi Arabia: a new species, new records and a key to the reported species. Acarologia. 57: 275–294.

Alatawi FJ, Basahih J and Kamran M, 2018. Suitability of date palm pollen as an alternative food source for the predatory mite *Cydnoseius negevi* (Swirski and Amitai) (Acari: Phytoseiidae) at a low relative humidity. Acarologia. 58(2): 357–365.

Alatawi FJ and Kamran M, 2018. Spider mites (Acari: Tetranychidae) of Saudi Arabia: two new species, new records and a key to all known species. J. Nat. Hist. 52(7): 429–455.

Audenaert JA, Vangansbeke D, Verhoeven R, De Clercq P, Tirry L and Gobin B, 2014. Variability in greenhouse temperature and humidity affects predation efficiency of *Phytoseiulus persimilis* and *Neoseiulus californicus* on *Tetranychus urticae*. IOBC-WPRS Bull. in the press.

Berryman AA, 1992. On choosing models for describing and analyzing ecological time series. Ecology. 73: 694–698.

Casas J and Hulliger B, 1994. Statistical analysis of functional response experiments. Biocont. Sci. Technol. 4: 133–145.

Dick JTA, Alexander ME, Jeschke JM, Ricciardi A, MacIsaac HJ, Robinson TB, Kumschick S, Weyl OL, Dunn AM, Hatcher MJ and Paterson RA, 2014. Advancing impact prediction and hypothesis testing in invasion ecology using a comparative functional response approach. Biol. Invasions. 16: 735–753.

Döker İ, Kazak C and Karut K, 2016. Functional response and fecundity of a native *Neoseiulus californicus* population to *Tetranychus urticae* (Acari: Phytoseiidae, Tetranychidae) at extreme humidity conditions. Syst. Appl. Acarol. 21: 1463–1472.

Elbadry EA, Afifi AM, Issa GI and Elbenhawy EM, 1968. Effectiveness of the predaceous mite *Amblyseius gossipi* as a predator of three tetranychid mites (Acarina: Phytoseiidae). J. Appl. Entomol. 62(4): 189-194.

El-Banhawy EM, Hafez SM and Saber SA, 1999. Effect of the nymph prey density of the two spotted spider mite *Tetranychus urticae* Koch (Acari: Phytoseiidae) on the consumption and reproduction rates of the predacious mite *Cydnoseius negevi* (Swirski and Amitai) in absence and presence of nymphs of the white fly *Bemisia tabaci* (Genn.). Anz. Schaedlingskd. Pflanzenschutz. 72: 55–56.

Ganjisaffar F and Perring TM, 2015. Prey stage preference and functional response of the predatory mite *Galendromus flumenis* to *Oligonychus pratensis*. Biol. Control. 82: 40–45.

Helle W and Sabelis MW, 1985. Spider mites: their biology natural enemies and control, pp. 103–129. In W. Helle and M. W. Sabelis (eds.), World Crop Pests, vol. 1B. Elsevier, Amsterdam.

Holling CS, 1959. The components of predation as revealed by a study of small-mammal predation of the European pine sawfly. Can. Entomol. 91: 293–320.

Holling CS, 1961. Principles of insect predation. Annu. Rev. Entomol. 6: 163–182.

Hoy MA, 2011. Agricultural acarology: introduction to integrated mite management, vol. 7. CRC Press, Florida.

Juliano SA, 2001. Non-linear curve fitting: predation and functional response curve, pp. 178 – 196. In: S. M. Scheiner and J. Gurevitch (eds.), Design and Analysis of Ecological Experiment. Oxford University press, New York.

Khoualdia O, 1997. Premières observations sur Oryctes agamemnon, ravageur du palmier dattier en Tunisie. Fruits. 52: 111–115.

Marafeli PDP, Reis PR, Da Silveira EC, De Toledo MA and Souza-Pimentel GC, 2011. *Neoseiulus californicus* (McGregor, 1954) preying in different life stages of *Tetranychus urticae* Koch, 1836 (Acari: Phytoseiidae, Tetranychidae). Acarologia. 51: 499–506.

Meyer JS, Ingersoll CG, McDonald LL and Boyce MS, 1986. Estimating uncertainty in population growth rates: jackknife vs. bootstrap techniques. Ecology. 67: 1156–1166.

Montserrat M, Albajes R and Castañé C, 2000. Functional response of four heteropteran predators preying on greenhouse whitefly (Homoptera: Aleyrodidae) and western flower thrips (Thysanoptera: Thripidae). Env. Entomol. 29: 1075–1082.

Murdoch WW and Oaten A, 1975. Predation and population stability. Adv. Ecol. Res. 9: 1–131.

Negm MW, Alatawi FJ and Aldryhim YN, 2012. A new species of *Neoseiulus* Hughes, with records of seven species of predatory mites associated with date palm in Saudi Arabia (Acari, Phytoseiidae). Zootaxa. 3356: 57–64.

Negm MW, Alatawi FJ and Aldryhim YN, 2014. Biology, predation and life table of *Cydnoseius negevi* and *Neoseiulus barkeri* on the old world date mite, *Oligonychus afrasiaticus*. J. Insect Sci. 14: 1–6.

Omkar P and Pervez A, 2004. Functional and numerical responses of *Propylea dissecta* (Mulsant) (Col., Coccinellidae). J. Appl. Entomol. 128: 140–146.

Palevsky E, Borochov-Neori H and Gerson U, 2005. Population dynamics of *Oligonychus afrasiaticus* in the southern Arava Valley of Israel in relation to date fruit characteristics and climatic conditions. Agri. Forest Entomol. 7: 283–290.

Pastorok RA, 1981. Prey vulnerability and size selection by *Chaoborus* larvae. Ecology 62: 1311–1324.

Pervez A and Omkar P, 2005. Functional responses of coccinellid predators: an illustration of a logistic approach. J. Insect Sci. 5: 1–6.

Rasmy AH, Abou-Elella GM and Osman MA, 2014. Functional response of the phytoseiid mite *Typhlodromus negevi* Swirski and Amitai to the two-spotted spider mite *Tetranychus urticae* (Acari: Tetranychidae). Arch. Phytopathol. Plant Prot. 47: 1327–1334.

Rogers D, 1972. Random search and insect population models. J. Anim. Ecol. 41: 353–360.

Sabelis MW, 1985. Reproductive strategies. In W. Helle and M. W. Sabelis, (eds) Spider mites. Their biology, natural enemies and control, pp 265–278, vol. 1A. Elsevier, Amsterdam.

Sabelis MW, 1986. The functional response of predatory mites to the density of two-spotted spider mites. In Dynamics of Physiologically Structured Populations, pp. 298–321. Springer, Berlin.

Sabelis MW, 1992. Predatory arthropods. In Natural Enemies: The population biology of predators, parasites and diseases, pp. 225–264.

SAS Institute, 2012. SAS Enterprise Guide 5.1. SAS Institute Inc., Cary, North Carolina.

Sepúlveda F and Carrillo R, 2008. Functional response of the predatory mite *Chileseius camposi* (Acarina: Phytoseiidae) on densities of it prey, *Panonychus ulmi* (Acarina: Tetranychidae). Rev. Biol. Tropical. 56: 1255–1260.

Shah MA and Khan AA, 2013. Functional response-a function of predator and prey species. Bioscan. 8: 751–758.

SPSS. 2008. SPSS Base 16.0 for Windows User's Guide. SPSS Inc.

Sigma plot®, 12. Systat Software, San Jose, CA.

Tello-Mercado VE, Zarzar-Maza ME and Suarez-Pantoja AM, 2017. Functional response of *Cydnodromus picanus* (Acari: Phytoseiidae) on two-spotted spider mite, *Tetranychus urticae* (Acari: Tetranychidae). Acta Agronómica. 66: 275–281.

Trexler JC, Charles EM and Travis J, 1988. How can the functional response best be determined? Oecologia. 76: 206–214.

Vangansbeke D, De Schrijver L, Spranghers T, Audenaert J, Verhoeven R, Nguyen DT, Gobin B, Tirry L and De Clercq P, 2013. Alternating temperatures affect life table parameters of *Phytoseiulus persimilis*, *Neoseiulus californicus* (Acari: Phytoseiidae) and their prey *Tetranychus urticae* (Acari: Tetranychidae). Exp. Appl. Acarol. 61: 285–298.

Veeravel R and Baskaran P, 1997. Searching Behaviour of Two Coccinellid Predators, *Coccinella transversalis* Fab. and *Cheilomenes sexmaculatus* Fab., on Egg Plant Infested with *Aphis gossypii* Glov. Int. J. Trop. Insect Sci. 17: 363–368.

Xia JY, Rabbinge R and Van Der Werf W, 2003. Multistage Functional Responses in a Ladybeetle-Aphid System: Scaling Up from the Laboratory to the Field. Env. Entomol. 32: 151–162

17

# Glyphosat and Paraquat herbicides weed control and yield effect after emergence in cotton

**Raheel Atif Hameed*[1], Shazia Ajum[1], Muhammad Naveed Afzal[2]**
[1]Cholistan Institute of Desert Studies, The Islamia University of Bahawalpur
[2]Central Cotton research Institute Multan. Punjab, Pakistan

*Corresponding author email: raheel.atif@iub.edu.pk

## Abstract

Response of various post emergence herbicides at different levels i.e. round up was applied as 4.75 lha$^{-1}$, 2.75 L ha$^{-1}$ and 1.75 L ha$^{-1}$ (Glyphosat), Gramaxone 20EC (Paraquat) as 2.55 lha$^{-1}$ and untreated (no spray) were field experimented against cotton cultivar CIM-473 under field condition at CCRI cotton research institute, Multan. Significant Control of weeds and increase in yield and yield contributing factors were observed. It was indicated that the highest significant yield, number of bolls, fresh weed biomass, dry weed biomass, plant height and weed control were obtained by using Round up (Glyphosate) @ 4.75 L ha$^{-1}$ as compared to other treatments including untreated (control). Average boll weight was not significant among treatments but significant against control. The highest net profit was obtained by the Round up 490 G/L when treated @ 4.75 L ha$^{-1}$ than all other treatments.

**Keywords:** Cost benefit analysis, Cotton growth, Glyphosate, Paraquat, Yield components

## Introduction

Cotton is a major cash crop cultivated in Pakistan and is an important source of foreign exchange. The cotton has 1% share in GDP and 5.1% in agriculture. It has been cultivated an area of 2917 thousand hectares with 10074 thousand bales and yield as 587 kegs ha$^{-1}$. At present, the average seed cotton yield of Pakistan is much lower than other advanced countries i.e. UK, China, and India (Anonymous, 2016).

Besides many other factors like cultivar selection, irrigation techniques, fertilizer application rates and methods etc, the low yield per hectare is caused by serious weed infestation in the crop. Weeds compete in several ways with crop plants for space, nutrients, water, sunlight and many other basic requirements. These are the host and provide shelter for many insect/pests diseases. These can reduce average yield 33.50% to 55% or even result in complete crop failure (Ali *et al.*, 2013).

Weeding by cultural practices is laborious, tedious and difficult in contrast herbicide weed control in cotton, the method is easy, time saving and effective. Ali *et al.*, 2013, Alves *et al.*, 2011, Chaudhry *et al.*, 2011, Johnson *et al.*, 2009, Holloway *et al.*, 2008, Oad *et al.*, 2007, Deshpande *et al.*, 2006, Sheikh *et al.*, 2006 and Ali *et al.*, 2005, experimented in field and stated that weeds were controlled and cotton yield was maximized by the application of chemicals at applied on various rates and did not have negative effects on fibre quality characteristics. The herbicides Round up 490 G/L @ 4.75 lha$^{-1}$, 2.75 lha$^{-1}$ and 1.75 lha$^{-1}$ and Gramaxone 20EC @ 2.55 lha$^{-1}$ were applied against untreated control after emergence of cotton plants, herbicides significant controlled all weeds and increased yield and yield components.

The chemical weed control appeared more beneficial and effective that was the objective of this research.

## Materials and Methods

The investigations were carried out at the field area of CCRI Institute, Multan, during 2011 and 2012 on silt loam soil. Experiment was laid out in randomized complete block (R.C.B.D) design with three repeats against five treatments. Round up 490 G/L @ 4.75 lha$^{-1}$, 2.75 lha$^{-1}$ and 1.75 lha$^{-1}$ and Gramaxone 20EC @ 2.25 lha$^{-1}$ and untreated control for cv CIM-473 by using net plot size 20ft x 50ft with 75cm R x R and 25cm P x P distance. The herbicides were applied after emergence of cotton plants. Each herbicide was mixed thoroughly in a spray volume of 250 L ha$^{-1}$ and sprayed uniformly with knapsack sprayer fitted with fiat fan nozzle.

Uniform and normal field operations were applied for all the treatments. Weed control and yields component characteristics were investigated like number of weeds per m$^2$, Fresh weed biomass gm per m$^2$, weeds dry weight gm per m$^2$, bolls count per plant, Boll weight (g), Final plant height and yield of seed cotton kg ha$^{-1}$. Particular crop husbandry practices were adopted and insect/pests were controlled through regular insecticidal sprays. Data on weed control collected after 10, 20 and 30 days of spray and on yield and yield components at maturity were statistically analyzed by analysis of variance techniques and statistically significant differences among the means of the treatments were analyzed by Duncan's test of multiple range at 5% probability level as described by (Steel and Torrie, 1986).

## Results and Discussion

Tested herbicides at different levels gave statistically significant decrease of weed population over untreated control as indicated in Table-1. Results were highly significant for lowest number of weeds (40.0 and 42) were found in plot treated with Round up 490 G/L @ 4.75 lha$^{-1}$ against untreated control (274.5and 275) after 20 DAS (days after spray) respectively during 2011-12. It is the quality of Round up 490 G/L that it gives good results after 20 DAS. These results are supported by (Ali *et al.*, 2013 and Deshpande *et al.*, 2006). Data also represented that application of Round up 490 G/L @ 4.75 L ha$^{-1}$produced the lowest fresh weed biomass (228.6 and 229.6 g) against untreated control (4489.0 and 4491 g) after 20 DAS during both the years according to its quality then weed fresh biomass started to increase. These findings are in line with that of (Chaudhry *et al.*, 2011 and Johnson *et al.*, 2009). (Table-1)

Table-2 showed that the lowest dry weed biomass was produced by Round up 490 G/L @ 4.75 L ha$^{-1}$ (177.4 and 179.6 g) against untreated control (645.0 and 646.5 g) after 20 DAS, then it started to increase. Ali *et al.*, 2013, Holloway *et al.*, 2008 and Ali *et al.*, 2005 were reported the same results. The highest significant bolls count per plant (19.17 and 20.01) was produced by Round up, when applied as 4.75 lha$^{-1}$ against untreated control (10.40 and 11.30). These results are supported by (Chaudhry *et al.*, 2011, Sheikh *et al.*, 2006 and Ali *et al.*, 2005. 16).

The data in Table-3 also presented that statistically the highest boll weight was obtained by Round up 490 G/L applied @ 4.75 lha$^{-1}$ (2.77 and 2.78 g) as compared with untreated control (2.16 and 2.18 g) These results are supported by Chaudhry *et al.*, 2011, Sheikh *et al.*, 2006 and Ali *et al.*, 2005. The tallest plant height was found in Round up 490 G/L treated plots when it was applied @ 4.75 L ha$^{-1}$ (91.00 and 93.40 cm) against untreated control (62.03 and 63.70 cm) These results were supported by Ali *et al.*, 2013, Chaudhry *et al.*, 2011, Johnson *et al.*, 2009, Oad *et al.*, 2007 and Sheikh *et al.*, 2006. Data also showed that application of Round up 490 G/L @ 4.75 lha$^{-1}$ produced significantly the highest yield of seed cotton (2076 and 2085 kilo grams ha$^{-1}$) against untreated control (870 and 891 kg ha$^{-1}$) and other treatments. It was occurred due to better growth of cotton plants as a result of minimum competition with weeds for moisture, nutrients, space etc. which attributed to yield of cotton. These results are in accordance with Ali *et al.*, 2013, Chaudhry *et al.*, 2011, Johnson *et al.*, 2009, Holloway *et al.*, 2008, Oad *et al.*, 2007 and Sheikh *et al.*, 2006.

Economics of new technology (inputs) was the basic consideration in this study, data indicated that maximum net profit was obtained by Round up 490 G/L when applied @ 4.75 lha$^{-1}$ (Rs.12552.25) with less expenditures against other treatments including untreated control. On the basis of this evaluation, we can conclude that Round up 490 G/L @ 4.75 lha$^{-1}$ may be sprayed for obtaining maximum return.

**Table - 1: Effect of herbicides on weeds count and fresh weed biomass $m^{-2}$**

| Treatments | No. of Weeds $m^{-2}$ 2011 | | | No. of Weeds $m^{-2}$ 2012 | | | Fresh Weed Weight ($gm^{-2}$) 2011 | | | Fresh Weed Weight ($gm^{-2}$) 2012 | | |
|---|---|---|---|---|---|---|---|---|---|---|---|---|
| | 10 DAS | 20 DAS | 30 DAS | 10 DAS | 20 DAS | 30 DAS | 10 DAS | 20 DAS | 30 DAS | 10 DAS | 20 DAS | 30 DAS |
| Round up 4.75 L $ha^{-1}$ | 48.7d | 40.0c | 78.0d | 49.6d | 42c | 79.8d | 341.2c | 228.6c | 382.2c | 343.4c | 229.6c | 381c |
| Round up 2.75 L $ha^{-1}$ | 52.8c | 86.5d | 116.0c | 52.4c | 88.2d | 116.c | 619.5c | 790.6d | 1112.0d | 620.5c | 793.6d | 1114.3d |
| Round up 1.75 L $ha^{-1}$ | 84.10b | 113.5b | 184.6b | 85.4b | 117.5b | 185.3b | 1368b | 2171.0b | 2860.0b | 1371b | 2173b | 2863.0b |
| Gramaxone 2.55 L $ha^{-1}$ | 37.4c | 100.5c | 124.6c | 38.0c | 102.5c | 126.5c | 593.4d | 1051.0c | 1579.0c | 596.4d | 1053c | 1582c |
| Control | 240.3a | 274.5a | 290.1a | 241.5a | 274.5a | 292a | 3209a | 4489.0a | 5472.0a | 3209a | 4491a | 5474a |

DAS: Days after spray

**Table - 2: Effect of herbicides on dry weed biomass $gm^{-2}$ and bolls $plant^{-1}$**

| Treatment | Boll Weight 2011 | Boll Weight 2012 | Plant Height 2011 | Plant Height 2012 | Seed Cotton Yield 2011 | Seed Cotton Yield 2012 |
|---|---|---|---|---|---|---|
| Round up 4.75 L $ha^{-1}$ | 2.77a | 2.78a | 91.00a | 93.40a | 2076a | 2085a |
| Round up 2.75 L $ha^{-1}$ | 2.67a | 2.69a | 85.00ab | 87.30b | 1579b | 1587b |
| Round up 1.75 L $ha^{-1}$ | 2.53a | 2.55a | 76.67c | 78.00c | 1349b | 1365b |
| Gramaxone 2.55 L $ha^{-1}$ | 2.60a | 2.63a | 83.00b | 84.80b | 1512b | 1526b |
| Control | 2.16b | 2.18b | 62.03d | 63.70d | 870c | 891c |

**Table - 3: Effect of herbicides on boll weight (g), plant height (cm) and seed cotton yield (kg $ha^{-1}$)**

| Treatments | Dry Weed Weight ($gm^{-2}$) 2011 | | | Dry Biomass of weed ($gm^{-2}$) 2012 | | | Bolls $plant^{-1}$ 2011 | Bolls $plant^{-1}$ 2012 |
|---|---|---|---|---|---|---|---|---|
| | 10 DAS | 20 DAS | 30 DAS | 10 DAS | 20 DAS | 30 DAS | | |
| Round up 4.75 L $ha^{-1}$ | 138.3d | 177.4c | 182.6c | 140.3d | 179.6c | 184.3c | 19.17a | 20.01a |
| Round up 2.75 L $ha^{-1}$ | 175.2c | 219.3d | 314.9d | 176.5c | 220.3d | 316.7d | 15.13b | 16.30b |
| Round up 1.75 L $ha^{-1}$ | 230.5b | 267.7b | 483.8b | 233.2b | 269.3b | 486.0b | 13.00bc | 13.62bc |
| Gramaxone 2.55 L $ha^{-1}$ | 129.1e | 225.8c | 341.0c | 130.2e | 227.7c | 343.2c | 15.03b | 16.50b |
| Control | 461.0a | 645.0a | 793.0a | 463.3a | 646.5a | 795.3a | 10.40c | 11.30c |

**Table - 4: Cost benefit analysis for Post-Emergence Herbicides**

| Treatment | Total Herbicide Cost | Ave. Yield kg $ha^{-1}$ | Cotton Sticks Value $ha^{-1}$ | Gross Benefit | Total cost of Production | Net Benefit Obtained |
|---|---|---|---|---|---|---|
| **Round up 4.75 L $ha^{-1}$** | 1927.00 | 2076 | 1500 | 43020 | 30467.75 | 12552.25 |
| **Round up 2.75 L $ha^{-1}$** | 1107.00 | 1579 | 1500 | 33080 | 29026.5 | 4053.5 |
| **Round up 1.75 L $ha^{-1}$** | 615.00 | 1349 | 1500 | 28480 | 28247 | 233.0 |
| **Gramaxone 2.55 L $ha^{-1}$** | 1100.00 | 1512 | 1500 | 31740 | 28935.75 | 2804.25 |
| **Control** | - | 870 | 1500 | 18900 | 27033.25 | -8133.25 |

Seed Cotton Value (Rs) =800 / 40 kg; Cotton Sticks Value =600/ acre, Round up 490 G/L =410/L; Gramaxone 20EC =440/L

## References


Ali H, Abid SA, Ahmad S, Sarwar N, Arooj M, Mahmood A and Shahzad AN, 2013. Impact of Integrated Weed management on Flat-Sown Cotton (*Gossypium hirsutum* L.) J. Anim. Plant. Sci. 23 (4): 1185-1192.

Ali H, Muhammad D and Abid SA, 2005. Weed control practices in cotton (*Gossypium hirsutum* L.) planted on bed and furrow. Pakistan J. Weed Sci. Res. 11(1-2): 43-48.

Anonymous, 2016. Agricultural Statistics of Pakistan. Government of Pakistan. Ministry of Food, Agriculture and Livestock. Economic Wing, Islamabad. 27-28.

Alves, Severino LS and Vale LS, 2011. Critical periods of weed control in naturally green colored cotton BRS Verde. Indust. Crop. Prod. 34: 1198-1202.

Chaudhary SU, Iqbal J, Hussain M and Wajid A, 2011. Economical weed control in lentils crop. J. Anim. Plant. Sci. 21(4): 734-737.

Deshpande RM, Pawar WS, Mankar PS, Bobde PN and Chimote AN, 2006. Integrated weed management in rainfed cotton (*Gossypium hirsutum* L.). Ind. J. Agron. 51(1): 68-69.

Holloway J, Trolinder L, Ellis JM and Baker S, 2008. New herbicide tolerance technology for glyphosate resistant weed management in cotton. Proc. Beltwide Cotton Prod. Res. Conf., Nashville, TN. 8-11 Jan. 2008. Natl. Cotton Counc. Am., Memphis, TN. p. 1726.

Johnson WG, Davis VM, Kruger GR and Weller SC, 2009. Influence of glyphosate-resistant cropping systems on weed species shifts and glyphosate-resistant weed populations. Eur. J. Agron. 31: 162-172.

Khan NU and Khan SU, 2003. Integrated weed management in upland cotton. Pakistan J. Weed Sci. Res. 9(3&4): 185-192.

Koger CH, Price AJ and Reddy KN, 2005.Weed Control and Cotton Response to Combinations of Glyphosate and Trifloxysulfuron. Weed Tech. 19: 113-121.

Maqbool M, Tanveer MA, Ali A and Ahmed R, 2001. Effect of sowing methods and herbicides on weeds and yield of cotton (*Gossypium hirsutum* L.). Pakistan J. Bot. 33(4): 383-387.

Oad FC, Siddique MH, Buriro UA and Solangi GS, 2007. Weed management practices in cotton crop. Asian J. Plant Sci. 6(2): 344-348.

Salgado TP, Alves P, Mattos ED and Martins JF, 2002. Periodos de interferencia das plantas daninhas na cultura do algodoeiro (*Gossypium hirsutum*). Planta Daninha. 20: 373-379.

Steel RGD and Torrie JH, 1986. Principles and procedures of statistics, McGraw Hill Book Inc. New York, USA.

Sheikh MA, Saleem A and Malik NA, 2006. Integrated weed management and its effect on the seed cotton yield in cotton. Pakistan J. Weed Sci. Res. 12(1-2): 111-117.

Tanveer A, Chaudhry NH, Ayub M and Ahmad R, 2003. Effect of cultural and chemical weed control methods on weed population and yield of cotton. Pakistan J. Bot. 35(2): 161-166.

# Pathogenic activity of *Fusarium equiseti* from plantation of citrus plants (*Citrus nobilis*) in the village Tegal Wangi, Jember Umbulsari, East Java, Indonesia

**Dalia Sukmawati* and Mieke Miarsyah**
Department of Biology, Faculty of Mathematics and Natural Sciences, Universitas Negeri Jakarta, Campus A, Jl. Rawamangun Muka East Java. Hasyim Ashari Building, 9th floor, Indonesia.

*Corresponding author email: Dalia-Sukmawati@unj.ac.id

## Abstract

Some fungi associate with fruit and dead or dying plant tissues as pathogen on a wide range of agricultural plants. This work comprised the isolation, identification and pathogenic assay from citrus fruit plantations (*Citrus nobilis*), Tegal Wangi, Jember, Jawa Timur, Indonesia with 34 mold isolates obtained. Color of 7-day-old colonies cultures on PDA was dominated by white while the reverse was whitish to pale yellow. Based on the pathogenicity test, four representative mold isolates were identified as pathogenic fungi using the sequence of internal transcribed regions Spacer (ITS) in the region of ribosomal DNA selected. Molds were identified as UNJCC (D5) D5K3A (*Fusarium equiseti* with 98% homology bootstrap value 100%), UNJCC (D6) D6. K3.B (*F. equiseti* with 99% homology bootstrap value of 100%), and UNJCC (D7) D7.K2.B (*F. equiseti* with 99% homology bootstrap value 66%) and UNJCC (D8) D3.K2.B (*F. equiseti* with 99% homology bootstrap value of 55%). *F. equiseti* is a main source of trichothecenes, zearalenone and other mycotoxins which can cause serious disease in humans and animals. Present information regarding the *Fusarium equiseti* damage to citrus leaves can be used help identify the occurrence of pathogenic fungi in citrus fruit plantations.

**Keywords:** *Citrus nobilis*, *Fusarium equiseti,* Pathogenicity, ITS rDNA region

## Introduction

Orange is one of the main crop in the village of Tegal Wangi, Indonesia which can help improve the well-being of its citizens in terms of economic development. Constraints faced by the farmers is a decrease in quality of citrus caused by mold destroyer (Sangwanich, 2013). Many citrus plants around agricultural land had suffered damage of tree trunks as brownish, mongering, leaves and fruits having wrinkles with black spots around on their surface. Mold cause serious losses annually on citrus fruit (Ghuffar et al. 2017). *Aspergillus, Penicillium, Rhizopus, Fusarium, Alternaria* and *Mucor* species are major disease source in citrus (Akhtar et al., 2013). Semangun (2007) reported *Fusarium* and *Aspergillus* in citrus fruits. Moscoso-Ramirez et al. (2013) reported mold *Penicillium digitatum* which resulted in whole green fruit and damaged stem with rotten fruit harmful to human health (Sangwanich et al., 2013; Sperandio et al., 2015. Plant diseases are of interest due to wide range of pathogens present in the rhizosphere especially fungi such as *Colletotrichum* sp. (Than et al. 2008; Cannon et al. 2012), *Fusarium* spp. (Tewoldemedhin et al. 2011), *Fusarium culmorum, F. Oxysporum, F. Sporo-trichioides, Alternaria alternata, A. tenuissima, A. arborescens, A. Infectoria* (Lee et al. 2005), *Alternaria* spp. (Serdani et al. 2002).

Apple trees are susceptible to wide variety of pathogens such as *Fusarium equiseti* (Alonso et al., 2015). Johnston (2008) isolated and identified a wide range of *Penicillium* molds from *Litchi chinensis* Sonn in South Africa using rDNA ITS regions. Wani (2011) reported *Colletotrichum coccodes, C. dermatum,* and *C. gleosporoides* cusing anthracnose in tomato fruits. Damage inflicted blackish-colored black wounds with concave pink colored mycelium growing (Rodrigues and Menezes 2005; Merr et al., 2013). Thiyam and Sharma (2013) showed fungal diseases from local fruits containing *Aspergillus, Acremonium, Alternaria, Aspergillus, Chalaropsis, Cladosporium, Curvularia, Fusariumm, Mucor, Penicillium, Rhizopus* and *Trichoderma* in all fruits during storage. In rainy season, the loss of production in infected fruits by molds can reach 100% (Arauz, 2000). Mold attacking citrus has become one of the limiting production factor in world citrus production (Timmer et al., 2000; Poppe et al., 2003). Fungi a universal pathogen that causes diseases on many fruits such as mango, papaya, and apple and especially in wither caused on citrus due to this pathogen. In this study isolation, identification and testing of highly pathogenic mold of citrus plantation of Siam in The Tegal Wangi Village, the control efficiency of yeasts isolate against pathogenic fungi from citrus leaves was investigated.

## Materials and Methods

### Sampling location

Location of the sampling site was Desa Krangkongan, Village Tegal Sari, Jember Umbul Wangi, East Java, Indonesia (Fig. 1). Samples were collected from seven citrus trees located in the four corners and center of the total area from orchards (Bagyaraj and Rangaswami, 2007).

### Isolation of fungi

For isolation of fungi from citrus plants, the agar method was applied. Fungal pathogens responsible for disease were isolated from leaf surface collected from sampled trees. From the samples obtained, serial dilutions (1:10) were prepared in test tubes with 9 ml sterile water, adding 1 g of leaf samples (previously

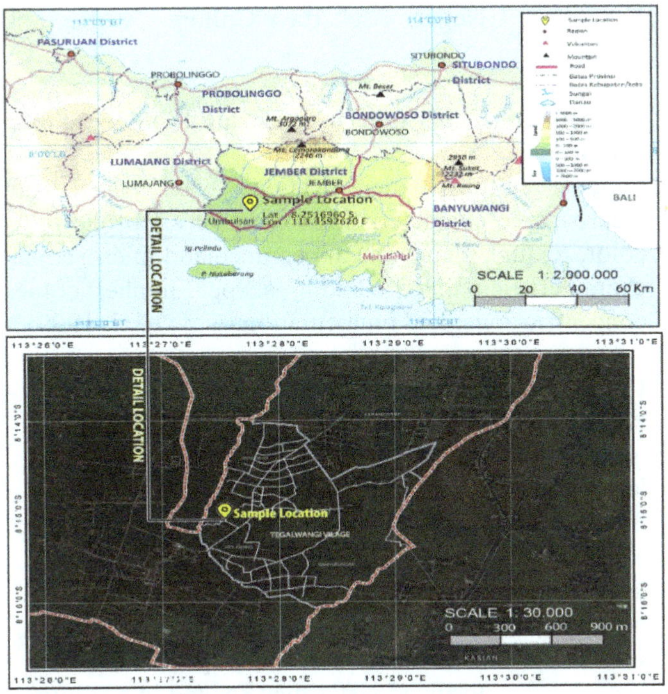


**Fig. - 1: Sampling location**

sieved) until dilutions $10^{-5}$, $10^{-6}$ and $10^{-7}$, of which 50 μL and shaked at 200 rpm for 3 hours. Samples collected were seeded by duplicate through diffusion technique on Petri dishes of 90 mm diameter containing one of the following artificial Potato Dextrose Aga (PDA) and Czapecks agar medium. After incubation (25°C for 7 days) the number for each fungi was calculated. All fungi were cultured using a single spore technique on PDA and Czapecks medium. After incubation, fungi were identified by their macroscopic characters such as colony color, pigment production and mycelium characteristics, and through their microscopic characteristics like the presence of spores and arrangement of sporulation structures examined with a compound microscope (Carl Zeiss) at 1000X. All fungi were subcultured on PDA medium and stored at 4°C. Other isolation methods were also performed based on Farrag (2011) with modifications. Isolation was performed with mold using agar on infected leaves. We used a modified dental needle to the hilt. Each piece of medium was incubated on PDA medium at room temperature (± 30°C).

### Purification of Pathogens

Pathogen fungal cultures obtained were purified by the single spore isolation method (Choi et al. 1999). Pure cultures were maintained on PDA slants for further study and preserved with L-drying method in the

University Negeri Jakarta Culture Collection (UNJCC).

**Identification of macroscopic and microscopic fungi**

Fungal cultures identification was based on macroscopic characteristics like colony morphology, color, texture, shape and appearance and microscopic characteristics like conidia shape, hyphe color, concentric zone, and pigmentation (Navi et al. 1999).

**Pathogenicity test**

Pathogenicity test used the agar smear method based on the principle of Koch's postulates. Parameter in testing was based on the measurement of disease incidence and severity using the formula described earlier (Embaby et al., 2013). Stages of the pathogenic test included sterilization of leaf surface and inoculation the pathogenic fungi on leaves. Leaf surface sterilization was performed by washing citrus leaves using sterilized water then soaked in a solution of sodium hypochlorite (NaOCl) 0.5% for one minute and further soaked in 70% alcohol for one minute before rinsing with sterilized water. Inoculation of the pathogenic fungi was done with an agar smear method based on Chutia et al. (2009). Test fungi were inoculated with 5mm mycelium plugs from 7-days-old cultures and observation was for 10 days at a temperature of 25-27°C. Growth of fungal species was recorded after one week of incubation and the percentage inhibition was computed after comparison with the control. Lime leaves were placed with 99% moisture and placed in the plastic tubs containing fruits before incubation. Observation was recorded after 10 days when kept at 25-27°C (Agrios, 2005).

**Identification of fungi using rDNA sequence**

Identification of pathogenic fungi was done using the rDNA on ITS region described by White et al. (1990). Reaction mixture contained specific primers for ITS (Internal Transcribed Spacer region) rDNA region with primer ITS4 (5'-TCCTCCGCTTATTGATATGC-3') and primer ITS5 (5'-GGAAGTAAAAGTCGTAACAAGG-3'). PCR reaction using PuReTaq™ Ready-To-Go™ PCR Beads (GE Healthcare) total reaction 25 μL, each reaction contained: 15 μL nuclease free water (NFW) dilution in PuReTaq™ Ready-To-Go (RTG) PCR beads (GE Healthcare), 10 pmol primer ITS4 and ITS5 (100 ng DNA template). PCR condition: denaturation at 95°C for 2 min (1 cycle); post denaturation at 94°C for 15 sec, annealing at 56°C for 30 sec, extension at 68°C for 1 min (40 cycles); and 70°C for 10 min in final extension (1 cycle) (Sukmawati et al. 2015). All PCR results were visualized using UV transilluminator after electrophoresis through a 1% agarose gel and ethidium bromide staining. PCR products were sent to 1stBASE (Malaysia) for sequencing.

**Phylogenetic analysis**

Nucleotide sequence datasets were automatically aligned using the MUSCULE program. Multiple alignments were carried out in MEGA6 (Molecular Evolutionary Genetics Analysis Version 6.0) (Tamura et al. 2007) and sequences retrieved from NCBI (http://www.ncbi.nlm.nih.gov). Phylogenetic analysis was conducted using the maximum likelihood (ML) method in MEGA6. ML analysis was tested by bootstrap (BS) analysis using 1000 replications. BS values of 50% or higher were shown and NR 130661 *Candida orthopsilosis* ATCC 96139 were used as outgroups.

## Result and Discussion

**Isolation of fungi**

Physically observation of citrus fruits showed light brown color with bark peeling and brittle. Plant leaves were developing mold pathogen with blackish leaf spots blackish, speckled blotch, freckle spot, hard spot (shot-hole spot), yellowish, blackish sooty mold developed on the leaves or fruit fouled. Leaves turned yellowish but larger veins remained slightly green which easily fell (Fig. 2). Citrus spp. often got affected by fungal pathogens causing heavy fruit losses (Stammler et al. 2013). General symptoms of citrus plant infected by pathogenic fungi included leaf spots and chlorosis (Teixeira et al. 2005). Five plant pathogenic fungi such as *Alternaria alternata, Rhizoctonia solani, Curvularia lunata, Fusarium oxysporum* and *Helminthosporium oryzae* can infect citrus with black spots on infected leaves (Chutia et al., 2009). Plant pathogen included *Zygomycetes, Ascomycetes* and *Basidiomycetes* (Sukmawati, 2016; Teixeira et al. 2005). Fungi need nutrient from plant for their metabolism.

A total of 34 fungal isolates from citrus leaves was obtained showed diverse morphology of molds on stems and leaves. Mold isolates from the leaves were white (38.2%); light ochre-flesh (26.5%); flesh (8.8%); ochre (5.9%) and others (20.6%) (Table 1).

Mold isolated from leaves had sporulation and pigments (Sukmawati, 2016). Citrus is affected by several mold colors like green affecting fruit quality responsible for major postharvest problems like market losses. Their colors helped in preliminary identification like green and blue mold infections were caused by *Penicillium* spp. (Akhtar et al. 2013) and brown by *Colletotrichum gloeosporioides* Penz (Chung et al. 2002)

**Fig. – 2: Citrus leaves infected with pathogen fungi a: leaves from first plant; b: leaves from second plant and c: leaves from third plant**

**Table – 1: Morphology of molds isolates on Potato Dextrose Agar (PDA), 3 days' incubation at 27--28$^0$C.**

| No. | Code of isolate (UNJCC) | Morphology colony of molds | | | | |
|---|---|---|---|---|---|---|
| | | **Colours** | **Margin** | **Sporulation** | **Reverse of colony** | **Diameter (L/W) cm** |
| 1 | D2. K1. B | White | White | Olive Green | Light Flesh | 19.29 / 19.24 |
| 2 | D3.K2. B | White | White | - | Cinnamon | 25.97 / 23.72 |
| 3 | D4. K1. A | White | White | Olive Green | Cinnamon | 26.78 / 30. 46 |
| 4 | D4. K1. B | White | White | Olive Green | Burnt ochre | 22.33 / 26.84 |
| 5 | D5. K1. A | White | White | Cinnamon | Ochre | 31.06 / 32.70 |
| 6 | D6. K3 | white | White | - | light flesh | 26.06 / 24.84 |
| 7 | D7.K2. A | White | White | Cedar Green | white | 28.15 / 28.32 |
| 8 | D7.K2. B | White | White | - | Light Flesh | 26.25 / 27.54 |
| 9 | D7. K3 | White | White | - | Light Ochre | 77.69/60.30 |
| 10 | D8. SP. K1. B | White | White | Grey green | White | 37.44 / 39.29 |
| 11 | D8.SP.K2. A | White | White | - | White | 27.69 / 25.60 |
| 12 | D8.SP.K2. B | White | White | Cedar green | White | 46.66 / 52.96 |
| 13 | D8. K1 | White | White | - | Cream | 82.68/45.00 |
| 14 | D1. K3 | Light Ochre | Light Ochre | - | Gold Ochre | 42.25 / 62.77 |
| 15 | D2. K1. A | Light Ochre | Warm Grey I | Cedar Green | Cinnamon | 18.74 / 20.32 |
| 16 | D3. K1 | Light ochre | White | - | Light Ochre | 72.84 / 82.36 |
| 17 | D2. K2 | Light Flesh | Cinnamon | - | Light Ochre | 62.85/55.00 |
| 18 | D5. K1. B | Light Flesh | White | - | Light Ochre | 31.21 / 29. 78 |

| | | | | | | |
|---|---|---|---|---|---|---|
| 19 | D5. K2 | Light flesh | White | Olive Green | Cinnamon | 37.35 / 46. 86 |
| 20 | D7. K1 | Light Flesh | White | - | Light Ochre | 81.62/46.00 |
| 21 | D8.K2. A | Light Flesh | White | Light ochre | Ochre | 45.47 / 79.17 |
| 22 | D8.K2. B | Light Flesh | White | Ivory | Light Ochre | 46.43 / 81.84 |
| 23 | D6. K1. A | Medium flesh | Dark flesh | Juniper Green | Light Flesh | 46.99 / 76.46 |
| 24 | D6. K1. B | Medium flesh | Dark flesh | - | Ochre | 39.59 / 60.07 |
| 25 | D6. K2 | Medium flesh | Light flesh | - | Light ochre | 42.45 / 72.42 |
| 26 | D3. K3 | Ochre | Light Ochre | - | Gold ochre | 44.07 / 75.75 |
| 27 | D8. K3 | Ochre | White | - | Gold Ochre | 82.5/45.67 |
| 28 | D1. K1. A | Cinnamon | Cold grey I | - | Light Flesh | 21.45 / 21.26 |
| 29 | D1. K1. B | Caput Mortum | Silver | - | Cold Grey II | 18.09 / 19.42 |
| 30 | D1.K2. A | Burnt Ochre | Brown Ochre | Warm Grey II | Raw Umber | 51.81 / 58.22 |
| 31 | D2. K3 | Dark Flesh | White | - | Light Flesh | 27.06 / 25.92 |
| 32 | D3.K2. A | Gold ochre | White | - | Cinnamon | 26.54 / 24.64 |
| 33 | D5. K3 | Cedar green | White | - | Olive green | 27.96 / 18.74 |
| 34 | D8.SP. K1.A | Olive green | White | Soft black | Juniper green | 55.33 / 65.60 |

Mold isolated from citrus leaves showed 14 isolates (42%) with color variations among others; green, brown, and black (Table 1). Leaf is one of the source of nutrients and living place of mold isolates. Molds of citrus plants had been characterized by the colony with sporulation such as black, brown, green and yellow greenish (Chutia et al. 2009; Mohammed et al. 2013; Nasiru et al. 2015). Spores are asexual structures in the mold which is useful in deployments to the host (Akhtar et al. 2013; Sperandio et al. 2015). Mold pathogen had various colors like white, light, dark fleash, medium flesh ochre ivory, gold, cadmium and Hyalin (Akhtar et al. 2013; Chutia et al. 2009; Mohammed et al. 2013; Nasiru et al. 2015).

**Pathogenicity test**

Selection of 8 representative isolates was based on ability of sporulation. The representative mold consisted UNJCC (D1) D8.SP.K1.B; UNJCC (D2) D5. K1. A; UNJCC (D3) D4.K1.B; UNJCC (D4) D8.SP.K1.A; UNJCC (D5) D5.K3.A; UNJCC (D6) D6.K3.B ; UNJCC (D7) D7.K2.B; and UNJCC (D8) D3.K2.B. According to postulant Koch the pathogenicity test showed that four molds isolate were pathogenic (Table 2; Fig. 3). The value of the disease incidence and disease severity was indicated the highest by the isolate with code (D8) D3.K2. B (85%; 100%) (Fig. 3).

Test results proved that the four isolates with original mold causing damage in citrus leaves by highly pathogenic in accordance of Koch's postulates. According to Carla and Renata (2012), Koch's postulates can be used as a criterion of highly pathogenic isolates of a mold. The principle of Koch's postulates consists of: 1) Isolates can be isolated from the diseased host; 2) Isolates can be grown in the laboratory; 3) Isolate the results of isolation will give the same symptoms of the disease on the host, if re-inoculation; 4) Isolates will have the same morphology.

**Table - 2: The pathogenicity test results 8 representatives of pathogen mold incubation 10 day at 30 $^0$C**

| Isolate Code (UNJCC) | Characteristic of observation (day) | | | | |
|---|---|---|---|---|---|
| | d0 | d1 | d3 | d5 | d10 |
| (D1)D8. SP. K1. B | Green leaves | Green leaves | - | - | - |
| (D2)D5. K1. A | Green leaves | Green leaves | - | - | - |
| (D3) D4. K1. B | Green leaves | Green leaves | - | - | - |
| (D4) D8. SP. K1. A | Green leaves | There are several causes of brown spots | The color of the leaves becomes brown to black | The color of the leaves becomes brown to black | The color of the leaves becomes brown to black |
| (D5) D5. K3. A | Green leaves | There are several causes of brown spots | The tips of leaves first, but gradually the dark coloring | The tips of leaves first, but gradually the dark coloring | The tips of leaves first, but gradually the dark coloring |
| (D6)D6. K3. B | Green leaves | There are several causes of brown spots | The tips of leaves first, but gradually the dark coloring with growing mycelium = 7.28 mm | The tips of leaves gradually the dark coloring with growing mycelium = 8.45 mm | The leaves first, but gradually the dark coloring |
| (D7) D7.K2. B | Green leaves | There are several causes of brown spots | The leaves become brown to dark with growing mycelium length = 11.59 mm; width = 2.54 mm | The leaves become brown to dark with growing mycelium with white in the margin, with length = 13,61 mm; width = 3,83 mm | The leaves become brown to dark with growing mycelium with white in the margin, with length = 15.61 mm; width = 4,83 mm |
| (D8) D3.K2. B | Green leaves | There are several causes of brown spots | The leaves become brown to dark with growing mycelium length = 14.29 mm; width = 4.32 mm | The leaves become brown to dark with growing mycelium length = 15.72 mm; width = 4.39 mm | The leaves become brown to dark with growing mycelium length = 17.72 mm; width = 6.39 mm |

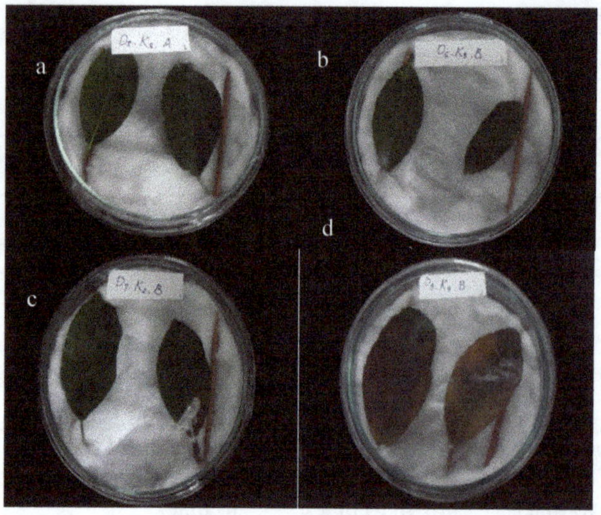


**Fig. – 3: Pathogenicity test a: UNJCC (D5) D5.K3.A; b: UNJCC (D6) D6.K3.B ; c: UNJCC (D7) D7.K2.B and d: UNJCC (D8) D3.K2.B incubation 10 day.**

**Macroscopic observation and identification of molecular ITS region**

Identification of molecular sequence analysis done using ITS rDNA region. PCR results on isolates obtained mold band with long bases 600 bp (Fig. 4).

Based on ITS regions of rDNA sequence data, four pathogenic isolates mold consisted following species as UNJCC (D5) D5K3A (*F. equiseti*) with similarity values 98%; UNJCC (D6) D6. K3. B (*F. equiseti*); UNJCC (D7) D7.K2.B (*F. equiseti*); and UNJCC (D8) (D3.K2.B) (*F. equiseti* and *F. oxysporum* sp. *fragariae*) with similar values 99%, which indicated high similarity to their closest species (Fig. 5, Table 4). Based on phylogenetic analysis, four isolates were found for *F. equiseti.*

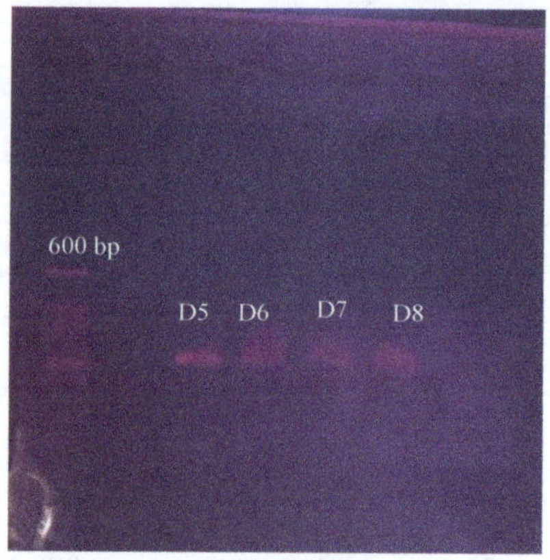


**Fig. – 4: Electrophoresis results UNJCC (D5) D5.K3.A; UNJCC (D6) D6.K3.B ; UNJCC (D7) D7.K2.B and UNJCC (D8) D3.K2.B**

**Table – 3: Identification result of mold isolates from the leaves of citrus from Tegal Wangi Village based on ITS region of rDNA.**

| Isolate code (UNJCC) | Closely related species | Max scores | Total score | Query score | E-value | Similarity (%) | Accession number |
|---|---|---|---|---|---|---|---|
| (D5) D5K3A | *F. equiseti* | 953 | 953 | 97% | 0 | 98% | AY928409. |
| (D6) D6K3B | *F. equiseti* | 1018 | 1018 | 99% | 0 | 99% | KX588103. |
| (D7) D7K2.B | *F. equiseti* | 987 | 987 | 98% | 0 | 99% | KX588103. |
| (D8) D3.K2B | *F. equiseti* | 1050 | 1050 | 99% | 0 | 99% | KR364600. |

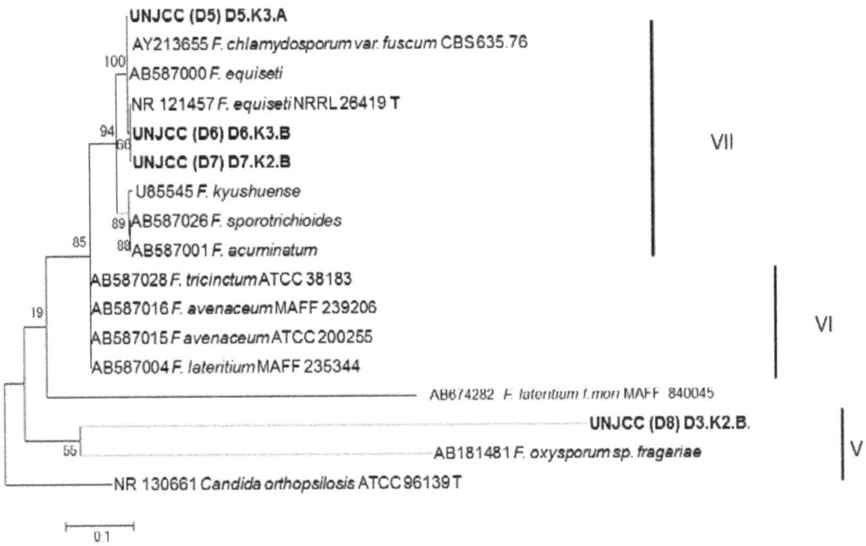


**Fig. - 5: Maximum likelihood tree showing taxonomic position of *Fusarium* strains isolated from orange leaves: UNJCC (D5) D5.K3.A; UNJCC (D6) D6.K3.B; UNJCC (D7) D7.K2.B and UNJCC (D8) D3.K2.B. The tree was rooted to NR 130661 *Candida orthopsilosis* ATCC 96139**

The *Fusarium* genome was first described by Link in 1809 (Aoki et al., 2014). Based on the phylogenetic tree, isolates (D5) D5.K3.A; (D6) D6.K3.B; (D7) D7.K2.B and (D8) D3.K2.B were identified as one clade with *F. equiseti* with bootstrap value of 94% (Fig. 5). The low value of the need for this data described the bootstrap analyzed by using a gene other than ITS rDNA. These regions have high success rates to identify a molecular approach (Schoch et al. 2012). But not all isolates of *Fusarium* can be identified are accurate based on a single gene. Isolate fungi with code (D8) D3.K2.B are identified as *F. equiseti*. While according to phylogenetic analysis isolate (D8) D3.K2.B was one clade with AB181481 *F. oxysporum* sp. *fragariae*. Watanabe et al. (2011) reported seven clade in *Fusarium*. Our riset consist of clade VII, clade VI and Clade V. Clade VI consists of *F. lateritium, F. avenaceum,* and *F. tricinctum*, which belong to different "sections", namely, *Lateritium, Roseum,* and *Sporotrichiella* respectively. The paraphyly of *F. avenaceum and F. lateritium* was supported by all the genes. Clade VII contains 4 "sections" with 9 species: Eupionnotes consisting of *F. incarnatum, Gibbosum consisting of F. equiseti* and *F. acuminatum, F. graminearum and F. culmorum,* and *Sporotrichiella* consisting of *F. poae, F. kyusyuense, F. sporotrichioides,* and *F. langsethiae*. Clade V consisting *F. oxysporum sp. fragariae.* (Watanabe et al. 2011). *Fusarium* species is known one of the most difficult species to be identified based on morphological markers among fungal species. One of the main reasons for this difficulty is that genetic and morphological characters vary among strains in a species and the ranges of character diversity are often overlapped among closely-related species. Although all fore gene trees supported the classification of *Fusarium* species into 7 major clades, I to VII.

According to Tunarsih et al. (2015) suggested to use suitable marker for the identification of *Fusarium* members as *Fusarium* genome has possibly unique evolutionary history. Watanabe et al. (2011) used multigene analysis for *Fusarium* genome (18S rDNA gene, ITS1, 5.8S rDNA, 28S rDNA, $\beta$-tubulin gene, and aminoadipate reductase gene (*lys2*) for inter-species identification of *Fusarium*. Their results showed that sequence has homology with bootstrap value of 65–100%. *F. equiseti* mold can cause damage to various crops, including corn, rice and wheat in field and storage (Hasem et al. 2010). Palmero et al. (2011) reported that all the tested *Fusarium* isolates were pathogenic on tomato and melon. Regarding Bakar et al. (2013), *Fusarium* species are one of the common pathogens of post-harvest disease to cause rot on tomato and other perishable vegetable fruits. A total of 180 *Fusarium* isolates were obtained from 13 locations throughout Selangor. *Fusarium solani* was the most abundantly isolated (34%) followed by *F. semitectum* (31%) and *F. oxysporum* (31%), *F. subglutinans* (3%) while the last was *F. equiseti* (1%).

*Fusarium equiseti* is a plant pathogenic fungus and produce secondary metabolites form toxins that can be pathogenic on the various plants on agricultural land (Kosiak et al. 2005). Secondary metabolites produced vary in amount and toxicity. This species produces a variety of toxins, such as trichothecenes type A, for example, neosolaniol (NEO), diacetoxy-scirpenol (DAS), the type of T-2 toxin and HT-2, type B trichothecenes, for example, nivalenol (NIV), and non-essential compounds such as trichothecene zearalenon (ZEA), equisetin and fusarochromanone (Barros et al. 2012). In addition, it can also be profitable. This fungi have potential infection in rooting plants (Macia-Vicente et al. 2008) and the special nature of belonging so that could make these fungi as a candidate for biological control of nematodes (Nitao et al. 2001; Horinouchi et al. 2007). Other potential owned by mold, *F. equiseti* i.e. capable of producing the enzyme xyloglucanases (XG) (Rashmi & Siddalingamurty, 2016). These enzymes are known to have potential in processing waste plant, modifications to improve the nature of xyloglucans reologi in the food industry and the feed, fabric treatment to change the brightness and color, to remove the fuzz from the surface of textile materials in the textile industry and the paper industry (Sinitsyna, 2010). The enzyme is easily obtained so easily applied to help lower the cost of production.

## Conclusion

Sampling mold pathogen in citrus plant plantation was done in isolation of pathogenic mold by direct and washing methods. Isolation retrieved as many as 34 isolates derived from leaf mold- Mold colonies from the leaves and stems were dominated by white colonists with mold. Testing of highly pathogenic samples was made from leaves of citrus plantation in Jember. Testing was conducted on eight isolates of highly pathogenic representative molds. Four potentials isolate mold caused the same damage when symptoms mold isolates from diseased leaves. These isolates were UNJCC (D5) D5. K3. A; UNJCC (D6) D6. K3. B; UNJCC (D7) D7. K2. B; and UNJCC (D8) D3. K2. B. Based on their phylogenetic analysis, all isolates were identified as *F. equiseti* (UNJCC (D5) D5K3A with 98% homology bootstrap values 64%, isolate UNJCC (D6) D6. K3. B 99% homology with bootstrap values 100%, isolate UNJCC (D7) D7. K2. B 99% homology with bootstrap values 99%, and UNJCC (D8) D8 D3. K2. B 99% homology with bootstrap values of 88%.

## Acknowledgments

This research was supported by Fund from FMIPA DIPA 2016 and DIKTI 2016 which provided financial assistance, the facilities, and materials required. The researcher would also like to thank to students (Agustiningsih, and Sherly) for the technical assistance during the course of the experiment. We thank Avinash Sharma, Retno Widowati and Marina S. for reading through the manuscript. A special thank you to get the sample to Sudarno Lulung and Agung Adiputra, S.Si. MSi from PT. BIP (Gramedia Group) for drawing the sampling map.

## References

Agrios NG, 2005. Plant Pathology. 5th Ed. Department of Plant Pathology. University of Florida, USA.

Akhtar N, Anjum T and Jabeen R, 2013. Isolation and identification of storage fungi from citrus sampled from major growing areas of Punjab, Pakistan. Int. J. Agri. Biol. 15(6): 1283–1288.

Altschul SF, Madden TL, Alejandro A, Schäffer Zhang J, Zhang Z, Miller W and Lipman DJ, 1997. Gapped BLAST and PSI-BLAST: A new generation of protein database search programs. Nucleic Acids Res. 25(17): 3389–3402.

Alonso D, Pérez C, Nancy Y, García G, Gabriel GM, María F, Ruiz C, David I, Berlanga R and Claudio RV, 2015. Isolation of actinomycetes associated to apple trees rhizosphere antagonistic to *Fusarium equiseti*. Mex. Cienc. Agríc. 6(7): 1630-1638.

Arauz LF, 2000. Mango anthracnose economic impact and current options for integrated management. Plant Dis. 84: 600–609.

Aoki T, O'Donnell K and Geiser DM, 2014. Systematic of key phytopathogenic *Fusarium* species: current status and future challenges. J. Gen. Plant Pathol. 80: 189–201. Doi: 10.1007/s10327-014-0509-3.

Barros G, Zanon MSA, Palazzini JM, Haidukowski M, Pascale M and Chulze S, 2012. Trichothecenes and zearalenone production by *Fusarium equiseti* and *Fusarium semitectum* species isolated from Argentinean soybean. Food Additives & Contaminants. Part A; Chemistry, Analysis, Control, Exposure & Risk Assessment. 29(9): 1436–42.

Bakar AIA, Nur Ain Izzati MZ and Umi Kalsom Y, 2013. Diversity of *Fusarium* Species Associated with Post-Harvest Fruit Rot Disease of Tomato. Sains Malaysiana. 42(7): 911–920.

Bagyaraj DJ and Rangaswarni G, 2007. Agricultural microbiology. PHI Learning. 53-59.

Carla E and Renata A, 2012. From Koch's postulates to biofilm theory. The lesson of Bill Costerton Garth D. Int. J. Artif. Organs. 35(10): 695-699.

Cannon PF, Damm U, Johnston PR and Weir BS, 2012. *Colletotrichum*: current status and future directions. Studies in Mycol. 73: 181-213.

Chutia MP, Deka B, Pathak MG, Sarma TC and Boruah P, 2009. Antifungal activity and chemical composition of *Citrus reticulata* blanco essential oil against phytopathogens from North East India. Food Sci. Technol. 42:777–780.

Chung KR, Turksen S and Wei Li LW, 2002. Timmer engineering a genetic transformation system for *Colletotrichum acutatum*, the causal fungus of lime anthracnose and postbloom fruit drop of citrus. FEMS Microbiol. Letters. 213: 33-39.

Chimbekujwo IB, 2000. Frequency and pathogenicity of *Fusarium* wilts (*Fusarium solani* and *Fusarium equiseti*) of cotton (*Gossypium hirsutum*) in Adamawa in Nigeria. Rev. Biol. Trop. 48: 1–5.

Choi YW, Hyde KD and Ho WH, 1999. Single spore isolation of fungi. Fungal Diversity. 3: 29-38.

Embaby EM, Hazaa M, Hagag LF, Ibrahim TE and El-Azem FS, 2013. Decay of some citrus fruit quality caused by fungi. J. Appl Sci. Res. 9(11): 5920-5929.

Farrag ESH, 2011. First record of *Cercospora* leaf spot disease on okra plants and its control in Egypt. Plant Pathol. J. 10 (4): 175-180.

Ghuffar S, Muhammad ZA, Muhammad FA, Luqman A and Sajjad H, 2017. Physiological studies on *Colletotrichum gloeosporiodes* associated with wither tip disease of citrus and its chemical control. Asian J. Agri. Biol. 5(2):77-82.

Hashem M, Moharam AM, Zaied AA and Saleh FEM, 2010. Efficacy of essential oils in the control of cumin root rot disease caused by *Fusarium* spp. Crop Protection. 29: 111-117.

Horinouchi H, Muslim A, Suzuki T and Hyakumachi M, 2007. *Fusarium equiseti* gf191 as an effective biocontrol agent against *Fusarium* crown and root rot tomato in rock wool systems. Crop Protection. 26: 1514-1523.

Johnston CL, 2008. Identification of *Penicillium* species in the South African litchi export chain. Dissertation, Department of Microbiology, University of Pretoria, South Africa.

Lee HB, Park JY and Jung HS, 2005. First report of leaf anthracnose caused by *Colletotrichum boninense* on spindle trees. Plant Pathol. 54: 254.

Lahlali R, Hamadi Y, El guilli M and Jijakli H, 2011. Efficacy assessment of *Pichia guilliermondii* strain Z1, a new biocontrol agent, against citrus blue mold in Morocco under the influence of temperature and relative humidity. Biological Control. 56: 217-224.

Macia-Vicente JG, Jansson HB, Abdullah SK, Descals E, Salinas J and Lopez-Llorca LV, 2008. Fungal root endophytes from natural vegetation in Mediterranean environments with special reference to *Fusarium* spp. FEMS Microbiol. Ecol. 64: 90-115.

Moscoso-Ramírez PA, Montesinos-Herrero C and Palou L, 2013. Control of citrus postharvest *Penicillium* molds with sodium ethylparaben. Crop. Prot. 46: 44–51.

Meer H, Iram S, Ahmad I, Fateh FS and Kazmi MR, 2013. Identification and characterization of post-harvest fungal pathogens of mango from domestic markets of Punjab. Int. J. Agri. Plant Prod. 4: 650-658.

Mohammed SSD, Kumar N, Damisa D, Bala E, Muhammad IL, Muhammad RG and Yunusa A, 2013. Fungi associated with decayed sweet oranges *(Citrus sinensis)* collected from Lapai, Nigeria State, Nigeria. Indian J. Life Sci. 2 (2): 123-131.

Nasiru AM, Salau IA and Yakubu M, 2015. Fungi associated with spoilage of *Citrus sinensis* in fruits and vegetables market, Sokoto, Nigeria. Global Adv. Res. J. Agri. Sci. 4(12): 919-922.

Navi SS, Bandyopadhyay R, Hall AJ and Bramel-Cox PJ, 1999. A pictorial guide for the identification of mold fungi on sorghum grain. International Crops Research Institute for the Semi-Arid Tropics. p. 118.

Nitao JK, Meyer SLF, Schmidt WF, Fettinger JC and Chitwood DJ, 2001. Nematode-antagonistic trichothecenes from *Fusarium equiseti*. J. Chem. Ecol. 27: 859-869.

Poppe L, Vanhoutte S and Hofte M, 2003. Modes of action of *Pantoea agglomerans* CPS-2, an antagonist of postharvest pathogens on fruits. Eur. J. Plant. Pathol. 109: 963-973.

Palmero D, de Cara M, Iglesias C, Gálvez L and Tello JC, 2011. Comparative study of the pathogenicity of seabed isolates of *Fusarium equiseti* and the effect of the composition of the mineral salt medium and temperature on mycelial growth. Brazilian J. Microbiol. 42(3), 948–953.

Rashmi R and Siddalingamurty KR, 2016. Production of xyloglucanases from three species of filamentous fungi *Fusarium equiseti*, *Aspergillus terreus* and *Cephalosporium* sp. Res. J. Biotechnol. 7(1): 55-59.

Rodrigues AAC and Menezes M, 2005. Identification and pathogenic characterization of endophytic *Fusarium* species from cowpea seeds. Mycopathologia. 159: 79–85.

Stammler G, Schutte GC, Speakman J, Miessner S and Crous PW, 2013. Phyllosticta species on citrus: Risk estimation of resistance to QoI fungicides and identification of species with cytochrome b gene sequences. Crop Protection. 48: 6-12.

Sangwanich S, Sangchote S and Leelasuphakul W, 2013. Biocontrol of citrus green mold and postharvest quality parameters. Int. Food Res. J. 20(6): 3381-3386.

Semangun H, 2007. Penyakit-penyakit tanaman holtikultura. Gajah Mada University Press. Yogyakarta, Indonesia.

Serdani M, Kang JC, Andersen B and Crous PW, 2002. Characterisation of *Alternaria* species-groups associated with core rot of apples in South Africa. Mycol. Res. 106:561- 569.

Schoch CL, Seifert K and Huhndorf S, 2012. Nuclear ribosomal internal transcribed spacer (ITS) region as a universal DNA barcode marker for Fungi. In Proceedings of the National Academy of Sciences of the United States of America. 109: 6241–6246. Doi: 10.1073/pnas.1117018109.

Sinitsyna OA, 2010 Isolation and Properties of xyloglucanases of *Penicillium* sp. Biochem. 75(1): 41-49.

Sperandio EM, Martins do Vale HM and Moreira GAM, 2015. Yeasts from native Brazilian Cerrado plants: Occurrence, diversity and use in the biocontrol of citrus green mold. Fungal Biol. 119(11): 984–993.

Sukmawati D, 2016. Antagonism Mechanism of Fungal Contaminant on Animal Feed using Phylloplane Yeasts Isolated from the Bintaro Plant (*Cerbera manghas*) Bekasi in Java, Indonesia. Int. J. Curr. Microbiol. Appl. Sci. 5(5):63-74.

Sukmawati D, Oetari A, Hendrayanti D, Atria M and Wellyzar S, 2015. Identification of phylloplane yeasts from paper mulberry (*Broussonetia papyrifera* (L.) L'Her.ex Vent.) in Java, Indonesia. Malaysian J. Microbiol.11(4): 324-340.

Tamura K, Dudley J, Nei M and Kumar S, 2007. MEGA4: Molecular Evolutionary Genetics Analysis (MEGA) software version 4.0. Mol. Biol. Evolu. 24(8):1596-1599.

Teixeira DC, Saillard C, Eveillard S, Danet JL, da Costa PI, Ayres AJ and Bove J, 2005. Candidatus *liberibacter americanus*, associated with citrus Huanglongbing (greening disease) in Sao Paulo State, Brazil. Int. J. Syst. Evol. Microbiol. 55: 1857-1862.

Tewoldemedhin YT, Mazzola M, Botha WJ, Spies CFJ and McLeod A, 2011. Characterization of fungi (*Fusarium* and *Rhizoctonia*) and oomycetes (*Phytophthora* and *Pythium*) associated with apple orchards in South Africa. Eur. J. Plant Pathol. 130 (2):215-229.

Timmer LW, Solel Z and Orozco SM, 2000. *Alternaria* brown spot of Mandarins. In: Compendium of citrus diseases, Timmer L. W., Garnsey S. M. & Graham J. H., Eds. American Phytopathological Society Press. pp. 19-21.

Than P, Prihastuti H, Phoulivong S, Taylor PWJ and Hyde KD, 2008. Chilli anthracnose disease caused by Colletotrichum species. J. Zhejiang University Sci. 9 (10): 764-778.

Thiyam AB and Sharma GD, 2013. Isolation and Identification of fungi associated with local fruits of Barak valley. Curr. World Environ. 8(2): 319-322.

Tunarsih, R. and Hidayat. 2015. Molecular Phylogenetic Analysis of Indonesian *Fusarium* Isolates from Different Lifestyles, based on ITS Sequence Data. *Plant Pathology & Quarantine*. 5(2): 63–72.

Kosiak EB, Holst-Jensen A, Rundberget T, Gonzalez Jaen MT and Torp M, 2005. Morphological, chemical and molecular differentiation of *Fusarium equiseti* isolated from Norwegian cereals. Int. J. Food. Microbiol. 99(2), 195–206.

Watanabe M, Yonezawa T, Lee K, Kumagai S, Sugita–Konishi Y, Goto K and Hara–Kudo Y, 2011. Evaluation of genetic markers for identifying isolates of the species of the genus *Fusarium*. J. Sci. *Food*. Agri. 91: 2500–2504. Doi: 10.1002/jsfa.4507.

Wani AH, 2011. An overview of fungal rot of tomato. Mycopath. 9: 33-38.

White TJ, Bruns T, Lee S and Taylor J, 1990. Amplification and direct sequencing of fungal ribosomal RNA genes for phylogenetics. *In:* M.A. Innis, D.H. Gelfand, J.J. Sninsky & T.J. White, PCR Protocols: Guide to Methods and Applications. Academic Press Incorporation, New York. pp. 315-322.

19

# First record of the crown rot fungus *Fusarium equiseti* affecting *Triticum aestivum* L and *Aptenia cordifolia* in Iraq

**Adnan A. Lahuf[1], Ola H. Jaafar[1], Muhassen Al-mosoy[1], Zainab L. Hameed[2], Junman Li[3]**
[1]Department of Plant protection, College of Agriculture, University of Kerbala, Kerbala, Iraq
[2]Department of Field crops, College of Agriculture, University of Kerbala, Kerbala, Iraq
[3] Institute of Virology and Biotechnology, Zhejiang Academy of Agricultural Sciences, Hangzhou, Zhejiang, China

*Corresponding author email: adnan.lahuf@yahoo.com

## Abstract

During growing seasons of 2015 and 2016, severe rot symptoms were noticed epidemiologically on root and crown of *Triticum aestivum* and *Aptenia cordifolia* seedlings in Kerbala Province, Iraq. Thus, the aims of this research were to isolate, identify and assess pathogenicity of the causal agent of the disease. Several fungal isolates of *Fusarium* genus were consistently isolated from the diseased root and crown of the seedlings. The pathogen was identified as *Fusarium equiseti* based on its cultural and morphological characteristics, pathogenicity and sequence of the ITS-rDNA region. This is first record of crown rot disease caused by *F. equiseti* on *T. aestivum* and *A. cordifolia* plants in Iraq.

**Keywords**: Crown rot disease, *Fusarium equiseti*, Morphological and molecular identification, Iraq

## Introduction

Wheat (*Triticum aestivum*), which belongs to Poaceae family, is considered one of the most economically essential food for approximately more than 1/3 of the global population (Bockus et al., 2010). It has been ranked as the fourth most important food source in the world, with global production more than 700 million tons (FAO, 2015). Wheat has multiple beneficial characteristics and properties such as its agronomic adaptation, simplicity of grain storage and easy processing into flour used to produce numerous edible and healthy food products (Orth and Shellenberger, 1988). It is also one of the main diet components due to its high content of carbohydrate, easily digested proteins, a variety of different lipids, fibers, minerals and vitamins (Khalil and Shrewry, 2009). Additionally, the low quality of wheat grains are a good source for animal fodder and produce some industrial products such as alcohol and paper additives (Zarrin et al., 2009; Bockus et al., 2010).

Heartleaf ice plant (*Aptenia cordifolia*) is from the family Aizoaceae. It is also called baby sun rose and red aptenia that is native to South Africa (Court, 2000). It is a perennial evergreen succulent plant with thick roots and rounded or angled stems, which grow in horizontal clusters on the soil (Sajeva and Costanzo, 2000). The leaves are glossy green in heart or oval shape-like structures arranging singly or in pairs. The flowers are red to magenta, emerging singularly or in clusters on the leaf axils and open throughout the day. They are self-fertilized leading to produce fruits in a capsule shape with multi-chambers containing brown to black seeds (Rowley, 1980; Smith and Vanwyk, 2008). It is widely used as an anti-inflammatory and an ornamental plant cultivated in different regions of Iraq (Vanwyk et al., 1997; Smith, 2005).

Crown rot disease is one of the common diseases that appears on different plant hosts causing a significant reduction in yield and quality (Lazreg et al., 2013; Gebremariam et al., 2017). Diseased seedlings of the host plants show light brown necrotic lesions covering the roots and extending to crown regions, which latterly develop to be darker and larger. Subsequently, the diseased seedlings collapse and die (Paulitz et al., 2002; Moya-Elizondoand Jacobsen, 2016). Although this disease has not been found to be recorded on heartleaf ice plants (*A. cordifolia*) worldwide, it was reported on wheat plants (*T. aestivum*) in different countries including the USA (Gonzalez and Trevathan, 2000), Norway (Kosiak et al., 2003), Iran (Hajieghrari, 2009), Tunisia ( Fakhfakh et al., 2011), Korea (Kim et al., 2016), Turkey (Gebremariam et al., 2017). Additionally, in Iraq, this disease was reported on wheat caused by *Fusarium avenaceum*, *F. solani*, *Rhizoctonia solani* and *Stemphylium* sp. (Salih et al., 2009) and *F. culmorum* (Khalifah and Matny, 2013). In fact, diverse pathogenic species of *Fusarium* genus including *F. equiseti* were reported to cause this disease (Rossi et al., 1995; Gonzalez and Trevathan, 2000; Gebremariam et al., 2017). However, so far in Iraq *F. equiseti* has not been identified neither on wheat *T. aestivum* nor on heartleaf ice *A. cordifolia* plants. Therefore, the purpose of this work is to investigate possibility of association the species *F. equiseti* with the crown rot disease on wheat and heartleaf plants in Iraq.

## Material and Methods

### Samples collection and pathogen isolation

To isolate the pathogen, seedling samples of *T. aestivum* and *A. cordifolia* showing symptoms of the crown rot disease were collected during a survey undertaken from December 2015 to April 2016 from several fields and plantations located in different regions of Kerbala Province/Iraq. The symptoms appeared on seedlings were brownish to black necrotic lesions accompanied with softness covering the roots and the basal stems. Later, the seedlings wilted, collapsed, rotted and died. The collected seedlings were transported to plant disease laboratory in Agriculture college/University of Kerbala and stored in fridge at 4 °C until further processing(Gebremariam et al., 2017; Hwang et al.,2017).

Symptomatic tissues from roots and basal stems of diseased seedlings were cut into 1-2 cm long pieces after washed gently with tap water and surface-sterilized in sodium hypochlorite (NaOCl) 1 % (w/v). These symptomatic tissues were then dried using sterilized filter paper and placed on water agar (WA) plates, followed by incubation at 25°C in darkness. Hyphal tip from each emerging colony was cut and transferred into potato dextrose agar (PDA) plates containing amoxicillin (200 µg/ml). Subsequently, all plates were incubated at 25 °C under darkness for 5-6 days (Gebremariam et al., 2017; Maridueña-Zavala et al., 2016; Hwang et al., 2017).

### Morphological identification

To determine the morphological characteristics including mycelial structure, colour and growth ratio of all pure fungal cultures obtained were monitored and measured after sub-culturing them on synthetic low-nutrient agar (SNA).The conidia and chlamydospores of the pure isolates were also observed after 7 and 14 days using a compound microscope (HumaScopePremiumLED, Germany) that was equipped with an ocular micrometer measurement. Subsequently, shapes and measurements of the fungal conidia and chlamydospores were compared with former descriptions (Burgess et al., 1994; Leslie and Summerell, 2006).

### Pathogenicity

To assess pathogenicity of the *Fusarium* sp. isolates, healthy seedlings at two-true leaf stage of *T. aestivum* and *A. cordifolia* grown on water agar media were transplanted into autoclaved and inoculated compost with the isolates of *Fusarium* sp. The inoculated compost was made by mixing spore suspension of a pure 7 days isolate culture in one plate with 0.5 kg of autoclaved commercial compost. The mixture was incubated for three days at 25 °C in darkness with shaking every day. The mixture was then distributed equally into surface-sterilized plastic pots. In each full pot, three seedlings of each plant were transplanted. In addition, healthy plants of each host were planted in sterile compost as a control. Subsequently, all pots were placed in a growth cabinet under conditions 16 h lightat 25°C and 8 h dark at 15°C with 70% humidity (Mitter et al., 2006). Watering of seedlings was performed as needed and monitored daily until symptoms of crown rot disease were observed. Re-isolation of *Fusarium* sp. was made from symptomatic seedlings of both plants as previously described. The morphological characteristics of the re-isolated *Fusarium* sp. isolates were compared with the corresponding *Fusarium* isolates to accomplish the requirement of Koch`s postulates. No *Fusarium* sp. infection was seen in all control plants of both hosts.

**Molecular identification**

The total genomic DNA from pure *Fusarium* isolates was extracted using the DNeasy Plant Mini kit (Qiagen, Germany) following the manufacturer`s instructions. Polymerase chain reaction (PCR) test was applied using the universal primer set (ITS1/ITS4) to amplify the internal transcribed spacer (ITS) region of fungi (White et al., 1990). The PCR amplicons were sequenced at Macrogen in Seoul, South Korea (ABI 3730xI automated Sequencer; Applied Biosystems). Analysis of chromatograms and nucleotide sequences of sequencing resulting were conducted using Chromas software version2.6.4. Then, Basic Local Alignment Search Tool (BLAST) was operated for comparison the obtained sequences with sequences database of the GenBank at the National Center for Biotechnology Information National (NCBI). Another comparison was operated with database of the Fusarium MLST website for same purpose. Subsequently, the Molecular Evolutionary Genetics Analysis (MEGA) version6.06 was applied for phylogenetic analysis sequence data and operating the Neighbour joining method of the ITS-rDNA *Fusarium* sp. sequences of this study and others recorded in the NCBI to construct a phylogenetic tree (Tamura et al., 2011). Edited sequences of Fusarium isolates were deposited to NCBI-GenBank Database for recording.

## Results and Discussion

**Isolation and morphological identification**

A fungus was consistently isolated from almost all symptomatic seedlings of *T. aestivum* and *A. cordifolia*. The colony growth of this fungus on SNA was rapid, reached more than 5 cm diameter after 6-7 days. Initially, the appearance of colonies was white aerial mycelia that later became brown and bright orange representing the causal agent – sporodochia (Figure 1 A-B). The reverse surface of the colony was bronze to brown, with dark brown spots. After 2 weeks incubation at 25 °C in darkness, an enormous collection of colorless conidia and chlamydospores with various sizes and shapes were observed and measured. Although the microconidia were not seen, the macroconidia were distinctly curved and falcate with thick-wall and the majority had 3-5 septate. Their apical cells were acute and showed an extended enhanced curving of the macroconidia while the basal cells were distinctly foot shape. Conidial sizes were extremely variable with average length 20-70 μm and width 2.5-6.5 μm (Figure 1 C-D). Additionally, the chlamydospores were spherical, 6.8–9 μm in diameter, either as chains or clusters with very obvious thick walls. The conidiophores were noticed to be either unbranched or loosely branched mono to tri-phialides. Depending on the morphological appearances of the fungal isolates, they were initially identified as *Fusarium equiseti* (Corda) Sacc. (Leslie and Summerell, 2006; Burgess et al., 1994).

**Molecular identification**

The PCR amplicons size of the ITS-rDNA region amplified were more than 600 bp in both isolates. The sequences were recorded at NCBI-GenBank with accession numbers MF099868.1 of *T. aestivum* and MF099783.1 of *A. cordifolia*. BLAST analysis of ITS-rDNA sequences of the two isolates showed that the first sequence shared 99% similarity to the known sequences of *F. equiseti* deposited in GeneBank database with accession numbers KX343174.1, KR094440.1 and KY318493.1. However, the second sequence demonstrated 100% identity to those sequences of published *F. equiseti* isolates with accession numbers KY560313.1 and EF483926.1. The phylogenetic analysis revealed that the two isolates respectively grouped with several reference isolates of the fungus *F. equiseti* (Figures 2 and 3). Thus, based on these results the fungus was confirmed to be *F. equiseti*.

**Pathogenicity**

The crown rot disease incidence of the inoculated *T. aestivum* and *A. cordifolia* seedlings was approximately 100% whereas the non-inoculated plants of both hosts did not reveal any of the disease symptoms. After 3 weeks, the seedlings showed symptoms similar to those noticed on diseased seedlings in the infested fields and plantations. These symptoms included wilting, brown lesions of the basal stems and roots that eventually progressed to rotting and collapsing of the seedlings. It was also observed that the height of the inoculated seedlings was notably reduced compared with the control seedlings (Figure 1E-F).The causal agent, *F. equiseti*, was re-isolated from all inoculated seedlings of both hosts and its microscopic features were similar to the original isolates. Consequently, Koch`s postulates were achieved successfully and revealed that *F. equiseti* was the causative agent of the crown rot disease on both plant hosts.

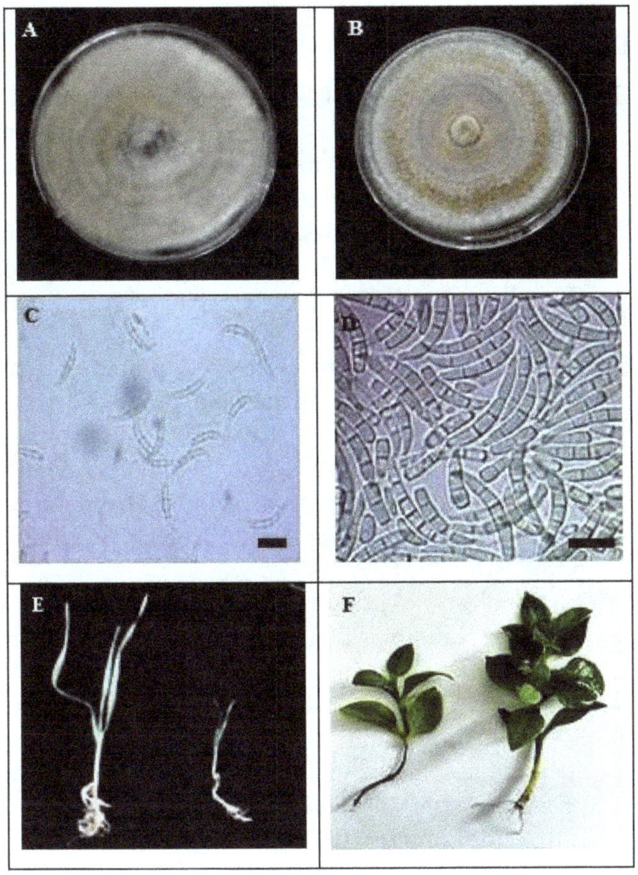


**Figure 1**: (Colour online) Cultural, morphological characteristics of *F.equiseti* and crown rot symptoms on *T. aestivum* and *A. cordifolia* seedlings. Colony of *F. equiseti* grown on PDA after 7 days (A) and 14 days (B). Macroconidia after 7 days (C) and 14 days (D). Crown rot symptoms appear on inoculated *T. aestivum* (E) the left plant is control while the right plant is diseased, on inoculated *A. cordifolia* plant (F) the left plant is diseased whereas the right is control. Note, in diseased plants roots and base stems (crown) are brown and rotted.
Bars: C–D = 20 µm.

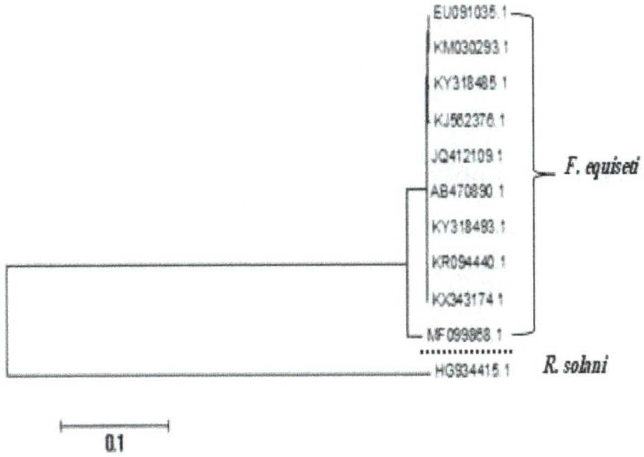


**Figure 2:** Phylogenetic tree created with the ITS-rDNA sequences of the *Fusarium equiseti* isolated from *T. aestivum* seedlings in this study (determined by dashed line) and other isolates of the same species retrieved from GenBank. Bootstrap values were based on 1000 replicates. The out-group species used was *Rhizoctonia solani*.

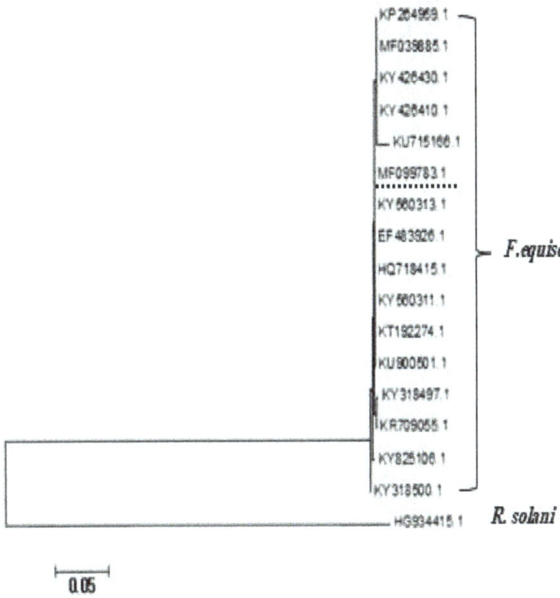


**Figure 3:** Phylogenetic tree constructed using the ITS-rDNA sequences of the *F. equiseti* obtained from the diseased *A. cordifolia* seedlings in this study (determined by dashed line) and other isolates of the same species retrieved from GenBank. Bootstrap values were depended on 1000 replicates. *R. solani* was the out-group species.

The current study has shown that *F. equiseti* is the causative agent of the crown rot disease on *T. aestivum* and *A. cordifolia* seedlings in Kerbala Province, Iraq. Identification of this pathogen was based on its morphological and molecular characteristics in addition to pathogenicity ability. The morphological features of the macroconidia, chlamydospores and conidiogenesis of both *F. equiseti* isolates were compatible with previous studies descriptions (Burgess et al., 1994; Leslie and Summerell, 2006; Li et al., 2017). It was observed clearly that the macroconidia were in diverse sizes and distinctively curved with more than 3 septate. Their apical cells were fusiform whereas the basal cells were in foot shape like-structure. Furthermore, the chlamydospores were globule shaped carried either in chains or clusters. However, the microconidia were not observed. Additionally, amplification of the ITS region yielded amplicons exceeding 600bp for both isolates and the phylogenetic analysis based on the ITS sequences demonstrated that both representative isolates grouped within a clade that included several known isolates of *F. equiseti* (Figures 2 and 3).
In the pathogenicity test, the *F. equiseti* was able to re-infect of *T. aestivum* and *A. cordifolia* seedlings producing the typical symptoms of crown rot that were similar to those observed in the infested fields and plantations previously.

Numerous studies were reported *F. equiseti* infecting *T. aestivum* and other plants of Poaceae family such as *Hordeum vulgare*, *Avena sativa* (Goswami et al., 2008) and *Zea mays* (Li et al., 2014). Furthermore, this pathogen was found to infect numerous members of Fabaceae family such as *Pisum sativum*, *Phaseolus vulgaris*, *Medicago sativa* and *Cicer arietinum* (Goswami et al., 2008). As well as, it infects *Vigna unguiculata* (Rodrigues and Menezes, 2005). Additionally, several other studies were revealed that *F. equiseti* infects other host plants belonging to different families for instance *Pinus halepensis* (Lazreg et al., 2013), *Cuminum cyminum* (Ramch and raand Bhatt, 2012), *Carya illinoinensis* (Lazarotto et al., 2014) and *Brassica oleracea* var. *botrytis* (Li et al., 2017). Although, *F. equiseti* is a plant pathogen prevalent worldwide infecting a wide range of plant hosts, it has not been reported in Iraq infecting *T. aestivum* and *A. cordifolia* plants. To the best of our knowledge, *F. equiseti* causing crown rot on *T. aestivum* and *A. cordifolia* plants is first report in Iraq.

## Acknowledgment

High appreciation is expressed to Dr. Louise Gamble for proofreading of the manuscript.

## References

Bockus WW, Bowden RL, Hunger RM, Morrill WL, Murray TD and Smiley RW, 2010. Compendium of wheat diseases and pests, 3rd ed. The American Phytopathological Society Press, Saint Paul, USA.

Burgess LW, Sumerell BA, Bullock S, Gott KP and Backhouse D, 1994. Laboratory manual for Fusarium Research. Sydney University Press, Sydney, Australia.

Court D, 2000. Succulent flora of southern Africa. Balkema, Rotterdam, Netherland.

Fakhfakh MM, Yahyaoui A, Rezgui S, Ilias EM and Daaloul A, 2011. Identification and pathogenicity assessment of *Fusarium* spp. sampled from durum wheat fields in Tunisia. Afr. J. Biotech. 10: 6529–6539.

FAO (Food and Agriculture Organization) of the United Nations, 2015. FAOSTAT statistics database. FAO, Rome, Italy.

Gebremariam SE, Sharma-Poudyal D and Paulitz TC, 2017. Identity and pathogenicity of Fusarium species associated with crown rot on wheat (*Triticum* spp.) in Turkey. Eur. J. Plant Pathol. (https://doi.org/10.1007/s10658-017-1285-7).

Gonzalez MS and Trevathan LE, 2000. Identity and pathogenicity of fungi associated with root and crown rot of soft red winter wheat grown on the upper coastal plain land resource area of Mississippi. J. Phytopathol. 148: 77–85.

Goswami RS, Dong Y and Punja ZK, 2008. Host range and mycotoxin production by *Fusarium equiseti* isolates originating from ginseng fields. Can. J. Plant Pathol. 30:155–160.

Hajieghrari B, 2009. Wheat crown and root rotting fungi in Moghan area, northwest of Iran. African J. Biotech. 8:6214–6219.

Hwang S, Strelkov SE, Ahmed HU, Zhou Q, Fu H, Fredua R and Turnbull GD, 2017. First report of *Verticillium dahlia* Kleb causing wilt symptoms in canola (*Brassica napus* L.) in North America. Can. J. Plant Pathol. 39: 514-526.

Khalifah MH and Matny ON, 2013. Pathogenicity evaluation of *Fusarium* spp. isolates, causal agent of crown rot disease in wheat. Iraqi J. Agric. Sci. 44: 480-489.

Khalil K and Shrewry PR, 2009. Wheat, chemistry and technology, 4th ed. Elsevier Inc, London, UK.

Kim DW, Kim GY, Kim HK, Kim J, Jeon SJ, Lee CW, Lee HB and Yun SH, 2016. Characterization of nivalenol-producing *Fusarium culmorum* isolates obtained from the air at a rice paddy field in Korea. Plant Pathol. J. 32: 182–189.

Kosiak B, Torp M, Skjerve E and Thrane U, 2003. The prevalence and distribution of *Fusarium* species in Norwegian cereals: a survey. Acta Agric. Scand. B. 53: 168-176.

Lazarotto M, Muniz MFB, Santos RFD, Blume E, Harakawa R and Hamann FA, 2014. First report of *Fusarium equiseti* associated on pecan (*Carya illinoinensis*) seeds in Brazil. Plant Dis. 98:847–848.

Lazreg F, Belabid L, Sanchez J, Gallego E, Garrido-Cardenas JA and Elhaitoum A, 2013. First report of *Fusarium equiseti* causing damping off disease on Aleppo pine in Algeria. Plant Dis. (dx.dio.org/10.1094/PDIS-02-13-0194-PDN).

Leslie JF and Summerell BA, 2006. The Fusarium laboratory manual. Blackwell Publishing, Ames, USA.

Li P, Shi Y, Guo M, Xie X, Chai A and Li B, 2017. Fusarium wilt of cauliflower caused by *Fusarium equiseti* in China. Can. J. Plant Pathol. 39: 77–82.

Li PP, Cao ZY, Wang K, Zhai H, Jia H, Liu N, Li SH, Hao ZM, Gu SQ and Dong JG, 2014. First report of *Fusarium equiseti* causing a sheath rot of corn in China. Plant Dis. 98:998.

Maridueña-Zavala MG, Villavicencio-Vásquez ME, Cevallos-Cevallos JM and Peralta EL, 2016. Molecular and morphological characterization of Moniliophthoraroreri isolates from cacao in Ecuador. Can. J. Plant Pathol. (http://dx.doi.org/10.1080/07060661.2016.1261372).

Mitter V, Zhang MC, Liu CJ, Ghosh R, Ghosh M and Chakraborty S, 2006. A high-throughput glasshouse bioassay to detect crown rot resistance in wheat germplasm. Plant Pathol. 55:433–441.

Moya-Elizondo EA and Jacobsen BJ, 2016. Integrated management of Fusarium crown rot of wheat using fungicide seed treatment, cultivar resistance, and induction of systemic acquired resistance (SAR). Biol. Cont. J. 92: 153–163.

Orth RA and Shellenberger JA, 1988. Origin, production, and utilization of wheat, pp. 173-198. In Y. Pomeranz (ed.), Wheat: chemistry and technology, vol. I. American Association, Saint Paul, USA.

Paulitz TC, Smiley RW and Cook RJ, 2002. Insights into the prevalence and management of soil borne cereal pathogens under direct seeding in the Pacific Northwest, USA. Can. J. Plant Pathol. 24:416–428.

Ramchandra S and Bhatt PN, 2012. First report of *Fusarium equiseti* causing vascular wilt of cumin in India. Plant Dis. 96:1821.

Rodrigues AAC, Menezes M, 2005. Identification and pathogenic characterization of endophytic Fusarium species from cowpea seeds. Mycopathologia. 159:79-85.

Rossi V, Cervi C, Chiusa G and Languasco L, 1995. Fungi associated with foot rots on winter wheat in northwest Italy. J. Phytopathol. 143:115–119.

Rowley GD, 1980. Name that succulent. Stanley Thornes, Cheltenham, UK.

Sajeva M and Costanzo M, 2000. Succulents, vol. II: the new illustrated dictionary. Timber Press, Oregon, USA.

Salih MM, Aboud HM and Kaream MK, 2009. Detection of wheat seedlings damping off and root rot diseases pathogens in Wasit and Anbar provinces of Iraq. Iraqi J. Agric. 14: 23-32.

Smith G, 2005. Gardening with succulents. Struik Press, Cape Town, South Africa.

Smith GF and Van Wyk BE, 2008. Guide to succulents. Briza Publications, Pretoria, South Africa.

Tamura K, Stecher G, Peterson D, Filipski A and Kumar S, 2013. MEGA6: Molecular Evolutionary Genetics Analysis, version 6th ed. Mole. Biol. Evol. 30: 2725–2729.

Vanwyk BE, Van Oudtshoorn B and Gericke N, 1997. Medicinal plants of South Africa. Briza Publications, Pretoria, South Africa.

White TJ, Bruns T, Lee S and Taylor J, 1990. Amplification and direct sequencing of fungal ribosomal RNA genes for phylogenetics, pp. 315–322. In M. A. Gelfand, D. H. Sninsky and T. J. White (eds.), PCR Protocols: A guide to methods and applications. Academic Press, San Diego, USA.

Zarrin F, Saleemi M, Zia M, Sultan T, Aslam M, Rehman R and Chandhary MF, 2009. Antifungal activity of plant growth promoting Rhizobactera isolates against *Rhizoctonia solani* in wheat. Afr. J. Biotech. 8: 219-225.

# 20

# Effectiveness of multi-strain biofertilizer in combination with organic sources for improving the productivity of Chickpea in Drought Ecology

**Maqshoof Ahamd[1], Azhar Hussain*[1], Muhammad Fakhar-U-Zaman Akhtar[1], Muhammad Zafar-Ul-Hye[2], Zafar Iqbal[1], Tayyaba Naz[3], Muhammad Mazhar Iqbal[4]**

[1]Department of Soil Science, University College of Agriculture and Environmental Sciences, The Islamia University of Bahawalpur, Bahawalpur, Pakistan

[2]Department of Soil Science, College of Agriculture, Bahauddin Zakariya University Multan

[3]Institute of Soil and Environmental Sciences, University of Agriculture, Faisalabad, Pakistan

[4]Soil and Water Testing Laboratory for Research, Chiniot, Department of Agriculture, Government of Punjab-Pakistan.

## Abstract

The present study was conducted in Bahawalpur, which is characterized by extremely dry and harsh desert conditions with very little rainfall and water availability. Due to less availability of water, the production of food crops is unpredictable that is not enough to feed the local population. To cope with the shortage of cereal grains and water, it is need of the hour to arrange some alternate food sources that can be produced with limited water. Chickpea can be successfully grown under water limited conditions. Biofertilizers have the ability to improve the growth and nodulation of chickpea enables it to withstand the periods of drought. The study was involved evaluation of multi-strain biofertilizer (developed from novel strains of *Mesorhizobium cicri*, *Pseudomonas* sp. and *Bacillus* sp.) effectiveness in combination with organic sources for improving the productivity of chickpea in Cholistan desert area of Bahawalpur. Four field trials were conducted in different villages of Bahawalpur. Biofertilizer along with enriched compost and farmyard manure were applied under field conditions with three replications. The combined application of farm yard manure and biofertilizer improved nodules formation, plant growth, yield and chemical parameters as compared to control. It can be concluded that the use of biofertilizer in combination with farm yard manure is significantly effective in improving the productivity and profitability of chickpea. So, the farmers of Cholistan area should be recommended to adopt the biofertilizer technology along with organic manures that will not only enhance the grain yield but also rejuvenate the soil health.

*Corresponding author email: azharhaseen@gmail.com

**Keywords:** Biofertilizer, ACC-deaminase activity, Water scarcity, Food security

## Introduction

The Cholistan is a part of the world's seventh largest desert, stretched along the south border of Punjab province. The Cholistan has a fertile soil, but the non-availability of irrigation water turned it into a desert. Moreover, frequent droughts resulting in a severe shortage of food in the Cholistan that is the main cause of hundreds of deaths in this area due to hunger. Pakistan is facing a famine-like situation in its arid climate districts i.e. Tharparkar and (Cholistan)

Bahawalpur. Due to the shortage of wheat grains and water scarcity, there is a need to find out some alternate ways and means to feed the undernourished people of the area.

Drought is a polygenic stress and is considered as one of the most important factors limiting crop production in desert areas. It adversely affects plant growth and yields that result in a shortage of food supply and the onset of famines in desert areas (Arshad et al., 2008). It has been reported that even a short-term water stress can cause significant losses in crop production (Pinheiro et al., 2005). However, the deleterious effects of water shortage on crop yield may be more obvious at some particular growth stage depending upon the nature of crop and species (Arshad et al., 2008).

Ethylene is the most important growth hormone, produced by almost all the plants under stress, which mediates a wide range of plant responses (Arshad and Frankenberger, 2002). The production of ethylene may be inhibitory or stimulatory depending upon its concentration, the nature of the physiological process and the growth phase of the plant. When it is present in high levels, it can be damaging to plants, leading to epinasty, shorter root, and premature senescence. The onset of drought may result in the production of the excess amount of ethylene, which causes inhibition of seed germination, root elongation and nodulation in legumes.

Chickpea has the potential to grow successfully under water limited conditions. It contributes more than 20% of the world pulse production (FAO, 2016). The growth of chickpea can be improved by using biofertilizer containing bacteria, which have the ability to improve plant growth through reducing the harmful effects of stresses. Under natural conditions, plants make symbiotic relationships with soil inhibiting plant growth promoting rhizobacteria (PGPR), rhizobia and fungi (Srinivasan et al., 1996). These play important role in helping the plants to withstand unfavorable conditions, like salinity (Ahmad et al., 2013; Joe et al., 2016), drought (Hussain et al., 2016), diseases (Nadeem et al., 2014), etc. In modern agriculture, these microorganisms are being used as biofertilizers i.e. live formulations of PRPR (O'Callaghan et al., 2001). The PGPR under drought conditions increases plant drought tolerance through different mechanisms. These mechanisms include the improvement in moisture availability to roots, enhanced nutrient availability, supplying phytohormones, and reduction in ethylene production (Zaheer et al., 2016). Different plant growth promoting compounds are supplied by bacteria through roots which alleviate the effect of stress on plants (Ahmad et al., 2011). They also facilitate the growth of their host plant by fixing atmospheric nitrogen, and siderophores production (which may solubilize and sequester iron thereby increasing its availability for plant uptake), producing phytohormones, and solubilizing minerals, such as phosphate, thus increasing their availability (Ahmad et al., 2011).

Keeping in view the entire scenario, it is imperative to find a technology that can easily be adapted by the resource poor farmers of the drought areas like the Cholistan. In our earlier studies at the University of Agriculture Faisalabad, we have developed a multi-strain biofertilizer through a series of experiments under laboratory, green house, and field conditions. The biofertilizer developed from novel strains of *Mesorhizobium cicri* and PGPR (*Bacillus* and *Pseudomonas*) have the ability to improve the growth and nodulation of chickpea, thus enabling it to withstand the periods of drought. These fertilizers can be more effective for chickpea production in a desert climate. Using this biofertilizer is a practical, environment-friendly and sustainable approach which could consequently result in a better yield of chickpea. These biofertilizers can help in alleviating stress in plants grown in the arid and semi-arid environment (Ahmad et al., 2013). The present paper reports the results of experiments conducted on farmer's fields in Cholistan area of Bahawalpur to evaluate the use of biofertilizer for chickpea production under limited water resources.

## Materials and Methods

### Preparation of biofertilizer

A multi-strain biofertilizer containing novel strains of *Mesorhizobium cicri*, *Pseudomonas* sp. and *Bacillus* sp. was prepared at the University of Agriculture Faisalabad. The already installed biofertilizer production unit was used to prepare the bio-fertilizer, which was used to inoculate the seed for field trials.

### Sowing of crop in the field

Four sites at different villages in Cholistan area were selected and sowing of chickpea crop was conducted in November 2014. The treatments comprised of control, farmyard manure, P-enriched compost, bio fertilizer, farmyard manure + biofertilizer, and P-enriched compost + bio fertilizer. The experiments were conducted in a Randomized Complete Block design (RCBD) with three replications. Recommended dose of P and K @ 60 kg ha$^{-1}$ each plot while half of the recommended dose of nitrogen (20 kg ha$^{-1}$) was applied as basal dose and seeds of chickpea were inoculated with bio-fertilizer. Sowing was done by hand drill and row to row distance was maintained 30 cm, while plant to plant distance 5-7 cm. All agronomic practices were adopted during the growth period of the crop.

**Data recording and plant analysis**

Nodulation data were recorded at flowering by carefully uprooting the plants with enough surrounding soil, keeping in view that minimal damage to roots occurs. Five plants per experimental unit were uprooted along with rhizospheric soil, washed thoroughly under running water on a sieve of mesh size that doesn't allow nodules to pass through but the soil. Nodules were counted and fresh as well as dry weight was recorded. At harvesting, plant agronomic data were recorded and samples were collected for analyses. Dried ground plant material was digested according to the method of Wolf (1982) and N, P, and K concentration was determined as described by Ryan et al. (2001).

Physiological parameters i.e. chlorophyll-a, and chlorophyll-b contents in leaves were recorded at physiological maturity by extracting leaf material (0.05 g) in 10 mL dimethyl sulfoxide (Hiscox and Israelstam, 1979). The samples were heated at 65°C for 4 h and then the absorbance of the extract was recorded at 663 and 645 nm as described by Arnon (1949). Data were subjected to statistical analysis by using LSD technique at 5% level of significance (Steel et al., 1997).

## Results

Results of the experiments conducted at four locations in district Bahawalpur are summarized in this section. It has been observed that separate application of bio-fertilizer, farm yard manure, and compost efficiently improved the growth and yield in chickpea in all trials. However, the integrated use of biofertilizer with farm yard manure and enriched compost was more effective in improving all the parameters as compared to the individual use of these sources. It is obvious in *Figure 1*, where clear difference among treatments in the improvement of root growth of chickpea can be noticed.

The data given in table 1 depicts that combined use of biofertilizer with farmyard manure and enriched compost significantly increased the number of nodules. Among sole applications, the biofertilizer resulted in a maximum increase in a number of nodules followed by farmyard manure and enriched compost. The combined application of biofertilizer with farmyard manure gave 3 fold more nodules than control plots from an experiment at 138DB while the combination biofertilizer with enriched compost resulted in 227% increase in a number of nodules at experimental site Jageer bhati dari. The sole application of biofertilizer also gave better results as compared to the sole application of farmyard manure and enriched compost. The combination of biofertilizer with farm yard manure gave 195% more nodules fresh weight when compared with control from the experiment at Baghdad research farm.

The results regarding plant height (Table 2) showed that the combined use of biofertilizer with farmyard manure and enriched compost significantly increased plant height. However, there was no significant difference among the sole application of farmyard manure and enriched compost. The maximum increase in plant height occurred with combined application of enriched compost and biofertilizer followed by farmyard manure and biofertilizer. The results regarding shoot dry weight (Table 2) showed that the use of organic nutrient sources and biofertilizer, individually and in combination, significantly enhanced shoot dry weight of chickpea. The sole application of farmyard manure, enriched compost and biofertilizer significantly increased shoot dry weight in chickpea, though the combined application gave better results. The maximum increase in shoot dry weight was observed in the treatment where biofertilizer was integrated with farmyard manure.

The results given in Table-3 showed that the number of pods plant$^{-1}$ increased significantly by the application of biofertilizer, farmyard manure, and enriched compost. The sole application of biofertilizer showed significantly better results than sole application of farmyard manure and enriched compost and. While, combined application of biofertilizer with farmyard manure and enriched compost showed a maximum increase in a number of pods plant$^{-1}$, which is significantly higher than individual application of bio fertilizer, farmyard manure, and enriched compost. The results in Table-3 also showed that biofertilizer with farmyard manure and enriched compost significantly increased the number of grains $pod^{-1}$ of chickpea. A significant increase in grains $pod^{-1}$ was recorded with combined application of farmyard manure and biofertilizer followed by the combined use of enriched compost and bio fertilizer.

The data showed that the use of biofertilizer with farmyard manure and enriched compost significantly increased 1000 grains weight in chickpea (Table 4). The maximum increase in 1000 grains weight was recorded with combined application of farmyard manure and biofertilizer followed by enriched compost + biofertilizer and sole application of biofertilizer. The grain yield was also significantly increased with separate as well as combined application of biofertilizer and organic sources. At all experimental sites, maximum increase in grain yield was observed where biofertilizer and farmyard manure were applied in combination (Table 4).

The application of organic nutrient sources and biofertilizer, individually and in combination, significantly improved the chlorophyll-a and chlorophyll-b contents in leaves of chickpea (Table 5).The combined application of biofertilizer along with farmyard manure and enriched compost showed maximum increase in chlorophyll-a and chlorophyll-b contents in leaves of chickpea. The chemical analysis of grains showed that nitrogen concentration in grains significantly increased by the application of biofertilizer in combination with farmyard manure and enriched compost. The results showed that sole application of farmyard manure, biofertilizer, and enriched compost showed non-significant difference regarding nitrogen concentration in grains (Table 6). However, combined application of biofertilizer with farmyard manure and enriched compost resulted in significant increase in nitrogen concentration in grains of chickpea.

The data regarding phosphorus concentration (Table 6) showed that the integrated use of organic nutrients and biofertilizer significantly increased phosphorus concentration in grains of chickpea. Sole application of farmyard manure, enriched compost, and biofertilizer significantly increased phosphorus concentration in grains as well. It was observed that increase in phosphorus in grains was significantly higher when biofertilizer was applied solely, as compared to the sole application of farmyard manure and enriched compost. Maximum increase was recorded when biofertilizer was applied in combination with farmyard manure. The results showed that there was the non-significant effect of farmyard manure, enriched compost and biofertilizer on potassium uptake (Table 7). No significant increase in potassium concentration in grains of chickpea was observed among all the treatments at all the experimental sites.

**Table – 1: Effect of farmyard manure, enriched compost and biofertilizer on number of nodules $plant^{-1}$ and nodules fresh weight of chickpea**

| Treatment | No. of nodules $plant^{-1}$ | | | | Nodules fresh weight (g $plant^{-1}$) | | | |
|---|---|---|---|---|---|---|---|---|
| | **107 DB** | **Jageer bhati dari** | **138DB** | **Baghdad** | **107 DB** | **Jageer bhati dari** | **138DB** | **Baghdad** |
| **Control** | 31 e | 22 f | 19 f | 25 f | 0.65 e | 0.42 f | 0.41 f | 0.55 e |
| **FYM** | 49 d | 36 d | 32 d | 41 d | 0.94 c | 0.81 d | 0.78 d | 0.85 c |
| **Enriched compost (EC)** | 37 e | 27 e | 25 e | 31 e | 0.78 d | 0.62 e | 0.54 e | 0.67 d |
| **Biofertilizer (B)** | 55 c | 55 c | 51 c | 59 c | 1.27 b | 0.98 c | 0.91 c | 1.16 b |
| **FYM + B** | 76 a | 82 a | 79 a | 86 a | 1.77 a | 1.65 a | 1.59 a | 1.62 a |
| **EC + B** | 62 b | 72 b | 61 b | 77 b | 1.53 b | 1.26 b | 1.12 b | 1.41 b |
| **LSD** ($p < 0.05$) | **5.6139** | **3.5722** | **3.3724** | **5.1245** | **0.1448** | **0.0770** | **0.0976** | **0.1342** |

Means sharing same letters are statistically show insignificant results at par at 5 % level of probability (n = 3)

**Table – 2: Effect of farmyard manure, enriched compost and biofertilizer on plant height and shoot dry weight of chickpea**

| Treatment | Plant height (cm) | | | | Shoot dry weight (g) | | | |
|---|---|---|---|---|---|---|---|---|
| | 107 DB | Jageer bhati dari | 138DB | Baghdad | 107 DB | Jageer bhati dari | 138DB | Baghdad |
| Control | 45.58 c | 50.29 d | 38.98 d | 49.39 c | 12.30 e | 14.43 e | 11.03 d | 13.86 f |
| FYM | 49.85 b | 54.54 c | 44.54 c | 54.03 b | 16.16 c | 17.26 cd | 14.01 c | 16.99 d |
| Enriched compost (EC) | 48.88 b | 53.73 c | 43.14 c | 53.58 b | 15.02 b | 15.94 e | 12.97 c | 15.94 e |
| Biofertilizer (B) | 54.99 a | 58.69 b | 47.70 b | 57.78 a | 18.43 b | 18.19 c | 16.49 b | 18.19 c |
| FYM + B | 56.34 a | 61.34 a | 49.60 a | 59.50 a | 20.07 a | 20.26 a | 19.38 a | 20.26 a |
| EC + B | 56.99 a | 60.83 a | 50.53 a | 59.28 a | 18.52 b | 19.68 b | 18.14 a | 19.68 b |
| LSD ($p < 0.05$) | **1.8632** | **1.9390** | **1.6158** | **1.9182** | **0.7964** | **0.5786** | **1.4025** | **0.5786** |

Means sharing same letters are statistically show insignificant results at par at 5 % level of probability (n = 3)

**Table – 3: Effect of farmyard manure (FYM), enriched compost and biofertilizer on number of pods plant$^{-1}$ and number of grains pod$^{-1}$ of chickpea**

| Treatment | Number of pods plant$^{-1}$ | | | | Number of grains pod$^{-1}$ | | | |
|---|---|---|---|---|---|---|---|---|
| | 107 DB | Jageer bhati dari | 138DB | Baghdad | 107 DB | Jageer bhati dari | 138DB | Baghdad |
| Control | 29.69 e | 33.42 d | 31.71 e | 33.79 d | 1.42 c | 1.46 d | 1.16 d | 1.39 d |
| FYM | 35.58 c | 36.05 c | 37.16 c | 38.53 c | 1.47 b | 1.52 c | 1.27 bc | 1.48 c |
| Enriched compost (EC) | 33.69 d | 34.71 cd | 35.80 d | 37.94 c | 1.50 b | 1.49 cd | 1.24 c | 1.45 c |
| Biofertilizer (B) | 39.97 b | 41.25 b | 39.14 b | 43.98 b | 1.56 a | 1.56 d | 1.29 ab | 1.53 b |
| FYM + B | 44.12 a | 43.93 a | 42.54 a | 49.86 a | 1.59 a | 1.61 a | 1.32 a | 1.61 a |
| EC + B | 40.68 b | 42.96 a | 39.56 b | 45.05 b | 1.59 a | 1.60 ab | 1.25 c | 1.58 a |
| LSD ($p < 0.05$) | **1.3005** | **1.4836** | **0.83 56** | **1.3532** | **0.0354** | **0.0345** | **0.0348** | **0.0380** |

Means sharing same letters are statistically show insignificant results at par at 5 % level of probability (n = 3)

**Table – 4: Effect of farmyard manure (FYM), enriched compost and biofertilizer on 1000 grains weight and grain yield of chickpea**

| Treatment | 1000 grains weight (g) | | | | Grain yield (kg acre$^{-1}$) | | | |
|---|---|---|---|---|---|---|---|---|
| | 107 DB | Jageer bhati dari | 138DB | Baghdad | 107 DB | Jageer bhati dari | 138DB | Baghdad |
| Control | 255.00 e | 256.33 d | 230.00 d | 274.67 e | 881.70 d | 821.7 e | 848.3 c | 715.0 d |
| FYM | 259.00 d | 258.33 cd | 235.33 c | 278.00 d | 967.70 b-d | 1011.0 bc | 997.0 b | 1011.0 a-c |
| Enriched compost (EC) | 261.33 c | 260.67 bc | 237.67 b | 280.33 c | 930.30 cd | 911.3 b | 827.7 c | 897.0 c |
| Biofertilizer (B) | 263.67 b | 262.67 ab | 240.33 a | 283.67 b | 1047.0 a-c | 963.7 cd | 897.0 bc | 957.0 bc |
| FYM + B | 266.00 a | 265.00 a | 242.00 a | 285.67 ab | 1162.3 a | 1162.3 a | 1180.3 a | 1129.0 a |
| EC + B | 264.67 ab | 265.00 a | 241.33 a | 287.67 a | 1071.3 ab | 1022.7 b | 927.7 bc | 1046.0 ab |
| LSD ($p < 0.05$) | **2.1440** | **2.8379** | **1.6828** | **2.0204** | **118.35** | **58.968** | **128.51** | **138.41** |

Means sharing same letters are statistically show insignificant results at par at 5 % level of probability (n = 3)

**Table - 5: Effect of farmyard manure (FYM), enriched compost and biofertilizer on chlorophyll-a and chlorophyll-b contents in leaves of chickpea**

| Treatment | Chlorophyll-a (µg $g^{-1}$) | | | | Chlorophyll-b (µg $g^{-1}$) | | | |
|---|---|---|---|---|---|---|---|---|
| | 107 DB | Jageer bhati dari | 138DB | Baghdad | 107 DB | Jageer bhati dari | 138DB | Baghdad |
| Control | 1.326 c | 1.313 d | 1.306 c | 1.323 c | 0.710 c | 0.613 e | 0.676 c | 0.796 d |
| FYM | 1.373 b | 1.356 c | 1.343 b | 1.373 b | 0.756 a-c | 0.640 de | 0.713 b | 0.843 bc |
| Enriched compost (EC) | 1.406 b | 1.360 c | 1.320 bc | 1.370 b | 0.746 a-c | 0.656 cd | 0.696 a-c | 0.826 cd |
| Biofertilizer (B) | 1.466 a | 1.436 b | 1.390 a | 1.440 a | 0.786 ab | 0.680 bc | 0.753 a | 0.866 ab |
| FYM + B | 1.493 a | 1.500 a | 1.406 a | 1.446 a | 0.800 a | 0.723 a | 0.770 a | 0.890 a |
| EC + B | 1.496 a | 1.506 a | 1.390 a | 1.443 a | 0.730 bc | 0.706 ab | 0.763 a | 0.873 ab |
| LSD ($p < 0.05$) | **0.0354** | **0.0224** | **0.0289** | **0.0394** | **0.0671** | **0.0268** | **0.0248** | **0.0315** |

Means sharing same letters are statistically show insignificant results at par at 5 % level of probability (n = 3)

**Table - 6: Effect of farmyard manure (FYM), enriched compost and biofertilizer on nitrogen and phosphorus concentration in grains of chickpea**

| Treatment | Nitrogen (%) | | | | Phosphorus (%) | | | |
|---|---|---|---|---|---|---|---|---|
| | 107 DB | Jageer bhati dari | 138DB | Baghdad | 107 DB | Jageer bhati dari | 138DB | Baghdad |
| Control | 1.68 b | 1.68 c | 1.61 c | 1.48 c | 0.45 c | 0.38 c | 0.72 c | 0.59 c |
| FYM | 1.87 ab | 1.87 bc | 1.84 b | 1.77 bc | 0.50 bc | 0.40 bc | 0.77 bc | 0.61 bc |
| Enriched compost (EC) | 1.85 ab | 1.85 bc | 1.76 bc | 1.69 bc | 0.50 bc | 0.47 a | 0.77bc | 0.67 ab |
| Biofertilizer (B) | 1.99 ab | 1.99 b | 1.92 b | 1.86 b | 0.55 ab | 0.44 ab | 0.79 a-c | 0.69 a |
| FYM + B | 2.22 ab | 2.20 a | 2.15 a | 2.05 a | 0.51 bc | 0.49 a | 0.85 a | 0.71 a |
| EC + B | 2.11 a | 2.08 a | 1.97 ab | 1.84 b | 0.58 a | 0.48 a | 0.83 ab | 0.73 a |
| LSD ($p < 0.05$) | **0.4539** | **0.2473** | **0.2110** | **0.1911** | **0.0690** | **0.0586** | **0.0771** | **0.0721** |

Means sharing same letters are statistically show insignificant results at par at 5 % level of probability (n = 3)

**Table - 7: Effect of farmyard manure (FYM), enriched compost and biofertilizer on potassium concentration in grains and nodules dry weight of chickpea**

| Treatment | Potassium (%) | | | | Nodules dry weight (g $plant^{-1}$) | | | |
|---|---|---|---|---|---|---|---|---|
| | 107 DB | Jageer bhati dari | 138DB | Baghdad | 107 DB | Jageer bhati dari | 138DB | Baghdad |
| Control | 1.50 a | 1.53 a | 1.72 a | 1.78 b | 0.15 f | 0.12 f | 0.17 f | 0.11 f |
| FYM | 1.53 a | 1.56 a | 1.76 a | 1.80 ab | 0.19 d | 0.16 d | 0.20 d | 0.17 d |
| Enriched compost (EC) | 1.54 a | 1.57 a | 1.75 a | 1.82 ab | 0.16 e | 0.14 e | 0.18 e | 0.13 e |
| Biofertilizer (B) | 1.56 a | 1.61 a | 1.78 a | 1.84 a | 0.18 c | 0.21 c | 0.20 c | 0.19 c |
| FYM + B | 1.57 a | 1.63 a | 1.80 a | 1.86 a | 0.26 b | 0.24 a | 0.28 b | 0.23 a |
| EC + B | 1.59 a | 1.63 a | 1.79 a | 1.86 a | 0.23 a | 0.22 b | 0.25 a | 0.20 b |
| LSD ($p < 0.05$) | **0.0904** | **0.1069** | **0.0525** | **0.0912** | **0.0240** | **0.0393** | **0.0194** | **0.0263** |

Means sharing same letters are statistically show insignificant results at par at 5 % level of probability (n = 3)

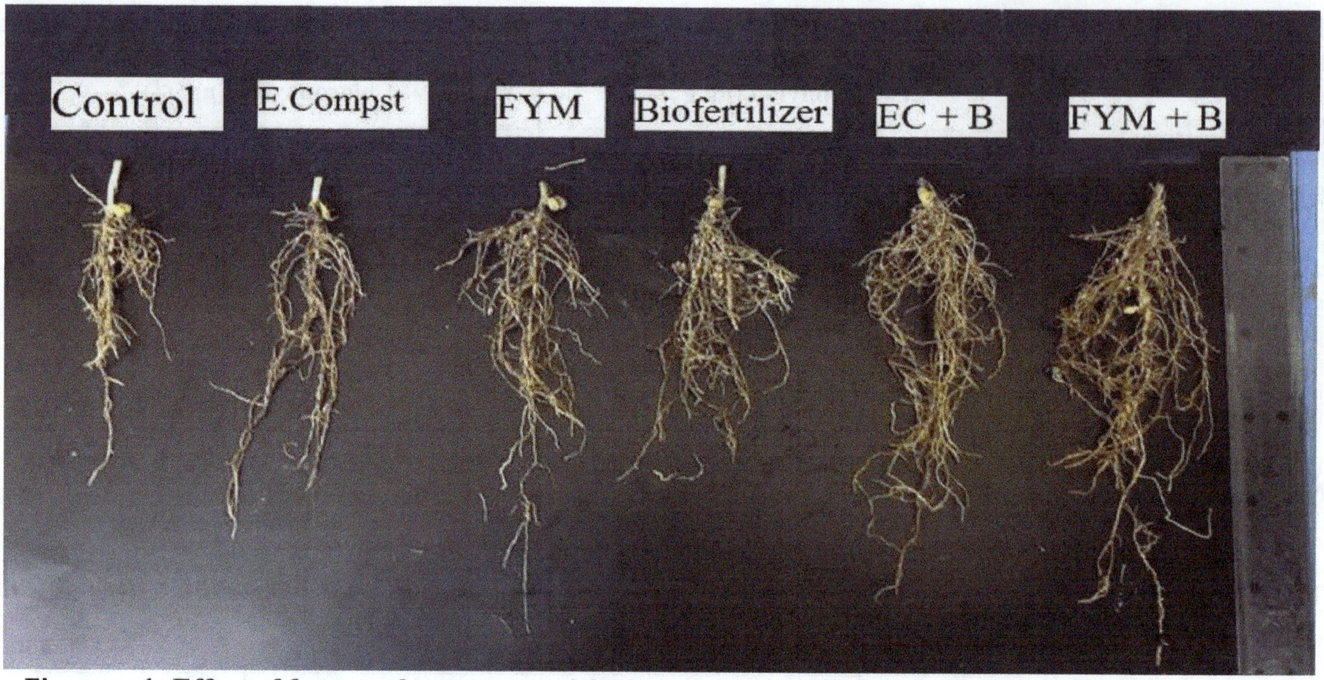


**Figure – 1: Effect of farmyard manure, enriched compost and biofertilizer on nodulation in chickpea E. Compost: P-enriched Compost, FYM: Farmyard Manure, B: Biofertilizer**

## Discussion

The results of our study showed that the inoculation with multi-strain biofertilizer containing *Mesorhizobium cicri*, *Pseudomonas* sp. and *Bacillus* sp. significantly improved the nodulation of chickpea at all experimental sites. It might be due to indirect stimulation in nodulation by rhizobacteria through increased root growth that provided more infection sites to rhizobia for nodulation. The improvement in nodule occupancy by *Bradyrhizobium* in soybean has also been reported by Mishra et al. (2009) due to co-inoculation with *P. fluorescens*. This kind of positive legume-Rhizobium association is well documented and recent reports confirmed its positive role on yield and nodulation of different legumes (Ahmad et al., 2013). The legume-Rhizobium association brought a number of changes in legumes but nodulation is the most important, as it has a central position in their developmental growth stages through $N_2$- fixation from the atmosphere.

Lowest nodulation in un-inoculated control elucidates decreased the activity of native rhizobia but the addition of organic nutrient sources (enriched compost and farmyard manure) and Rhizobium individually and in combination increased nodulation. It is advocated that the application of farmyard manure and enriched compost resulted in increased Rhizobium bioactivity in the rhizosphere and increased the colonization of bacteria that promoted the nodulation and also these organic sources enhanced the carbon and nutrient availability for the proliferation of rhizospheric Rhizobium (Zahran, 1999; Otieno et al., 2007). Organic amendments also boost nutrient availability (by solubilization and dissolution) ensuring plant uptake with enhancement of plant growth as well as nodulation (Bashan et al., 2013).

The results of the study under discussion showed that use of biofertilizer, enriched compost, and farmyard manure significantly increased yield attributes of chickpea. The findings were in line with the studies of Gharib et al. (2008) who reported that the use of biofertilizer and enriched compost enhanced crop yield significantly by improving the grain weight and their numbers. In addition to the positive attributes of organic sources, the application of biofertilizer enhanced nutrient uptake and growth leading to improved yield and grain nutritional quality. The increase in yield most likely due to the promotion of root growth by the decreased ethylene levels due to ACC-deaminase activity of rhizobacteria present in multi-strain biofertilizer (Ahmad et al., 2012) along with some other growth promoting attributes. The increase in growth and yield components of chickpea by combined application of biofertilizer and organic amendments may also be attributed to cumulative

effects, such as enhanced supply of nitrogen and phosphorus to the crop and other growth promoting substances produced by microorganisms present in biofertilizer (Ahmad et al., 2013). In addition to biocontrol activity against soil born fungal pathogens, PGPR also increased grain yield by suppressing the diseases (Vinale et al., 2008). So, the reduced activity of pathogens and disease incidences may also be attributed to increased grain yield to.
The increase in growth and yield of chickpea could also be related to nutrient supplementation by the bacteria present in multi-strain biofertilizer and the role of organic fertilizers in promoting the colonization of these microbes (Ahmad et al., 2014), which might have enhanced their efficiencies like $N_2$ fixation, P-solubilization, and effective pathogen suppression. Similarly, increasing nitrogen fixation due to phosphorus supplementation has been reported by Manjunath and Bagyaraj (1984).

The present study demonstrated the effectiveness of biofertilizer in combination with enriched compost and farmyard manure for improving growth, yield and nutritional quality (N, P, and K) of chickpea both individually and in combination at different experimental sites in Cholistan villages. The increase in grain quality could be attributed to the enhanced nutrient use efficiency in the presence of organic fertilizers and bio-inoculants. Studies have shown that the composted organic materials release nutrients slowly and may reduce the leaching losses, particularly of N (Muneshwar et al., 2001) and thus enhance nutrient use efficiency (Rudresh et al., 2005).This premise is further supported by the fact that total N and P uptakes in chickpea were significantly increased in response to combined application of biofertilizer and organic fertilizers as P-enriched compost and farmyard manure increase the solution P, and also phosphate solubilizing action of inoculants and nitrogen fixing ability of bacterial strains enhancing the nutrients uptake (Ahmad et al., 2014). All these factors resulted in an enhanced nutrient uptake leading to better quality grains (Nevens and Reheul, 2003). The increase in nutrient uptake may be due to an increase in available N and P contents in the soil, and improved soil structure due to increased organic matter that increased nutrients availability and hence enhanced growth and quality of the crop (Manna et al., 2001).

## Conclusion

In the present study, the application of biofertilizer, farmyard manure, and enriched compost significantly enhanced the nodulation, growth, yield, grain quality and physiology of chickpea under field conditions. It is concluded that the use of biofertilizer in combination with organic nutrient sources is significantly effective in improving the productivity and profitability of chickpea. The combined use of biofertilizer in combination with farm yard manure was the most effective treatment. So, the farmers of the desert and drought areas like the Cholistan should be recommended to adopt the biofertilizer technology along with organic manures to get a higher yield of chickpea.

## Acknowledgement

The authors acknowledge the support provided by the research team during the execution of the project activities. We also acknowledge the facilities/place provided by the Islamia University of Bahawalpur, and the University of Agriculture Faisalabad for project activities. Financial support was received from the research Competitive Grants Program, a joint initiative of the Planning Commission of Pakistan and the Pakistan Strategy Support Program, International Food Policy Research Institute (IFPRI), funded by USAID.

## References

Ahmad M, Zahir ZA, Jamil M, Nazli F, Latif M and Akhtar MF, 2014. Integrated use of plant growth promoting rhizobacteria, biogas slurry and chemical nitrogen for sustainable production of maize under salt-affected conditions. Pak. J. Bot. 46(1): 375-382.

Ahmad I, Akhtar MJ, Asghar HN and Khalid M, 2013. Influence of Rhizobium applied in combination with micro nutrients on mung bean. Pak. J. Life Social Sci. 11:53-59.

Ahmad M, Zahir ZA, Nadeem SM, Nazli F, Jamil M and Khalid M, 2013. Field evaluation of *Rhizobium* and *Pseudomonas* strains to improve growth, nodulation, and yield of mung bean under salt-affected conditions. Soil Environ. 32(2):

Ahmad M, Zahir ZA, Asghar HN and Arshad M, 2012. The combined application of rhizobial strains and plant growth promoting rhizobacteria improves growth and productivity of mung bean (*Vigna radiata* L.) under salt-stressed conditions. Ann. Microbiol. 62: 1321–1330

Ahmad M, Zahir ZA, Asghar HN and Asghar M, 2011. Inducing salt tolerance in mung bean through co-inoculation with *Rhizobium* and PGPR containing ACC-deaminase. Can. J. Microbiol. 57(7): 578-589

Arnon DI, 1949. Copper enzymes in isolated chloroplasts, polyphenoxidase in *Beta vulgaris*. Plant Physiol. 24: 1-15.

Arshad M and Frankenberger Jr. WT, 2002. Ethylene: Agriculture Sources and Applications. Kluwer Academic Publishers, New York, USA.

Arshad M, Shaharoona B and Mahmood T, 2008. Inoculation with *Pseudomonas* spp. containing ACC-deaminase partially eliminates the effects of drought stress on growth, yield, and ripening of pea (*Pisum sativum* L.). Pedosphere. 18: 611-620.

Bashan Y, Kamnev AA and de-Bashan LE, 2013. A proposal for isolating and testing phosphate-solubilizing bacteria that enhance plant growth. Biol. Fertil. Soils. 49: 1-2.

FAO, 2015. FAOSTAT. Food and Agriculture Organization of the United Nations. http://faostat.fao.org/site/291/default.aspx.

Gharib FA, Moussa LA, and Massoud ON, 2008. Effect of compost and bio-fertilizers on growth, yield and essential oil of sweet marjoram (*Majorana hortensis*) plant. Int. J. Agri. & Biol. 10: 381-387.

Hiscox JD and Israelstam GF, 1979. A method for the extraction of chlorophyll from leaf tissue without maceration. Can. J. Microbiol. 57: 1332-1334

Hussain MB, Zahir ZA, Asghar HN, Mubaraka R and Naveed M, 2016. Efficacy of rhizobia for improving photosynthesis, productivity, and mineral nutrition of maize. Clean Soil, Air, Water. 44(11): 1564-1571

Joe MM, Devaraj, S, Benson A and Sa T, 2016. Isolation of phosphate solubilizing endophytic bacteria from *Phyllanthus amarus* Schum & Thonn: Evaluation of plant growth promotion and antioxidant activity under salt stress. J. Appl. Res. Medicin. Arom. Plants. 3: 71-77.

Manjunath A and Bagyaraj DJ, 1984. The response of pigeon pea and cowpea to phosphorus and dual inoculation with vesicular arbuscular Mycorrhiza and Rhizobium. Tropical Agriculture. 61: 48-52.

Manna MC, Ghosh PK, Ghosh BN and Singh KN, 2001. Comparative effectiveness of phosphate-enriched compost and single super- phosphate on yield, uptake of nutrients and soil quality under soybean–wheat rotation. J. Agri. Sci. 137: 45–54.

Mishra PK, Mishra S, Selvakumar G, Kundu S and Shankar GH, 2009. Enhanced soybean (*Glycine max* L.) plant growth and nodulation by *Bradyrhizobium japonicum*-SB1 in presence of *Bacillus Thuringian is*. Acta. Agr. Scand. B-S. P. 59:189-196.

Singh M, Singh VP, and Reddy KS, 2001. Effect of integrated use of fertilizer nitrogen and farmyard manure or green manure on the transformation of N, K and S and productivity of rice-wheat system on a vertisol. J. Society Soil Sci. 49: 430-435

Nadeem SM, Ahmad M, Zahir Z A, Javaid A and Ashraf M, 2014. The role of mycorrhizae and plant growth promoting rhizobacteria (PGPR) in improving crop productivity under stressful environments. Biotechnol. Adv. 32: 429-448.

Nevens F and Reheul D, 2003. The application of vegetable, fruit, and garden waste (VFG) compost in addition to cattle slurry in a silage maize monoculture: nitrogen availability and use. Eur. J. Agron. 19: 189-203.

O'Callaghan KJ, Dixon RA and Cocking EC, 2001. *Arabidopsis thaliana*: a model for studies of colonization by non-pathogenic and plant-growth-promoting rhizobacteria. Funct. Plant Biol. 28: 975-982.

Otieno PE, Muthomi JW, Cheminingwa GN and Nderitu JH, 2007. Effect of rhizobia inoculation, farmyard manure and nitrogen fertilizer on growth, nodulation, and yield of selected food grain legumes. In African Crop Science Conference Proceedings. 8:305-312.

Rudresh DL, Shivaprakash MK and Prasad RD, 2005. Effect of combined application of Rhizobium, phosphate solubilizing bacterium and *Trichoderma* spp. on growth, nutrient uptake and yield of chickpea (*Cicer aritenium* L.). Appl. Soil Ecol. 28: 139-146

Ryan J, Estefan G and Rashid A, 2001. Soil and Plant Analysis Laboratory Manual, 2nd Ed. International Center for Agriculture in Dry Areas (ICARDA), Syria

Srinivasan M, Holl FB and Petersen DJ, 1996. Influence of indole acetic-acid-producing *Bacillus* isolates on the nodulation of *Phaseolus vulgaris* by *Rhizobiumetli* under gnotobiotic conditions. Can. J. Microbiol. 42(10): 1006-1014

Steel RGD, Torrie JH and Dickey DA, 1997. Principles and Procedures of Statistics: A Biometrical Approach. 3rd Ed. WCB/McGraw-Hill, Boston, Mass., USA

Vinale F, Sivasithamparam K, Ghisalberti EL, Marra R, Woo SL and Lorito M, 2008. Trichoderma-plant-pathogen interactions. Soil Biol. Biochem. 40: 1-10

Wolf B, 1982. The comprehensive system of leaf analysis and its use for diagnosing crop nutrient status. Commun. Soil Sci. Plan. 13:1035-1059

Zaheer A, Mirza BS, Mclean JE, Yasmin S, Shah TM, Malik KA and Mirza MS, 2016. Association of plant growth-promoting Serratia spp. with the root nodules of chickpea. Res. Microbiol. 167: 510-52.

# 21

# Comparative Toxicity of Insecticides against Two Important Insect Pests of Cauliflower Crop

**Muhammad Imran[1*], Kanwal Hanif[2], Munir Ahmad[2], Muhammad Nasir[2], Umer Ayyaz Aslam Sheikh[1]**
[1]Department of Entomology, Faculty of Agriculture, The University of Poonch, Rawalakot, AJ&K, Pakistan.
[2]Department of Entomology, Pir Mehr Ali Shah, Arid Agriculture University, Rawalpindi, Pakistan.

*Corresponding author email: muhammadimran@upr.edu.pk

## Abstract

Study was conducted to test four insecticides, profenofos, emamectin benzoate, λ-cyhalothrin and lufenuron against two field populations of *Plutella xylostella and Spodoptera litura* at different larval instars during 2010-11. The field populations collected from Rawalpindi and Taxila cauliflower fields was tested using leaf dip bioassay method under laboratory conditions. Results showed that emamectin benzoate was the most toxic insecticide for both *P. xylostella* and *S. litura*. However, profenofos was the second most toxic insecticides to different larval instars of *S. litura* and λ-cyhalothrin to *P. xylostella.* High $LC_{50}$ values for lufenuron for this limited time exposure might be due to its slow acting as chitin synthesis inhibitor. Emamectin benzoate can be suggested as the most effective insecticides against both field populations along with profenofos and λ-cyhalothrin.

**Keywords:** *Spodoptera litura, Plutella xylostella,* comparative toxicity, mode of action, age-dependent

## Introduction

Cruciferous family especially cabbages and cauliflowers are the most important winter vegetables grown extensively in temperate and tropical areas of the world (Liu et al., 2003). Two lepidopteran insects, Diamondback moth diamondback moth (DBM) (*Plutella xylostella* L.) and tobacco caterpillar (*Spodoptera litura* Fab.) are the most destructive insect pest of cruciferous plants mostly for cabbage and cauliflower over the world (Shankar et al., 1996; MalliKarjuna et al., 2004). They feed mainly on leaves resulting in reduced superiority and quantity of food. Its serious attack can cause 30-100% crop failure if no insecticide is used for its management (Verkerk and Wright, 1996). Frequently usage of insecticides for long period, they have developed resistance against many insecticides which make it prominent insect pest in one of the 20 resistant insect species (Shelton et al., 2000; Mota-Sanchez et al., 2002). Variation in detoxification enzymes activity among *S. litura* strains could be attributed due to insecticide usage pattern (Karuppaiah et al., 2017).

Selection of insecticide for the management of diamondback moth (DBM) and tobacco caterpillar required continuous testing under both field and laboratory condition, so it is important to select appropriate insecticide according to insect pest. Many insecticides having longer residual action on plants or insects like prothiophos, cartap and fenvalerate mixture are suitable for their management (Nakagome and Kato, 1981). Organophosphates have been considered as the most important group of insecticide due to variations in chemical structures (Liu et al., 2003). In Pakistan farmers relay only on insecticides to control insect pests and due to indiscriminate use of

insecticides against *S. litura* resistance problems occurred (Tong et. al., 2013; Basit et al., 2013).
Commonly used insecticides against *S. litura* showed that resistance ratios were high for endosulfan, cypermethrin, profenofos, chlorpyriphos, quinalphos, phoxim, triazophos, methomyl or thiodicarb and low to medium for deltamethrin and β-cyfluthrin (Mushtaq et al., 2008). One of the major reasons for the development of resistance to insecticides by DBM is the increase in number of sprays. Farmer still uses the broad spectrum pyrethroids, organophosphates, organochlorines and many other conventional insecticides against DBM (Kumar, 1995). Eco-friendly and less toxic new insecticides are also available in the market. (Vastrad et al., 2003) reported that thiodicarb, fipronil, lufenuron, spinosad, carbosulfan and indoxacarb are still performing well as compared to malathion. In Malaysia, high uses of abamectin in crucifer crops against diamondback moth have developed serious problem of resistance (Verkerk and Wright, 1996). To find out most effective insecticide and information about insecticide availability in future this study was planned by using conventional insecticides at different stages of diamondback moth and tobacco caterpillar.

## Materials and Methods

### Field collection

Two field populations of *S. litura* and *P. xylostella* were collected from Rawalpindi and Taxila cauliflower/cabbage fields during 2010-11. Larvae and eggs were collected and kept in plastic jars along with host plant leaves for their safe transport to laboratory and kept separately from previously tested for initial susceptibility levels.

### Laboratory rearing of DBM

The larvae were kept at 25±2°C, 65%RH and 16 hour photophase in plastic jars (1 kg) on cauliflower leaves. Pupae were collected daily and provided with 10% honey solution at adult stage. Two to three fresh leaves of cauliflower were provided on daily basis for eggs laying. Bioassays were performed after completion of one generation laboratory rearing to reduce field residual effect and to get sufficient number of larvae for bioassays.

### Laboratory rearing of tobacco caterpillar

Fields collected larvae were reared on semi-synthetic gram-based diet in the laboratory at 25±2°C and 60-65% relative humidity with a light (16 h): dark (8 h) photoperiod (Ahmad et al., 2007a). Small pieces of diet were put in six hole petri dishes and larvae were released on it, one larvae in each hole was placed under controlled laboratory condition. Diet was changed after 24 hour and pupae were collected on alternative days. Cells of petri dishes were cleaned for rearing of larvae till pupation. Mature pupae were collected and kept in separate plastic box lined with tissue paper. Emerged adults were kept in plastic jars and fed on a solution containing sucrose, vitamin solution in a soaked cotton wool ball (Ahmad et al., 2007b). Egg batches were collected daily and larval instars in next generation were used for bioassays.

### Bioassays

Four commercial insecticides, λ-cyhalothrin (Karate® 2.5EC; Syngenta (Pvt) Ltd, Pakistan), profenofos (Curacron® 50EC; Syngenta (Pvt) Ltd, Pakistan), emamectin benzoate (Proclaim® 1.9EC, Syngenta (Pvt) Ltd, Pakistan) and lufenuron (Match® 05EC, Syngenta (Pvt) Ltd, Pakistan) were used. Top Film® (Helb Pesticides (Pvt) Ltd, Pakistan) as a surfactant was used at 5 ppm for increased adhesiveness to leaf surface in preparation of insecticide solutions and also in control. Standard leaf disc bioassay method (Sayyed et al., 2008: Ahmad et al., 2007a) was used during bioassay. Leaves of cauliflower crop collected from unsprayed fields, washed with water, dried and immersed in a test solution for 10-15 second and then allowed to dry at room temperature for one hour. After drying, the leaf discs were placed in petri dishes containing moistened filter paper.

### Data analysis

Mortality rate of *S. litura* and *P. xylostella* were examined after 48 and 72 hours of insecticide exposure. Insects were considered as dead when they failed to show any movement with gentle touch with blunt needle. Abbott's formula was used to calculate the corrected mortality (Abbott, 1925) and analyzed by Probit analysis (Finney, 1971) using POLO-PC software (LeOra software, 1987). These values were compared from significance difference for these insecticides at particular age level and for different populations under study of *p. xylostella* and *S. litura* (Travis and Rick, 2000).

## Results and Discussion

Comparison of $LC_{50}$ values of four insecticides, profenofos, emamectin benzoate, λ-cyhalothrin and lufenuron against first instar larvae of *P. xylostella* population collected from Taxila revealed that emamectin benzoate to be the most effective insecticide with least $LC_{50}$(0.79 and 0.59) value after 48 and 72 hours, respectively follow by λ-cyhalothrin, profenofos and lufenuron. Similarly at 2nd and 3rd larval instars of Taxila population, lufenuron was most toxic insecticide with least $LC_{50}$ (0.73, 0.40 and 0.60, 0.48) respectively after 48 and 72 hours. At 3rd instars larvae of Taxila population profenofos was most effective with least $LC_{50}$ (0.82 and 0.63) respectively (Table 1).

The results at Rawalpindi populations showed that profenofos was most effective insecticides against 1st and 3rd instars larvae with least $LC_{50}$ (0.41, 0.36 and 0.74, 0.57 respectively) and lufenuron was less effective with high $LC_{50}$ values. Similarly at 2nd instars larvae emamectin benzoate was most effective (Table 1). Overall results revealed that among these four insecticides against *P. xylostella* profenofos was most effective with least CF1 and CF2 values (Table 1).

Comparison of $LC_{50}$ values of four insecticides against first instar larvae of *S. litura* population collected from Taxila revealed that emamectin benzoate was most effective against 1st (0.03 and 0.02), 2nd (0.23 and 0.16), 3rd (0.89 and 0.66), and 4th(2.38 and 1.83) instars larvae after 48hr and 72hr respectively with least $LC_{50}$. Similar result was found from Rawalpindi population that emamectin was most effect against 1st, 2nd, 3rd, and 4th (0.05 and 0.04, 0.13 and 0.09, 1.57 and 0.93, 2.01 and 1.79 after 48hr and 72hr respectively) instars larvae (Table 2). Overall results showed that emamectin benzoate was most effective on both field collected population of Rawalpindi and Taxila after 48hr and 72hr with least $LC_{50}$ and CF1 and CF2 values (Table 2).

Overall finding of these four insecticides on both tested insect *Plutella xylostella* and *Spodoptera litura* showed that emamectin benzoate was most effective with least CF3 values (Table 2). Comparison of these four insecticides revealed that lufenuron was most effective at Taxila population and profenofos at Rawalpindi population of *P. xylostella*, similarly emamectin benzoate was most effective at both population of *S. litura* collected from Taxila and Rawalpindi (Table 1, 2).

Susceptibility level of the tested populations was comparatively more than that used by as local reference strain (Rafiq, 2005). This might be the variation in the less use of these insecticides due to less incidence of *P. xylostella* for the last couple of years. The number of application may have decreased on cauliflower and cabbages due to less population pressure, and less exposure may have resulted in number of susceptible individuals in the field population. This could be the possible reason for the increase in susceptibility to these insecticides, and the variation in different larval instars with non-significant variation might be due to this susceptibility (Mazlan and Mumford, 2005; Ronald et al., 2000)

Variation in susceptibility to insecticides is considered as an important factor in different population of insect pests (Mohan and Gujar, 2003; Rafiq, 2005). In most of these resistance cases, insecticide to insecticides is compared to the $LC_{50}$ or mortality values of these with a reference population either for laboratory or field studies (Sexena et al., 1989; Vastrad et al., 2003). This helps to identify the crop areas with susceptible to resistant populations. Information regarding less resistance to the tested *P. xylostella* in field populations can be utilized for area-wide management of this important insect pest. It also foresee the importance of insecticide monitoring especially at the second instar larvae of *P. xylostella* collected from different ecological and cropping zones for establishment of management practices as per different requirements of the cruciferous growers.

Comparison to the reference strains (Ahmad, 2008) proved the field population tested to be more susceptible as compared to that of cotton areas where this pest is under intense selection with indiscriminate use of insecticides (Khan and Mehmood, 1999). The reason may be less use of insecticides and small land holding of vegetable growers in the collection areas of Rawalpindi and Taxila cropping fields. Increase in susceptibility might be due to less population pressure of this insect pest on different vegetable grown in this area that may have resulted in decreased selection pressure with non-significant susceptibility variation (Sayyed et al., 2000; Ahmad, 2009). However, there existed quite variable response for different instars tested for these insecticides which might be their distinct variability in size, feeding potential and exposure rate.

Information of susceptibility to emamectin benzoate and profenofos suggests their use for management of this important insect pest in vegetables especially cauliflower and cabbages. This can also be utilized for other crop growing areas especially where cash crop like cotton are under intensive cultivation and need to be relied on chemical control. However, other insecticides like cyhalothrin and thiamethoxam were proved to be ineffective when used for this pest. Although increased dose rate may be helpful but it will be uneconomical in field situations and become source of more environmental pollution. Lufenuron may be considered either alone or in combination with these insecticides and binary combination of these effective insecticides with this insect growth regulator may also be tested for extraction of further possibilities of more options for pest management.

**Table 1:** Toxicity of four different insecticides to four larval instars of field collected diamondback moth, *Plutella xylostella* L. under laboratory conditions using leaf dip bioassay method

| Insecticide | strain | Instars | Time | $LC_{50}$ (FL at 95%) | $LC_{90}$ (FL at 95%) | Slope ± SE | CF1 | CF2 | CF3 |
|---|---|---|---|---|---|---|---|---|---|
| Profenofos | TXL | 1st | 48 | 1.57 (0.79-2.26) | 4.86 (3.33-10.5) | 2.60±0.65 | 3.83 | 3.83 | 52.3 |
| | | | 72 | 1.06 (0.44-1.53) | 2.78 (1.98-5.38) | 3.06±0.87 | 2.92 | 2.94 | 53 |
| | | 2nd | 48 | 0.85 (0.48-1.22) | 3.25 (2.21-6.26) | 2.19±0.44 | 2.07 | 2.07 | 28.3 |
| | | | 72 | 0.79 (0.46-1.09) | 2.34 (1.68-4.05) | 2.73±0.57 | 2.19 | 2.19 | 39.5 |
| | | 3rd | 48 | 0.80 (0.41-1.28) | 7.62 (4.16-24.0) | 1.31±0.25 | 1.95 | 1.95 | 26.6 |
| | | | 72 | 0.63 (0.30-1.00) | 4.55 (2.72-11.3) | 1.49±0.29 | 1.75 | 1.75 | 31.5 |
| | | 4th | 48 | 1.05 (0.64-1.95) | 27.1 (9.42-394) | 0.91±0.22 | 2.56 | 2.56 | 35 |
| | | | 72 | 0.76 (0.33-1.31) | 11.3 (5.23-59.7) | 1.09±0.24 | 2.11 | 2.11 | 38 |
| | RWP | 1st | 48 | 0.41 (0.14-0.75) | 5.24 (2.72-19.9) | 1.16±0.26 | 1 | 1 | 13.6 |
| | | | 72 | 0.36 (0.14-0.62) | 2.33 (1.41-5.49) | 1.59±0.34 | 1 | 1 | 18 |
| | | 2nd | 48 | 1.03 (0.41-1.75) | 9.15 (4.57-51.2) | 1.35±0.34 | 2.51 | 2.51 | 34.3 |
| | | | 72 | 0.82 (0.32-1.36) | 6.63 (3.57-27.5) | 1.41±0.34 | 2.27 | 2.27 | 51.5 |
| | | 3rd | 48 | 0.74 (0.32-1.29) | 14.8 (6.21-108) | 0.99±0.22 | 1.80 | 1.80 | 24.6 |
| | | | 72 | 0.57 (0.18-1.05) | 8.37 (3.93-45.4) | 1.09±0.26 | 1.58 | 1.58 | 28.5 |
| | | 4th | 48 | 1.49 (0.85-2.65) | 29.8 (11.1-283) | 0.99±0.21 | 3.63 | 3.63 | 49.6 |
| | | | 72 | 0.94 (0.40-1.69) | 16.9 (6.98-135) | 1.02±0.23 | 2.61 | 2.61 | 47 |
| Emamectin Benzoate | TXL | 1st | 48 | 0.79 (0.31-1.32) | 6.23 (3.41-24.4) | 1.43±0.35 | 1.21 | 1.92 | 26.3 |
| | | | 72 | 0.59 (0.28-0.89) | 2.31 (1.56-4.63) | 2.19±0.49 | 1.11 | 1.64 | 29.5 |
| | | 2nd | 48 | 0.78 (0.43-1.15) | 3.89 (2.52-8.51) | 1.83±0.36 | 1.2 | 1.90 | 26 |

| | | | | | | | | | |
|---|---|---|---|---|---|---|---|---|---|
| | | | 72 | 0.55 (0.23-0.87) | 3.19 (1.99-7.79) | 1.68±0.37 | 1.03 | 1.53 | 27.5 |
| | | 3rd | 48 | 1.24 (0.62-2.20) | 23.0 (9.03-204) | 1.00±0.22 | 1.90 | 3.02 | 41.3 |
| | | | 72 | 0.88 (0.39-1.53) | 13.7 (6.09-84.2) | 1.07±0.24 | 1.66 | 2.44 | 44 |
| | | 4th | 48 | 1.63 (0.53-4.86) | 111 (19.4-113069) | 0.69±0.22 | 2.50 | 3.97 | 54.3 |
| | | | 72 | 1.11 (0.25-2.94) | 73.3 (14.3-52157) | 0.70±0.23 | 2.09 | 3.08 | 55.5 |
| | RWP | 1st | 48 | 0.89 (0.37-1.59) | 13.9 (6.06-89.9) | 1.07±0.24 | 1.37 | 2.17 | 29.6 |
| | | | 72 | 0.70 (0.27-1.26) | 7.82 (3.90-32.8) | 1.22±0.27 | 1.32 | 1.94 | 35 |
| | | 2nd | 48 | 0.65 (0.15-1.23) | 9.57 (4.19-113) | 1.09±0.31 | 1 | 1.58 | 21.6 |
| | | | 72 | 0.53 (0.12-0.99) | 6.67 (3.25-47.6) | 1.17±0.32 | 1 | 1.47 | 26.5 |
| | | 3rd | 48 | 0.94 (0.37-1.73) | 15.1 (6.40-114) | 1.06±0.25 | 1.44 | 2.29 | 31.3 |
| | | | 72 | 0.67 (0.24-1.21) | 8.20 (4.03-37.8) | 1.18±0.27 | 1.26 | 1.86 | 33.5 |
| | | 4th | 48 | 2.90 (1.46-9.65) | 145 (27.1-31505) | 0.76±0.21 | 4.46 | 7.07 | 96.6 |
| | | | 72 | 1.59 (0.63-4.19) | 98.3 (19.1-29894) | 0.72±0.21 | 3 | 4.41 | 79.5 |
| λ-cyhalothrin | TXL | 1st | 48 | 1.26 (0.71-1.89) | 7.08 (4.21-19.7) | 1.71±0.35 | 1.93 | 3.07 | 42 |
| | | | 72 | 0.81 (0.33-1.28) | 3.56 (2.27-84.6) | 1.99±0.48 | 1.45 | 2.25 | 40.5 |
| | | 2nd | 48 | 0.88 (0.46-1.35) | 4.51 (2.84-10.7) | 1.82±0.38 | 1.35 | 2.14 | 29.3 |
| | | | 72 | 0.78 (0.36-1.15) | 2.48 (1.71-4.77) | 2.54±0.61 | 1.39 | 2.16 | 39 |
| | | 3rd | 48 | 1.29 (0.55-2.39) | 19.4 (7.96-162) | 1.08±0.26 | 1.98 | 3.14 | 26 |
| | | | 72 | 1.00 (0.39-1.79) | 11.2 (5.36-56.3) | 1.22±0.28 | 1.78 | 2.78 | 50 |
| | | 4th | 48 | 1.19 (0.52-2.31) | 34.8 (11.0-738) | 0.87±0.22 | 1.83 | 2.90 | 39.6 |
| | | | 72 | 0.92 (0.35-1.74) | 17.6 (6.96-170) | 1.00±0.24 | 1.64 | 2.55 | 46 |
| | RWP | 1st | 48 | 0.65 (0.31-1.05) | 6.36 (3.51-19.6) | 1.29±0.25 | 1 | 1.58 | 21.6 |
| | | | 72 | 0.56 (0.25-0.90) | 3.40 (2.10-7.80) | 1.63±0.33 | 1 | 1.55 | 28 |
| | | 2nd | 48 | 1.63 (0.82-2.62) | 10.5 (5.63-46.4) | 1.59±0.38 | 2.51 | 3.97 | 54.3 |
| | | | 72 | 0.94 (0.31-1.68) | 12.1 (5.28-136) | 1.15±0.32 | 1.68 | 2.61 | 47 |
| | | 3rd | 48 | 1.08 (0.43-2.14) | 28.8 (9.58-548) | 0.89±0.23 | 1.66 | 2.63 | 36 |
| | | | 72 | 0.78 (0.30-1.40) | 11.4 (5.19-66.6) | 1.10±0.25 | 1.39 | 2.16 | 39 |

| | | | | | | | | | |
|---|---|---|---|---|---|---|---|---|---|
| | | 4th | 48 | 1.91 (0.73-6.27) | 152 (23.8-233875) | 0.67±0.21 | 2.94 | 4.65 | 63.6 |
| | | | 72 | 0.91 (0.27-1.98) | 57.7 (13.1-9231) | 0.71±0.21 | 1.62 | 2.53 | 45.5 |
| Lufenuron | TXL | 1st | 48 | 1.63 (0.87-2.40) | 6.00 (3.93-14.2) | 2.26±0.53 | 2.71 | 3.97 | 54.3 |
| | | | 72 | 1.34 (0.58-1.96) | 3.82 (2.65-7.88) | 2.81±0.76 | 3.35 | 3.72 | 67 |
| | | 2nd | 48 | 0.73 (0.24-1.28) | 5.33 (2.93-20.9) | 1.49±0.38 | 1.22 | 1.78 | 24.3 |
| | | | 72 | 0.40 (0.08-0.76) | 3.53 (1.94-14.6) | 1.36±0.38 | 1 | 1.11 | 20 |
| | | 3rd | 48 | 0.60 (0.20-1.13) | 19.5 (6.95-297) | 0.85±0.21 | 1 | 1.46 | 20 |
| | | | 72 | 0.48 (0.16-0.87) | 9.15 (4.17-54.6) | 1.00±0.23 | 1.2 | 1.33 | 24 |
| | | 4th | 48 | 1.34 (0.59-2.64) | 30.1 (10.3-475) | 0.95±0.23 | 2.23 | 3.26 | 44.6 |
| | | | 72 | 0.99 (0.38-1.81) | 12.6 (5.73-74.6) | 1.16±0.27 | 2.47 | 2.75 | 49.5 |
| | RWP | 1st | 48 | 0.68 (0.34-1.09) | 7.04 (3.82-22.3) | 1.27±0.24 | 1.13 | 1.66 | 22.6 |
| | | | 72 | 0.54 (0.29-0.83) | 2.97 (1.92-6.09) | 1.74±0.32 | 1.35 | 1.5 | 27 |
| | | 2nd | 48 | 0.91 (0.32-1.59) | 10.8 (4.92-92.7) | 1.19±0.32 | 1.51 | 2.22 | 30.3 |
| | | | 72 | 0.76 (0.26-1.28) | 7.99 (3.95-48.9) | 1.24±0.32 | 1.19 | 2.11 | 38 |
| | | 3rd | 48 | 1.35 (0.55-2.81) | 37.5 (11.5-1003) | 0.88±0.23 | 2.25 | 2.92 | 45 |
| | | | 72 | 0.76 (0.23-1.50) | 15.4 (6.02-167) | 0.98±0.25 | 1.19 | 2.11 | 38 |
| | | 4th | 48 | 2.92 (1.23-9.69) | 107 (22.1-26131) | 0.82±0.24 | 4.86 | 7.12 | 97.3 |
| | | | 72 | 1.77 (0.64-4.81) | 82.8 (17.5-20368) | 0.77±0.23 | 4.43 | 4.92 | 88.5 |

CF1, compared with least value of each insecticide separately for each test insect
CF2, compared with least value of all insecticides for each insect separately
CF3, compared with least value of all insecticides of both test insects
TXL= Taxila
RWP= Rawalpindi

**Table 2:** Toxicity of four different insecticides to four larval instars of field collected, *Spodoptera litura* L. under laboratory conditions using leaf dip bioassay method

| Insecticide | strain | Instars | Time | $LC_{50}$ (FL at 95%) | $LC_{90}$ (FL at 95%) | Slope ± SE | CF1 | CF2 | CF3 |
|---|---|---|---|---|---|---|---|---|---|
| Profenofos | TXL | 1st | 48 | 1.81 (0.63-1.79) | 28.3 (11.8-142) | 0.91±0.15 | 2.82 | 60.3 | 60.3 |
| | | | 72 | 0.57 (0.31-0.89) | 7.32 (4.15-18.5) | 1.15±0.18 | 1.63 | 28.5 | 28.5 |
| | | 2nd | 48 | 2.21 (1.38-3.27) | 32.1 (18.4-74.8) | 1.10±0.14 | 3.45 | 73.6 | 73.6 |
| | | | 72 | 1.51 (0.84-2.34) | 16.0 (9.91-32.4) | 1.25±0.17 | 4.31 | 75.5 | 75.5 |
| | | 3rd | 48 | 2.57 (1.51-3.84) | 22.5 (14.1-45.1) | 1.36±0.19 | 4.01 | 85.6 | 85.6 |
| | | | 72 | 2.02 (1.04-3.15) | 15.9 (10.2-31.0) | 1.43±0.22 | 5.77 | 101 | 101 |
| | | 4th | 48 | 16.2 (8.84-27.0) | 386 (167-1829) | 0.93±0.16 | 23.3 | 540 | 540 |
| | | | 72 | 12.1 (6.19-20.1) | 185 (95.9-587) | 1.08±0.18 | 34.6 | 605 | 605 |
| | RWP | 1st | 48 | 0.64 (0.38-0.95) | 6.14 (3.80-12.7) | 1.30±0.18 | 1 | 21.3 | 21.3 |
| | | | 72 | 0.35 (0.19-0.54) | 4.61 (2.74-10.0) | 1.14±0.16 | 1 | 17.5 | 17.5 |
| | | 2nd | 48 | 1.98 (1.15-3.10) | 42.1 (21.8-120) | 0.91±0.12 | 3.09 | 66 | 66 |
| | | | 72 | 0.57 (0.30-0.89) | 7.34 (4.59-14.5) | 1.16±0.16 | 1.63 | 19 | 19 |
| | | 3rd | 48 | 1.39 (0.86-2.04) | 16.9 (10.5-34.6) | 1.18±0.15 | 2.17 | 46.3 | 46.3 |
| | | | 72 | 0.62 (0.31-0.98) | 7.68 (4.77-15.3) | 1.17±0.17 | 1.77 | 31 | 31 |
| | | 4th | 48 | 10.5 (5.72-17.2) | 263 (123-928) | 0.92±0.14 | 16.4 | 350 | 350 |
| | | | 72 | 7.23 (3.81-11.7) | 120 (65.7-308.4) | 1.05±0.15 | 20.6 | 361 | 361 |
| Emamectin Benzoate | TXL | 1st | 48 | 0.03 (0.02-0.04) | 0.37 (0.22-0.81) | 1.20±0.15 | 1 | 1 | 1 |
| | | | 72 | 0.02 (0.01-0.04) | 0.23 (0.15-0.45) | 1.38±0.19 | 1 | 1 | 1 |
| | | 2nd | 48 | 0.23 (0.14-0.36) | 2.95 (1.62-8.07) | 1.17±0.18 | 7.66 | 7.66 | 7.66 |
| | | | 72 | 0.16 (0.08-0.24) | 1.59 (0.95-3.59) | 1.28±0.19 | 8 | 8 | 8 |
| | | 3rd | 48 | 0.89 (0.58-1.28) | 9.92 (5.75-23.5) | 1.22±0.16 | 29.6 | 29.6 | 29.6 |
| | | | 72 | 0.66 (0.39-0.97) | 4.69 (3.06-8.97) | 1.51±0.22 | 33 | 33 | 33 |
| | | 4th | 48 | 2.38 (1.50-3.52) | 30.5 (17.9-68.3) | 1.15±0.14 | 79.3 | 79.3 | 79.3 |
| | | | 72 | 1.83 (1.00-2.85) | 18.9 (11.6-38.9) | 1.27±0.18 | 91.5 | 91.5 | 91.5 |

| | | | | | | | | | |
|---|---|---|---|---|---|---|---|---|---|
| | RWP | 1st | 48 | 0.05 (0.02-0.06) | 0.61 (0.34-1.49) | 1.08±0.19 | 1.66 | 1.66 | 1.66 |
| | | | 72 | 0.04 (0.01-0.05) | 0.35 (0.21-0.77) | 1.38±0.23 | 2 | 2 | 2 |
| | | 2nd | 48 | 0.13 (0.07-0.20) | 2.22 (1.21-5.69) | 1.05±0.15 | 4.33 | 4.33 | 4.33 |
| | | | 72 | 0.09 (0.04-0.15) | 0.09 (0.78-3.29) | 1.09±0.16 | 4.5 | 4.5 | 4.5 |
| | | 3rd | 48 | 1.57 (0.86-2.49) | 31.7 (16.9-85.3) | 0.98±0.14 | 52.3 | 52.3 | 52.3 |
| | | | 72 | 0.93 (0.50-1.45) | 13.4 (8.00-28.7) | 1.10±0.15 | 46.5 | 46.5 | 46.5 |
| | | 4th | 48 | 2.01 (1.09-3.25) | 34.6 (18.6-91.5) | 1.04±0.15 | 67 | 67 | 67 |
| | | | 72 | 1.79 (0.92-2.87) | 18.0 (11.1-37.6) | 1.28±0.19 | 89.5 | 89.5 | 89.5 |
| λ-cyhalothrin | TXL | 1st | 48 | 5.49 (2.24-9.20) | 102 (49.7-488) | 1.01±0.21 | 1 | 183 | 183 |
| | | | 72 | 5.06 (2.60-7.87) | 46.2 (27.7-109) | 1.34±0.23 | 1.23 | 253 | 253 |
| | | 2nd | 48 | 10.8 (6.77-16.4) | 203 (106-560) | 1.00±0.13 | 1.97 | 360 | 360 |
| | | | 72 | 7.78 (4.09-12.5) | 113 (63.4-277) | 1.10±0.16 | 1.89 | 389 | 389 |
| | | 3rd | 48 | 66.8 (42.4-100) | 948 (499-2745) | 1.11±0.16 | 12.2 | 2226 | 2226 |
| | | | 72 | 46.7 (26.7-71.8) | 539 (308-1326) | 1.21±0.18 | 11.4 | 2335 | 2335 |
| | | 4th | 48 | 71.6 (41.1-113) | 985 (508-3089) | 1.12±0.18 | 13 | 2386 | 2386 |
| | | | 72 | 51.0 (27.0-80.9) | 528 (303-1309) | 1.26±0.21 | 12.4 | 2550 | 2550 |
| | RWP | 1st | 48 | 7.08 (2.87-12.5) | 147 (64.9-856.4) | 0.97±0.21 | 1.29 | 236 | 236 |
| | | | 72 | 4.11 (1.63-6.88) | 37.9 (22.4-98.8) | 1.33±0.27 | 1 | 205 | 205 |
| | | 2nd | 48 | 7.46 (3.98-11.8) | 87.3 (51.8-191) | 1.20±0.17 | 1.36 | 248 | 248 |
| | | | 72 | 7.58 (3.71-12.3) | 7.58 (42.1-140) | 1.32±0.21 | 1.84 | 379 | 379 |
| | | 3rd | 48 | 67.0 (43.5-113) | 1190 (572-477) | 1.02±0.17 | 12.2 | 2233 | 2233 |
| | | | 72 | 39.8 (17.5-68.4) | 500 (277-1327) | 1.16±0.20 | 9.68 | 1990 | 1990 |
| | | 4th | 48 | 69.8 (33.3-130) | 2192 (850-12895) | 0.86±0.15 | 12.7 | 2326 | 2326 |
| | | | 72 | 19.7 (8.74-34.7) | 405 (211-1146) | 0.98±0.15 | 4.79 | 985 | 985 |
| Lufenuron | TXL | 1st | 48 | 2.35 (1.35-3.71) | 31.1 (16.1-97.2) | 1.14±0.18 | 1.59 | 78.3 | 78.3 |
| | | | 72 | 1.68 (0.87-2.70) | 17.7 (10.0-45.4) | 1.25±0.21 | 1.69 | 84 | 84 |
| | | 2nd | 48 | 4.84 (2.95-7.25) | 56.8 (33.8-124) | 1.19±0.15 | 3.27 | 162 | 162 |

| | | | | | | | | | |
|---|---|---|---|---|---|---|---|---|---|
| | | | 72 | 4.17 (2.29-6.42) | 36.5 (22.9-72.5) | 1.36±0.20 | 4.21 | 208 | 208 |
| | | 3rd | 48 | 27.9 (15.9-43.9) | 453 (230-1454) | 1.05±0.16 | 18.9 | 930 | 930 |
| | | | 72 | 20.6 (11.2-32.2) | 214 (126-497) | 1.26±0.19 | 20.8 | 1030 | 1030 |
| | | 4th | 48 | 157 (86.5-248) | 1633 (900-4631) | 1.26±0.21 | 106 | 5233 | 5233 |
| | | | 72 | 94.1 (45.7-14) | 640 (407-1316) | 1.53±0.28 | 95.1 | 4705 | 4705 |
| | RWP | 1st | 48 | 1.48 (0.88-2.26) | 20.6 (11.5-51.3) | 1.12±0.15 | 1 | 49.3 | 49.3 |
| | | | 72 | 0.99 (0.49-1.58) | 8.98 (5.59-18.4) | 1.34±0.22 | 1 | 49.5 | 49.5 |
| | | 2nd | 48 | 10.7 (5.54-17.7) | 199 (101-611) | 1.01±0.16 | 7.22 | 356 | 356 |
| | | | 72 | 7.67 (3.79-12.3) | 66.2 (41.1-134) | 1.37±0.22 | 7.75 | 383 | 383 |
| | | 3rd | 48 | 22.4 (12.5-35.9) | 354 (189-963) | 1.07±0.15 | 15.1 | 746 | 746 |
| | | | 72 | 15.5 (8.48-24.5) | 201 (116-460) | 1.15±0.16 | 15.7 | 775 | 775 |
| | | 4th | 48 | 41.3 (18.9-69.5) | 435 (262-918) | 1.25±0.19 | 27.9 | 1376 | 1376 |
| | | | 72 | 56.4 (27.1-96.8) | 1231 (618-3793) | 0.95±0.14 | 56.9 | 2820 | 2820 |

CF1, compared with least value of each insecticide separately for each test insect
CF2, compared with least value of all insecticides for each insect separately
CF3, compared with least value of all insecticides of both test insects
TXL= Taxila
RWP= Rawalpindi


## Acknowledgment

We are thankful to the cauliflower growing farmers for their help in collection of *P xylostella* field populations and department to provide facilities to perform laboratory studies.


## References


Abbott WS, 1925. A method of computing the effectiveness of an insecticide. J. Econ. Entomol. 18: 265-267.

Ahmad M, 2008. Insecticide resistance, mechanisms, genetics and management of *Spodoptera litura* (Fab). In the Punjab, Pakistan. Ph.D thesis, University College of Agriculture, Bahauddin Zakariya University, Multan Pakistan. 120 p.

Ahmad M, 2009. Observed potentiation between pyrethroid and organophosphate insecticides for the management of *Spodoptera litura* (Lepidoptera: Noctuidae). Crop Protect. 28: 264-268.

Ahmad M, Arif I and Ahmad M, 2007a. Occurrence of insecticide resistance in field populations of *Spodoptera litura* (Lepidoptera: Noctuidae) in Pakistan. Crop Prot., 26: 807-809.

Ahmad M, Sayyed AH, Crickmore N and Saleem MA, 2007b. Genetics and mechanism of resistance to deltamethrin in a field population of *Spodoptera litura* (Lepidoptera: Noctuidae) in Pakistan. Pest Manag. Sci., 63: 1002-1010.

Basit MS, Saeed MA and Saleem MD, 2013. Shah. Detection of resistance, cross resistance and stability of resistance to new chemistry insecticides in Helicoverpa armigera. J. Econ. Entomol. 106:1414-1422.

Finney JD, 1971. Probit Aalysis (3rd eds.). Cambridge University Press. UK.

Karuppaiah V, Srivastava C and Subramanian, S, 2017. Variation in insecticide detoxification enzymes activity in *Spodoptera litura* (Fabricius) of different geographic origin. J. Entomol. Zool. Stud., 5: 770-773.

Khan AY and Mehmood RA, 1999. Cotton Crop Survey Report, 1998-99. Pakistan Central Cotton Committee. 67 p.

Kumar AVK, 1995. Externalities in the use of pesticides: An economic analysis in a cole crop. M.Sc. Agri thesis (unpublished), U.A.S. Bangalore, India.

LeOra Softwar 1987. POLO-PC. A user's guide to probit or logit analysis. Berkeley, California.

Liu TX, Hutchison WD, Chen W and Burkness EC, 2003. Comparative susceptibilities of diamondback moth (*Plutella xylostella*) (Lepidoptera: Plutellidae) and cabbage looper (Lepidopter: Noctuidae) from Minnesota and South Texas to λ-cyhalothrin and indoxacarb. J. Econ. Entomol. 94(4): 1230-1236.

Mallikarjuna N, Kranthi KR, Jadhav DR, Kranthi S and Chandra S, 2004. Influence of foliarchemical compounds on the development of *Spodoptera litura* (Fab.) in interspecific derivativesof groundnut. J. Appl. Entomol., 128: 321-328.

Mazlan N and Mumford J, 2005. Insecticide use in cabbage pest management in the Cameron Highlands, Malaysia. Crop Protect. 24: 31–39.

Mohan M and Gujar GT, 2003. Local variation in susceptibility of diamondback moth to insecticides and role of detoxification enzymes. J. Crop Protect. 22: 495-504.

Mota-Sanchez D, Bills PS, Whalon ME and Wheeler WB, 2002. Arthropod resistance to pesticides: status and overview. Pestic. Agric. Envir., 214-227.

Mushtaq AS, Ahmad M, Ahmad M, Aslam M and Sayyed AH, 2008. Resistance toselected organochlorin, organophosphate, carbamates and pyrethroid, in *Spodoptera litura* (Lepidoptera: Noctuidae) from Pakistan. J. Econ. Entomol., 101(5): 1667-1675.

Nakagome T and Kato K, 1981. Control of insects in cruciferous vegetables in Aichi Prefecture with special reference to diamondback moth. In: Insects in cruciferous vegetables and their control with special reference to diamondback moth (*Plutella xylostella*) Takeda Chemical Industries Ltd., Tokyo.79-92.

Rafiq MN, 2005. Insecticide resistance in diamondback moth, *Plutella xylostella* (L.) (Lepidoptera: Plutellidae) and strategies of its management. PhD thesis. Department of Entomology, university of Arid Agriculture, Rawalpindi, 165 p.

Ronald FLM, Dunbar DM, Minuto LG and Shimabuku RS, 2000. Management of diamondback moth with Emamectin benzoate and *Bacillus thuringiensis*, Proc. on the management of diamondback moth and other crucifer pests.178-183.

Sayyed AH, Ahmad M and Saleem MA, 2008. Cross-resistance and genetics of resistance to indoxacarb in *Spodoptera litura* (Lepidoptera: Noctuidae). J. Econ. Entomol. 101: 472-479.

Sayyed AH, Haward R, Herrero S, Ferre J and Wright DJ, 2000. Genetic and biochemical approach for characterisation of resistance to *Bacillus thuringiensis* toxin Cry1Ac in a field population of the diamondback moth. Appl. Environ. Microb., 66: 1509-1516.

Sexena JD, Rai S, Srivastava KM and Sinha SR, 1989. Resistance in the field population of the diamondback moth to some commonly used synthetic Pyrethroids and Local variation in susceptibility of diamond back moth. Indian J. Entomol., 51: 265 268.

Shankar G, Arjuna M, Bhatt UG, Kendappa GN and Miphyantha MS, 1996. *Bt* Berliner Subsp. kurstaki in the management of DBM in India. Proc. 3rd Int. Workshop, Kuala Lumpur, Malaysia. 29th Oct.-1st Nov. pp. 104-108.

Shelton AM, Sances FV, Hawley J and Tang JD, 2000. Assessment of insecticide resistance after the outbreak of diamondback moth (Lepidoptera: Plutellidae) in California in 1997.J. Econ. Entomol. 93: 931-936.

Tong HQ, SU X, Zhou and Bai L, 2013. Field resistance of *Spodoptera litura* (Lepidoptera: Noctuidae) to organophosphates, pyrethroids, carbamates and four newer chemistry insecticides in Hunan, China. J. Pest Sci., 86:599-609.

Travis A and Rick E, 2000. Effect of Insecticides on the Diamondback Moth (*Plutella xylostella*)

(Lepidoptera: Plutellidae) and Its Parasitoid *Diadegma insulare* (Hymenoptera: Ichneumonidae) J. Econ. Entomol. 93(3): 763-768.

Vastrad AS, Lingappa S and Basavanagoud K, 2003. Management of insecticide resistant populations of diamondback moth. J. Pest Managt. Hort. Ecosys., 1: 33-40.

Verkerk RHJ and Wright DJ, 1996. Multi-tropic interactions and management of the diamondback moth: a review. Bull. Entomol. Res., 86: 205-216.

# 22

# Determination of the critical period for weed control of sweet corn under tropical organic farming system

**Marulak Simarmata*, Uswatun Nurjanah, Nanik Setyowati**
Department of Agronomy, University of Bengkulu, Jalan W.R. Supratman Kandang Limun, Bengkulu 38371, Indonesia

## Abstract

An understanding of the critical period of the crop for weed control (CPWC) is needed before making a decision on weed management. A field experiment to study a CPWC of sweet corn was carried out at the highland of Bengkulu Province, Indonesia from October 2015 to January 2016. The objective was to determine the CPWC of sweet corn under the tropical organic farming system. Weed infestations in the research plots including 14, 28, 42, 56, 70, and 84 days after planting (DAP) of weedy and weed-free periods were arranged in a completely randomized block design (CRBD) with three replications. Results showed that the plant height, leaf area, and yield of sweet corn descended and ascended due to the increase of weedy and weed-free periods, respectively. The biomass and yield losses due to weed competition during the growing season reached 49.5 and 54.7 %, respectively. The relative yield descended or ascended in logistic equation curves due to the increase of weedy or weed-free periods, respectively. Based on the acceptable yield loss (AYL) of 5 %, the CPWC of sweet corn under organic farming system was determined from 2 to 77 DAP, and with the AYL of 10 %, the CPWC was determined from 3 to 53 DAP.

*Corresponding author email: marulak_simarmata@yahoo.com

**Keywords**: Sweet corn, CPWC, Tropical organic farming, Weed control

## Introduction

Sweet corn (*Zea mays saccharata*) is a horticultural crop cultivated over the world. The consumption of sweet corn in Indonesia continues to increase because of its nutrition and naturally sweet taste, but production remains low. Efforts to optimize the yields of sweet corn were done intensively using many synthetic inputs such as fertilizers and pesticides (Johnson et al., 2010; Akintoye and Olaniyan, 2012). But, using of synthetic inputs intensively may cause adverse impacts to the environment due to the deterioration of soil and water quality (Mulvaney et al., 2009; Savci, 2012; Ruark et al., 2012). Therefore, the organic farming system, known as an environmentally friendly practice without using synthetic materials, becomes a wise choice not only to save soil environment but also to produce healthy food (Hue and Silva, 2000; Taguling, 2013). An organic farming system is trending over the world today, especially for horticultural crops (Ruark et al., 2012; Muktamar et al., 2017).

Organic farming can be interpreted as a crop production system based on biological recycling of nutrients. Soil fertility can be improved not only by recycling the organic material in-situ but also by using the organic materials from the outside of the farming areas, such as composted forage plants and animal manures (Taguiling, 2013). However, some organic materials such as cattle manure and forage composts may carry weed propagules (Barberi, 2002). The number of weed populations in planting area may increase because of the emerging weeds both from animal manure and from soil seed bank (Carr et al., 2013). Weeds will be very detrimental to the crops if they are not controlled by the right measures at the appropriate time. Since organic farming does not use any synthetic chemicals such as herbicides, the choice of weed control practices may be limited to cultural, mechanical, biological, or integrated control practices by combining those methods (Barberi, 2002). An appropriate weed management should be carried out when the presence of weeds is harmful to the crops through a significant reduction of yield known as the CPWC (Knezevic at al., 2002; Johnson et al., 2010). According to Zimdahl (2004), CPWC is the growth stages of the crop where weeds must be controlled to prevent the apparent losses of crop yields. An understanding of the CPWC is very useful to make timely decisions for specific weed control on each plant species (Knezevic and Datta, 2015).

Determination of the CPWC can be approached by investigation of the weedy and weed-free periods on a specific crop. The limit of the acceptable yield loss (AYL) in general varies from 5 to 10 % (Knezeviz and Datta, 2015). The yield or relative yield of weedy and weed-free trials are described in logistic equation curves, where the intercept of the weedy and weed-free curves at the AYL on x-axis can determine the maximum of weedy periods and the minimum of weed-free periods of crop growth stages to weed competition, respectively (Juraimi et al., 2009; Mekonnen et al., 2017).

Some researchers have reported the CPWC among crops worldwide that included corn (Evans et al., 2003; Gantoli et al., 2013), soybeans (Knezevic et al., 2003), rice (Juraimi et al., 2009; Chauhan and Johnson, 2011; Mekonnen, 2017), peanuts (Everman et al, 2008), and sweet corn (Williams II, 2006). The CPWC varies due to climates factors, environmental conditions, cropping system, and cultivation technology such as row spacing, planting date, and fertilizer application (Williams II, 2006; Juraimi et al., 2009; Chauhan and Johnson, 2011). A field experiment was carried out at the highland of tropical areas to determine the critical periods of sweet corn to weed competition under a closed organic farming system.

## Material and Methods

The research was carried out at the Closed Agricultural Production System (CAPS) Research Station of the University of Bengkulu located at District of the Rejang Lebong, Bengkulu Province, Indonesia from October 2015 to January 2016. The field is positioned at 1020 36' 56" E, 30 27' 37" S, and the altitude of 1,054 meters above the sea level. The soil type is Andept and the soil texture is classified as sandy loam. The site was regularly cultivated for organic vegetable production since 2009.

Weed assessment on the research site was conducted in 3 blocks based on weeds stratification. Weeds were enumerated in 5 sampling plots for each block using a wooden square plot size of 0.5 m x 0.5 m each, following the methods of Simarmata et al. (2015). Data observed were density, frequency, and dry biomass weight of weed species in each sampling plot. The rank of dominant weed species was determined based on the values of summed dominance ratio (SDR) calculated from the average of relative density, frequency, and dry biomass weight as modified from Janiya and Moody (1989) (Eq. 1).

$$SDR = \frac{Dr+Fr+Br}{3} \qquad (1)$$

Where, SDR is summed dominance ratio; Dr is relative density calculated from density of one species divided by total density of all species; Fr is relative frequency calculated from frequency of one species divided by total of frequency of all species; and Br is relative biomass calculated from biomass of one species divided by total biomass of all species.

Land preparation was started by cutting the weeds and land was cultivated twice using hoes.The experimental site was formed for 36 plots (12 trials with 3 replications) with sizes of 3 m x 1.5 m each. Organic fertilizer was applied one week before planting using composted solid cow's manure of 10 ton $ha^{-1}$. The manures were mixed homogeneously within 20 cm depth of soil surface. Seeds of sweet corn var. Secada were planted in a hole of 3 cm depth with planting spaces of 75 cm x 25 cm.

Sweet corn plants were maintained regularly by watering and pest were controlled mechanically as needed. In addition to the manure fertilizers, liquid organic fertilizer (LOF) was sprayed on soil surface at 1 and 6 WAP at the rates of 5 ml $m^{-2}$. The LOF was produced at the University of Bengkulu, consisted of dairy cattle feces, dairy cattle urine, soil containing local microorganism, green leaves of *Tithonia diversifolia*, and solution of EM-4, diluted and fermented in water for 10, 10, 1, 2.5, 10 %, respectively (Muktamar et al., 2017). Weeds were controlled by physical control method (PCM) in accordance with weedy and weed-free periods (Table 1). The CPWC of sweet corn under tropical organic farming systems was evaluated by variations of weed infestations in the research plots including 14, 28, 42, 56, 70, and 84 DAP of weedy and weed-free periods.

The weed-free periods were maintained by manually removing weeds that appear in the specified weed-free period trial, likewise, the weedy periods were allowing weeds to grow within the specified period trial and after that period, the emerged weeds were manually removed from the plots. The experiment was arranged in a completely randomized block design (CRBD) with three replications.

**Table 1. Periods of weed infestations on sweet corn under tropical organic farming system.**

| No. | Treatment | 0-14 DAP | 0-28 DAP | 0-42 DAP | 0-56 DAP | 0-70 DAP | 0-84 DAP |
|---|---|---|---|---|---|---|---|
| 1. | Weed-free 14 DAP | Weed-free | Weedy | Weedy | Weedy | Weedy | Weedy |
| 2. | Weed-free 28 DAP | Weed-free | Weed-free | Weedy | Weedy | Weedy | Weedy |
| 3. | Weed-free 42 DAP | Weed-free | Weed-free | Weed-free | Weedy | Weedy | Weedy |
| 4. | Weed-free 56 DAP | Weed-free | Weed-free | Weed-free | Weed-free | Weedy | Weedy |
| 5. | Weed-free 70 DAP | Weed-free | Weed-free | Weed-free | Weed-free | Weed-free | Weedy |
| 6. | Weed-free 84 DAP =Weedy 0 DAP | Weed-free | Weed-free | Weed-free | Weed-free | Weed-free | Weed-free |
| 7. | Weedy 14 DAP | Weedy | Weed-free | Weed-free | Weed-free | Weed-free | Weed-free |
| 8. | Weedy 28 DAP | Weedy | Weedy | Weed-free | Weed-free | Weed-free | Weed-free |
| 9. | Weedy 42 DAP | Weedy | Weedy | Weedy | Weed-free | Weed-free | Weed-free |
| 10. | Weedy 56 DAP | Weedy | Weedy | Weedy | Weedy | Weed-free | Weed-free |
| 11. | Weedy 70 DAP | Weedy | Weedy | Weedy | Weedy | Weedy | Weed-free |
| 12. | Weedy 84 DAP =Weed-free 0 DAP | Weedy | Weedy | Weedy | Weedy | Weedy | Weedy |

DAP = Days After Planting; [hatched] = Weedy; [blank] = Weed-free

Ten sample plants from each plot were harvested at 84 DAP and data recorded for plant height, leaf area, and yield of unhusked cob's weight, cob's length, cob's diameter, and biomass weight. Residual of weed biomass were collected from 2 sampling plots (sizes of 0.5 m x 0.5 m each) of weed-free period trials. Biomass of sweet corn and weeds were oven-dried at 70 C for 72 hours. Data were subjected to one-way analysis of variances (ANOVA) and further separated by Duncan's multiple range test (DMRT) at 5 % level. Data of the yield were converted to relative yield as a percent of control and further analyzed using non-linear regression model. The equation with the highest determination factor ($R^2$) was judged as the most appropriate model to determine critical periods of sweet corn (Williams II, 2006; Juraimi et al., 2009). The intercept of weedy and weed-free curves on the x-axis at 5 % and 10 % of acceptable yield loss (AYL) were chosen arbitrarily to determine the CPWC of sweet corn under the tropical organic farming system (Knezeviz and Datta, 2015).

## Results and Discussion

### Weed analysis

Weed vegetation in the experimental site was identified before and after the experiment. Initial analysis was conducted in 3 blocks based on visual views of weed stratifications. Weed assessment in block I (Fig. 1A) identified 11 species of weeds in which 4 species have SDR > 10 which were *Ageratum conyzoides, Mimosa invisa, Echinochloa colona, Stachytarpeta jamaicensis*, respectively with SDR of 26.0, 18.0, 14.6, and 13.2 %. In block II (Fig. 1B) there were 7 species of weeds in which 5 species have SDR > 10 which were *Stachytarpeta jamaicensis, Cyperus kyllingia, Echinochloa colona, Ageratum conyzoides, and Mimosa invisa* with SDR of 29.4, 16.8, 16.4, 15.1 and 14.7 %, respectively. In Block III (Fig. 1C) there were 8 weed species but only 3 species have SDR > 10 which were *Echinochloa colona, Ageratum conyzoides, Stachytarpeta jamaicensis* with SDR of 40.9, 25.6, and 10.7 %, respectively. Based on the differences in weed distributions in the three blocks before the experiment, the study was designed in a randomized block design (RBD) with 3 replications as blocks.

Weeds that grew at the end of the study as residuals of the treatment were only observed in the plots of weed-free periods. There were 9 species that grew at the harvested time but only 3 species had SDR > 10 which were *Euphorbia prunifolia, Ageratum conyzoides, Echinochloa colona* with SDR of 24.0, 22.7, and 15.6 %, respectively. Compared to the initial analysis, two new weed species were *Amaranthus spinosus, Borreria latifolia* with SDR of 4.4 and 3.7 %, while five weed species were absent were *Cyperus kyllingia, Phyllanthus niruri, Syndrella nodiflora, Spilanthes acmella*, and *Stachytarpeta jamaicensis* (Fig. 1D). The presence or absence of weed species indicated the shifting in weed vegetation due to manipulated microenvironment such as the different periods of weed infestation (Simarmata et al., 2015). Newly emerging weeds may be carried away from the manure or emerged from dormant seed or from the soil seed bank (Barberi, 2002). Weed residue harvested at the end of the study as the total weed biomass was only observed in weed-free trials because there was no weed residue in the weedy trials (William II et al., 2008). The heaviest weed residue in the weed-free trials reached 656 g $m^{-2}$. The weed residue decreased with longer periods of weed-free and there was no weed residue if the plot was free from weeds during the season (Fig. 2).

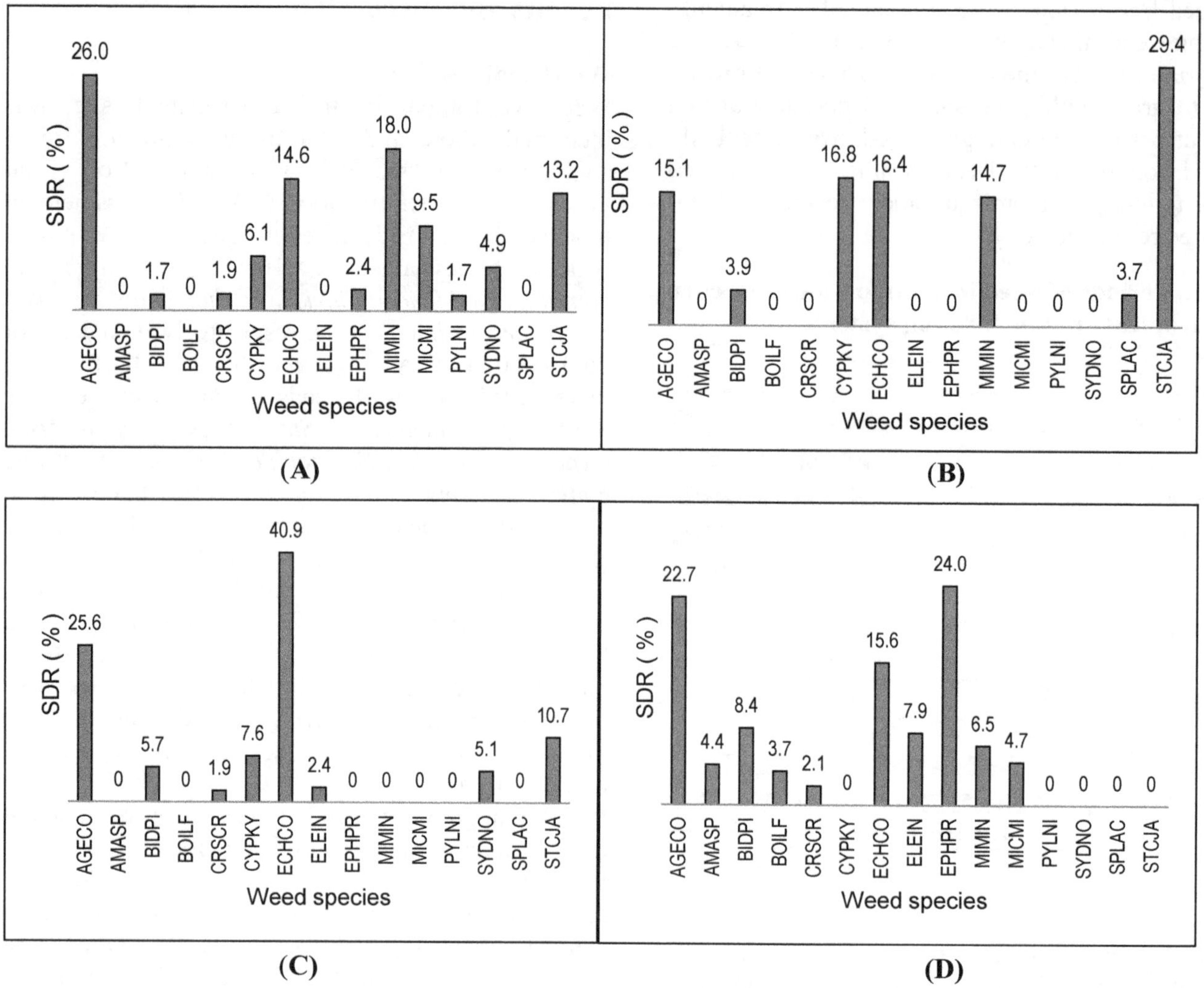


**Fig. 1. Initial assessment of weed vegetation in research sites (A, B, C) , and final assessment at the end of the growing season (D) counted as summed dominance ratio (SDR); AGECO = *Ageratum conyzoides*, AMASP = *Amaranthus spinosus*, BIDPI = *Bidens pilosa*, BOILF = *Borreria latifolia*, CRSCR = *Crassocephalum crepidioides*, CYPKY = *Cyperus kyllingia*, ECHCO = *Echinocloa colona*, ELEIN = *Eleusine indica*, EPHPR = *Euphorbia prunifolia*, MIMIN = *Mimosa invisa*, MICMI = *Mikania micrantha*, PYLNI = *Phyllanthus niruri*, SYDNO = *Synedrella nodiflora*, SPLAC = *Spilanthes acmella*, STCJA = *Stachytarpheta jamaicensis.***

## Growth, yield and biomass production

The sweet corn seeds germinated and grew at 99 %, so seedlings were thinned become one plant per planting hole. With low rain in October 2015, 35 mm in 5 days of rain categorized as a dry month, seedlings were watered every day. But in November, December 2015, and January 2016, the rainfalls were 355, 592, and 391 mm with rainy days of 23, 26, and 19 days, respectively. These rainfalls were optimum for sweet corn growth (Fig. 3). Overall, the crops grew well and there was no evidence of diseases and insects in the experimental plots.

Data on the growth variables, yield, and plant biomass are presented in Table (2). The period of weed infestations significantly affected plant height, leaf area, yield, and biomass production, but no effect was found on the yield components of cob's diameter and length. The longer the weedy period, the lower the height and the less the leaf area of sweet corn was. If the plots were weedy during agrowing season, the plant height and leaf area were depressed to 161.9 cm and 567.5 cm$^2$ compared to weed-free in a season, the plant height and leaf area were 230.9 cm and 786.6 cm$^2$, respectively. On the other hand, the opposite was found on the weed-free trial, where the longer the weed-free period the higher the plant height and the more the leaf area. If the plots were free from weeds during a season the plant height increased from 161.6 to 230.9 cm and the leaf area increased from 567.6 to 786.6 cm$^2$. The responses of plant growth to weed infestations can be explained by the competition periods between crop and weeds to the life necessities such as nutrition, growing space, water, and $CO_2$ (Zimdahl, 2004). If weeds were suppressed by increasing the weed-free periods then the crop growth and yield increases (Williams II et al., 2008).

The decrease and increase of the plant height and the leaf area on weedy and weed-free trials affected the yield and plant biomass production. The yield decreased when the weedy period increased and the yield increased when the weed-free periods increased (Table 2). If the plot was free from weeds during a season, the yield of unhusked cobs was 463.5 g plant$^{-1}$. But, when the plots were weedy during the season, the yield decreased to 209.9 g plant$^{-1}$. The decrease or increase of yield was not correlated to the yield components of the diameter and length of the cob, but it was suspected due to the size and the number of seeds. The more opportunities the crops free from weeds, the higher the growth and yield of crops that were harvested (Zimdahl, 2004; Williams II et al., 2008).

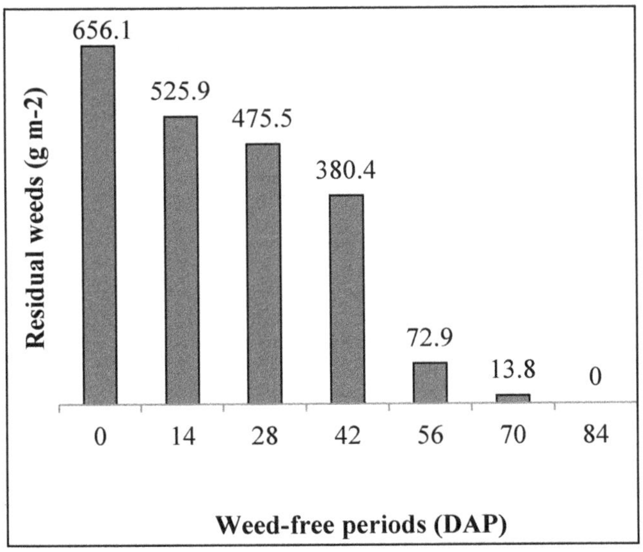


**Fig. 2. Residual weeds of in sweet corn plots harvested at the end of experiment.**

Biomass production of sweet corn also showed the same pattern with the growth parameters because biomass was the accumulation of the plant height and leaf areas of the plant. Biomass production decreased from 104.7 to 52.9 g plant$^{-1}$ if the plots were weedy during the growing season and vice versa occurred if the plots were weed-free during the season. The biomass production was also an important variable of sweet corn because it can be utilized for industries such as bioethanol or for local needs as fresh ruminant food (Barros-Rioss et al., 2015).

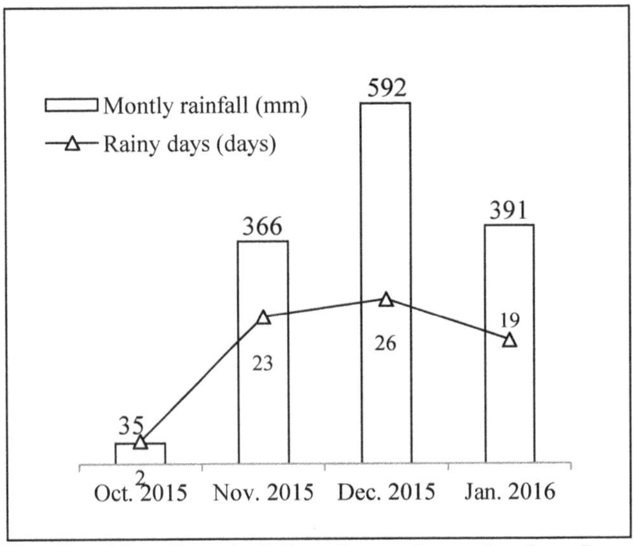


**Fig. 3. Montly rainfall and number of rainy days in the research location from October 2015 to January 2016.**

**Determination of the CPWC**

The yield and biomass relatives of sweet corn expressed in the percent of control are presented in Table 3. It appeared that the yield losses increased or decreased when the weedy or weed-free period increased, respectively. The highest yield loss of sweet corn grown under tropical organic farming reached 54.7 % if the weeds were not controlled during the season (84 days of weedy). Similarly, the highest biomass loss reached 49.5 % due to uncontrolled weeds during the season (84 days of weedy).

The logistic equation curves of relative yield of weedy and weed-free periods were used to determine the CPWC (Knezevic et al., 2002; Gantoli et al., 2013). The AYL due to adverse effects of weeds varies from 5 – 10 % (Knezevic and Datta, 2015). Determination of the critical period was judged by analysis of non-linear curves of the relative yields of weedy and weed-free treatments. In some publications, the critical periods of crops to weed control were fitted to Gompertz and logistic equation curves (Juraimi et al., 2009). In this study, a logistic equation was used to determine the CPWC with determination factors ($R^2$) reaching 94.9 and 85.8 % on the weedy and weed-free curves, respectively (Fig.4).

Based on the intercept of the curves on the x-axis with the AYL of 5 %, the maximum weedy period was 2 DAT, and the minimum of the weed-free period was 77 DAP. Thus, the CPWC of sweet corn under the tropical organic farming system with the AYL of 5 % was from 2 to 77 DAT (Table 4). If the AYL become 10 %, then the maximum weedy period was 3 DAT, and the minimum of the weed-free period was 53 DAP. Thus, the CPWC of sweet corn under the tropical organic farming system with the AYL of 10 % was from 3 to 53 DAT (Table 4).

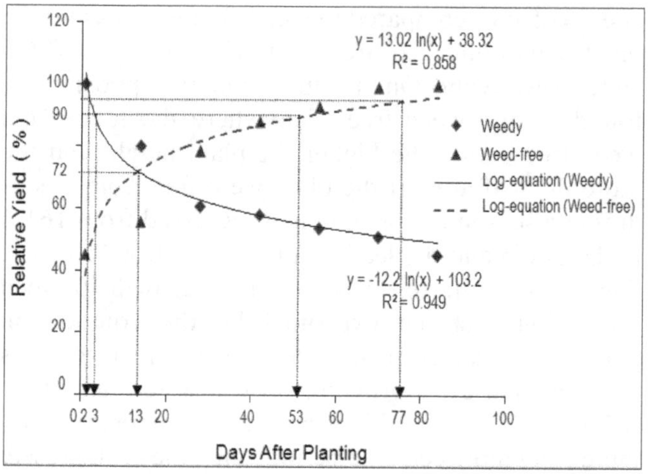


**Fig. 4. The critical periods for weed control (CPWC) of sweet corn under tropical organic farming system.**

**Table 2. Effect of weed infestation on growth, yield, yield components, and biomass of sweet corn.**

| Periods of weed infestation (DAP) | | Plant height (cm) | Leaf area (cm$^2$) | Unhusked - cob (g plant$^{-1}$) | Cob- diameter (cm) | Cob- lenght (cm) | Oven-dried biomass (g plant$^{-2}$) |
|---|---|---|---|---|---|---|---|
| Weedy | 0 | 230.9 a | 786.6 a | 463.5 a | 5.77 | 20.8 | 104.7 a |
| | 0-14 | 233.2 a | 775.8 a | 371.8 b | 5.88 | 20.8 | 93.0 ab |
| | 0-28 | 189.1 ab | 668.2 b | 281.9 c | 5.76 | 19.7 | 78.8 bcde |
| | 0-42 | 190.4 ab | 633.7 b | 269.2 c | 5.80 | 21.9 | 67.5 def |
| | 0-56 | 196.9 ab | 626.4 b | 250.2 cd | 5.54 | 19.2 | 60.9 ef |
| | 0-70 | 189.9 ab | 591.0 bc | 237.2 cd | 5.02 | 19.5 | 54.9 f |
| | 0-84 | 161.9 b | 567.5 c | 209.9 d | 5.00 | 21.5 | 52.9 f |
| Weed-free | 0 | 161.9 b | 567.5 c | 209.7 d | 5.00 | 21.5 | 52.9 f |
| | 0-14 | 206.3 a | 586.6 bc | 258.4 bc | 5.17 | 20.7 | 70.9 cdef |
| | 0-28 | 215.2 a | 742.8 a | 364.6 b | 5.13 | 21.4 | 83.9 bcd |
| | 0-42 | 221.2 a | 749.6 a | 405.0 b | 5.23 | 21.3 | 90.8 abc |
| | 0-56 | 226.8 a | 760.4 a | 431.1 a | 5.25 | 21.8 | 91.3 abc |
| | 0-70 | 222.6 a | 772.1 a | 460.4 a | 5.84 | 21.3 | 96.2 ab |
| | 0-84 | 230.9 a | 786.6 a | 463.5 a | 5.77 | 20.8 | 104.7 a |
| **ANOVA (P<0.05)** | | * | * | * | NS | NS | * |

*** = Significant effect; NS = Non-significant effect; Numbers followed by the same letter in one column are not significantly different by DMRT (P<0.05).**

**Table 3. Effect of the weed infestations on yield and biomass losses of sweet corn.**

| Period of weed infestation (DAP) | | Relative Yield ( % ) | Yield losses ( % ) | Relative biomass ( % ) | Biomass losses ( % ) |
|---|---|---|---|---|---|
| Weedy | 0-14 | 80.2 | 19.8 | 88.8 | 11.2 |
| | 0-28 | 60.8 | 39.2 | 75.3 | 24.7 |
| | 0-42 | 58.1 | 41.9 | 64.5 | 35.5 |
| | 0-56 | 53.9 | 46.1 | 58.2 | 41.8 |
| | 0-70 | 51.2 | 48.8 | 52.4 | 47.6 |
| | 0-84 | 45.3 | 54.7 | 50.5 | 49.5 |
| Weed-free | 0 | 45.3 | 44.7 | 50.5 | 49.5 |
| | 0-14 | 55.8 | 44.2 | 67.7 | 32.3 |
| | 0-28 | 78.7 | 21.3 | 80.1 | 19.9 |
| | 0-42 | 88.2 | 11.8 | 86.7 | 13.3 |
| | 0-56 | 93.0 | 7.0 | 87.2 | 12.8 |
| | 0-70 | 99.3 | 0.7 | 91.9 | 8.1 |
| | 0-84 | 100 | 0.0 | 100 | 0 |

**DAP = days after planting**

**Table 4. Determination of the CPWC of sweet corn based the AYL 5 and 10 % of weedy and weed-free curves.**

| Relative Yield (%) | Yield Loss (%) | Maximum weedyperiods (DAP) | Minimum weed-free periods (DAP) |
|---|---|---|---|
| 95 | 5 | 2 | 77 |
| 90 | 10 | 3 | 53 |

**CPWC = critical periods for weed control; AYL = acceptable yield loss; DAP = days after planting.**

## Conclusion

The periods of weed infestations influenced the growth and yield of sweet corn cultivated under tropical organic farming systems which ascended or descended with the non-linear curves of the logistic equations due to the increase of weedy or weed-free periods, respectively. Based on the AYL of 5 %, the CPWC of sweet corn under tropical organic farming was from 2 to 77 DAP and with the AYL of 10 %, the CPWC was from 3 to 53 DAP.

## Acknowledgment

Appreciation is expressed to staffs of Agriculture Research Center, the University of Bengkulu for cooperation to facilitate field experiments. Special thanks are also presented to the students who helped the works at the laboratory and for field experiments.

## References

Akintoye HA and Olaniyan AB, 2012. Yield of sweet corn in response to fertilizer sources. J. Agric. Sci. 1(5):110-116.

Barberi P, 2002. Weed management in organic agriculture: Are we addressing the right issue? Weed Res. 42(3):177-193.

Barros-Rios J, Romani A, Garrote G. and Ordas B, 2015. Biomass, sugar, and bioethanol potential of sweet corn. GCB Bioenergy 7:153-160.

Carr PM, Gramig, GG and Liebig MA, 2013. Impacts of organic zero tillage system on crops, weeds, and soil quality. Sustainability. 5:3172-3201.

Chauhan BS and Johnson DE, 2011. Row spacing and weed control timing affect yield of aerobic rice. Field Crop Res. 121(2):226-232.

Evans SP, Knezevic SZ, Lindquist JL, Shapiro CA and Blankenship EE, 2003. Nitrogen application influences the critical period for weed control in corn. Weed Sci. 51(3):408-417.

Everman WJ, Clewis SB, Thomas WE, Burke IC and Wilcut JW, 2008. Critical period of weed interference in peanut. Weed Tech. 22(1):63-67.

Gantoli G, Ayala VR and Gerhards R, 2013. Determination of the critical period for weed control in corn. Weed Tech. 27(1):63-71.

Hue NV and Silva JA, 2000. Organic Soil Amendments for Sustainable Agriculture: Organic Sources of Nitrogen, Phosphorous and Potassium. In: Silva, J.A. and Uchida, R. (eds). Plant Nutrient Management in Hawaii's Soils: Approaches for Tropical and Subtropical Agriculture. College of Tropical Agriculture and Human Resources, University of Hawaii at Manoa, Hawaii. pp. 133-144.

Janiya JD and Moody K, 1989. Weed populations in transplanted and wet-seeded rice as affected by weed control method. Tropical Pest. Manag. 35(1):8-11.

Johnson HJ, Colquhoun JB, Bussan AJ and Rittmeyer RA, 2010. Feasibility of organic weed management in sweet corn and snap bean for processing. Weed Tech. 24(4):544-550.

Juraimi AS, Najib MYM, Begum M, Anuar AR, Azmi M and Puteh A, 2009. Critical period of weed competition in direct seeded rice under saturated and flooded conditions. Pertanika J. Trop. Agric. Sci. 32(2):305-316.

Knezevic SZ, Evans SP and Blankenship EE, 2002. Critical period for weed control: the concept and data analysis. Weed Sci. 50(6):773-786.

Knezevic SZ, Evans SP and Mainz M, 2003. Row spacing influences the critical timing for weed removal in soybean (Glycine max). Weed Tech. 17(4):666-673.

Knezevic SZ and Datta A, 2015. The critical period for weed control: Revisiting data analysis. Weed Sci. Special Issue: 188-202.

Mekonnen G, Woldesenbet M and Yegezu E, 2017. Determination of critical period of weed-crop competition in rice (Oryza sativa L.) in Bench Maji and Kaffa Zone, South Western Ethiopia. J. Plant Sci. 5(3):90-98.

Muktamar Z, Sudjatmiko S, Chozin M, Setyowati N and Fahrurrozi, 2017. Sweet corn performance and its major nutrient uptake following application of vermicompost supplemented with liquid organic fertilizer. IJASEIT 7(5):602-608.

Mulvaney RL, Khan SA and Elsworth TR, 2009. Synthetic nitrogen fertilizers deplete soil nitrogen: A global dilemma for sustainable cereal production. J. Environ. Qual. 38(6):2295-2314.

Ruark M, Brundy L, Andraski T and Peterson A, 2012. Fifty years of continuous corn: Effects on soil fertility. Proc. Wisconsin Crop Manag. Conf. 51:127-132.

Savci S, 2012. An agricultural pollutant: Chemical fertilizer. Int. J. Environ. Sci. Dev. 3(1):77-80.

Simarmata M, Sitanggang CD and Djamilah D, 2015. Shifting weed compositions and biomass production in sweet corn field treated with organic composts and chemical weed controls. Agrivita J. Agric. Sci. 37(3):226-236.

Taguiling LG, 2013. Quality improvement of organic compost using green biomass. Eur. Sci. J. 9(36):319-341.

Williams II MM, 2006. Planting date influences critical period of weed control in sweet corn. Weed Sci. 54(3):928–933.

Williams II MM, Rabaey TL and Boerboom CM, 2008. Residual weeds of processing sweet corn in the North Central Region. Weed Tech. 22(4):646–653.

Zimdahl RL, 2004. Weed Crop Competition: A Review, 2nd edition. Blackwell, Oxford, UK. pp. 220.

23

# Modeling the potassium requirements of potato crop for yield and quality optimization

**Farheen Nazli[1]*, Bushra[2], Muhammad Mazhar Iqbal[3], Fatima Bibi[4], Zafar-ul-Hye[5], Muhammad Ramzan Kashif[1] and Maqshoof Ahmad[6]**

[1]Pesticide Quality Control Laboratory Bahawalpur-63100, Pakistan
[2]Institute of Soil and Environmental Sciences, University of Agriculture Faisalabad, Pakistan
[3]Soil and Water Testing Laboratory, Chiniot, Pakistan
[4]Mango Research Station, Multan, Pakistan
[5]Department of Soil Science, College of Agriculture, Bahauddin Zakariya University, Multan, Pakistan
[6]Department of Soil and Environmental Sciences, the Islamia University of Bahawalpur, Bahawalpur-63100, Pakistan

*Corresponding author email: farheenmaqshoof@gmail.com

## Abstract

The intensity, quantity as well as capacity factors are important to predict the amount of nutrient in soil required for maximum plant growth. Sorption isotherm considers these three factors so believed to be one of the most important techniques in soil which control the fate and mobility of nutrients. The field experiment was conducted to find out site-specific and crop-specific potassium requirement for potato crop. The potassium adsorption isotherm was constructed and Freundlich model was used to theoretically work out different soil solution K levels (0, 5, 10, 15, 20, 25, and 30 mg $L^{-1}$). The K fertilizer doses were calculated against these specific soil solution levels. Field experiment was conducted with seven model based K fertilizer treatments (0, 49, 94, 139, 183, 228, and 273 kg K $ha^{-1}$) and three replications in Randomized Complete Block Design (RCBD) using potato as test crop. The results showed that growth parameters like plant height, leaf area and chlorophyll significantly contributed to potato tuber yield. Different yield response models were tested and it was observed that linear plus plateau and quadratic plus plateau predicted equally well the optimum fertilizer K rate both for yield and quality attributes of potato. For maximum potato tuber yield i.e.34.41 Mg $ha^{-1}$ the economic optimum K was 100 kg $ha^{-1}$. Optimum fertilizer K rates (at 95 % relative yield) for potato tuber yield, dry matter percentage, protein, starch contents and vitamin C contents were 100, 103, 180, 230 and 200 kg K $ha^{-1}$, respectively. So, it is suggested that adsorption isotherm technique should be used to calculate site specific and crop specific fertilizer requirements of crops and 100 kg $ha^{-1}$ is recommended as optimum potassium fertilizer for potato crop. Moreover, the K fertilizer application would improve crop quality that would support the quality based marketing system in Pakistan.

**Keywords**: Potato, Quadratic plus plateau, Adsorption isotherm, Yield, Quality

## Introduction

Soil testing methods are essentially used as tool for monitoring soil fertility. However, they fail frequently, if used to predict optimal fertilizer requirements for crops. In Pakistan, the fertilizer recommendations are usually generalized and are given as a range (FAO, 2016) which are predicted from simple experiments in the fields and are extrapolated on all soils which is one of main reasons that crop responses to potassium fertilizers are very irregular and erratic (Mengel et al., 1998). Maximum profitability and environmental sustainability is only possible through fertilizer recommendations for maximizing the yield and quality (Thompson et al., 2017). These recommendations are developed through extensive field trials by studying the effect of different rates of fertilizers on crop yield and quality. To solve this problem, attempts should be made to predict fertilizer requirements using nutrient adsorption models. These models are based on different soil nutrient concentrations such as intensity, quantity and capacity factors. These all are important to predict the amount of soil nutrient requirements to adjust model based soil solution level required for maximum plant growth (Louison et al., 2015). The nutrient adsorption isotherms can be used for the estimation of fertilizers needed to adjust the soil solution nutrients to the level optimum for maximum yield (Kenyanya et al., 2014). The use of these models can only be possible through identification of critical solution level to get maximum plant growth which is specific for that particular soil and for that particular crop (Samadi, 2003, 2006). The adsorption characteristics are linked with type and quality of minerals in soil as well as other soil chemical properties, and vary greatly from soil to soil, in particular, the amount and type of clay minerals, cation exchange capacity and organic matter contents. So, the knowledge of the adsorption characteristics of soils can be an accurate and precise source of estimation for fertilizer requirement of potato crop.

In Pakistan, the use of potassium fertilizer for potato crop is not common (Hannan et al., 2011). The farmers either use generalized fertilizer for potato crop of even they do not use K fertilizer. So, the crop meets the potassium requirements from native soil K supply in tube well irrigated soils, and potassium from irrigation water in canal commanded areas. As potato is heavy K feeder and in most of the soils under potato crop the K is deficient so both sources cannot meet the K requirements of crop. The ratio of different nutrients in soil is affected by low or even no use of fertilizers for specific nutrients like P and K that restricts the crop growth and keeps it far below the genetic potential of the crop. Moreover, the potassium has pivotal role in the induction of resistance against different biotic and abiotic stresses including pests and disease resistance. The inadequate K supply to crop plants results in weakened plants which become more susceptible to various kinds of biotic and abiotic stresses such as disease stress, pests attack, heavy metal stress, frost, salinity, drought, nutritional stress, etc. These stresses affect the plant phenology and lower the ultimate yield and profitability.

Potato crop labeled as heavy feeder of K has high nutrient requirements and uptake of over 300 kg K $ha^{-1}$ is common under optimum K supply (Perrenoud, 1993). As potassium deficiency has been reported by Directorate of Soil Fertility, Pakistan in most of the soils under potato crop in Pakistan so there exists a great opportunity to increase yield and quality of potato crop by improving nutrient management. Therefore, it appears rational that soil test results and fertilizer recommendations must be site-specific, model based (Wang et al., 2016) and calibrated scientifically. Scanty information is available on determination of optimum potassium fertilizer rates for vegetables in Pakistan so it is imperative to optimize the potassium fertilizer rate for maximizing potato yield by comparing and evaluating different yield response models in potato cropping sequence. In view of this background, field experiment was conducted to see the effect of potassium fertilizer rate using adsorption isotherm technique for potato crop by improving nutrient management to work out the site specific fertilizer recommendations for potato crop in Pakistan.

## Material and Methods

A field study was conducted in potato growing tract of Pakistan (Farmer's field in Kassowal) to compare the effect of different potassium levels on the growth, yield and quality of potato by using sorption isotherm technique. The composite soil sample was collected and analyzed for the physical and chemical characteristics of the experimental field following the standard protocols (Ryan et al., 2001). The same was used to develop adsorption isotherm for the calculation of K fertilizer rates.

The potassium adsorption isotherm was constructed by using 2.50 g soil samples. These samples were equilibrated for 24 hours under shaking conditions at 25±1 °C with different K levels viz. 0, 25, 50, 75, 100, 125, 150, 175, 200, 225 and 250 µg $mL^{-1}$ in 25 mL $CaCl_2$ (0.01 M) solution. After achieving the steady state condition, the amount of K adsorbed was determined. Freundlich model was used to theoretically work out different soil solution K levels (0, 5, 10, 15, 20, 25, and 30 mg $L^{-1}$). The K fertilizer doses were calculated against these specific soil solution levels. Field experiment was conducted with seven model based K fertilizer treatments (0, 49, 94, 139, 183, 228, and 273 kg K $ha^{-1}$) and three replications in Randomized Complete Block Design (RCBD) using potato as test crop.

**Field studies**

The field experiment was conducted in Kassowal for the evaluation of different K fertilizer rates on the growth, yield and quality of potato tuber to optimize the fertilizer doze under field conditions. For this purpose, three replications were made and adsorption model based K fertilizer rates (0, 49, 94, 139, 183, 228, and 273 kg K $ha^{-1}$) were applied to potato crop as sulphate of potash using randomized complete design (RCBD). Potato seed tubers (cultivar Cardinal) were treated with fungicide Topsin-M and planted on ridges 75 cm apart at 20 cm spacing between plants. Plot size was 5m x 3m and there were four ridges per plot. All plots received basic application of 250 kg $P_2O_5$ (DAP) and 300 kg N $ha^{-1}$(Urea). Nitrogen was applied in three splits i.e. 1/3 at the time of sowing, second and third dose after 45 and 60 days of planting, respectively. An overall check (K=0) was also kept. The crop was irrigated with good quality irrigation water (canal water) with first irrigation after three days of planting. The subsequent irrigations were applied as and when required by the crop without any stress. The other plant protection measures and agronomic practices were carried out according to crop needs. Growth parameters like plant height, leaf area and chlorophyll contents were determined after 60 days of plant emergence. The crop was harvested at maturity on the development of potato tubers and data were recorded. Potato tubers were then graded manually into large (>75 g), medium (75-25 g) and small (< 25 g) sizes. Marketable tuber yield was determined and only marketable tuber data were reported in this manuscript. Internal and external K requirements of potato were worked out.

**Physiological parameters**

The physiological parameters of potato tubers were determined. Leaf area meter MK2 (Delta-T Devices Ltd, Cambridge, UK) was used for the measurement of leaf area. For the determination of chlorophyll contents, the chlorophyll meter (SPAD-502, Minolta Camera Co., Ltd, Japan) was used. The dry matter contents were determined by drying a known weight ($W_1$) of the sample in an oven at 105 °C to a constant weight ($W_2$), (AOAC, 1995).

**Plant quality parameters**

Total protein concentration was determined by following the method of Chapman and Parker (1961). For this purpose, one gram of well prepared (dried and ground sample) plant sample was digested in Kjeldahl flask by following the standard protocol. The digested material was distillated on micro Kjeldahl distillation apparatus and titrated against 0.1 N sulphuric acids. Data were used to calculate total protein concentration. The starch contents were determined by using the recommended method of Blankensh et al. (1993). Reducing sugars in the extract were estimated as described by Hortwitz (1960) while the vitamin C contents were estimated by following the method of Ruck (1961).

**Statistical analysis**

Data were subjected to statistical analyses using simple and multiple regression equations. Quadratic, square root, linear plus plateau, exponential and Quadratic plus plateau yield response models were tested using Graph pad ver.4.1. The treatment means were compared through analysis of variance techniques at 5% level of probability (Steel el al., 1997).

## Results

Field studies were conducted to see the effect of potassium fertilizer rate using adsorption isotherm technique for potato crop. The basic analysis of the soil indicated that it was normal ($EC_e$ = 0.75 dS $m^{-1}$), alkaline in reaction ($pH_s$ = 7.81), low in organic matter (0.72 %), deficient in available nutrients (N = 0.04 %, P = 5.80 ppm and K = 71.00 ppm), calcareous in nature ($CaCO_3$ = 8.71 %) and loam in texture (Table 1).

**Freundlich adsorption isotherm of the selected soil**

Results showed that Freundlich isotherm of equilibrium K concentration against adsorbed K gave a highly significant linear relationship (Fig. 1). The potassium sorption data were fitted in the Freundlich equation that gave good results with highest value of coefficient of determination (0.96**). The potassium adsorption isotherm was constructed and Freundlich model was used to theoretically work out different soil solution K levels (0, 5, 10, 15, 20, 25, and 30 mg $L^{-1}$). The K fertilizer doses were calculated against these specific soil solution levels. For the adjustment of same soil solution K levels in the field experiment, the equivalent K fertilizer rates were calculated which varied from 0 to 237 kg $ha^{-1}$ (Table 2). All potassium fertilizer rates were applied as basal dose of sulphate of potash during potato planting / at the time of sowing.

**Effect of K fertilizer on growth, physiological and yield parameters**

The results of the present study showed that plant height increased significantly with the application of K fertilizer (Table 3). It was improved from 40.67 cm with no K ($T_0$) to 56.33 cm with 139 kg K $ha^{-1}$ ($T_3$) against soil solution level of 15 mg K $L^{-1}$ which is 22.47 % increase as compared to control. Further increase in plant height with increasing K rate in $T_4$, $T_5$ and $T_6$ was statistically non-significant. The results of our study showed that potassium fertilization improved leaf area of potato crop up to 3438 $cm^2$ $plant^{-1}$ with K application @ 228 kg $ha^{-1}$ ($T_5$) as indicated in Table 3 and it was 39.44 % higher when compared with control. Adequate supply of K improved chlorophyll contents of potato plant significantly. It increased from 35.00 % with native K to a maximum level of 43.67 % with K supply @ 273 kg $ha^{-1}$ ($T_6$) equivalent soil solution level for this treatment was 30 mg K $L^{-1}$ (Table 3).

Data (Table 4) showed that tuber yield was increased with increasing rate of potassium fertilizer and the maximum potato tuber yield (34.05 Mg $ha^{-1}$) was observed in the treatment $T_3$ where 139 kg K $ha^{-1}$ was applied (Table 4). Further increase in K application rate ($T_5$ to $T_7$) could not bring about any significant change in potato tuber yield. The response of the crop to K application was due to low available status of K in soil before planting (Table 1).

The results showed that tuber dry matter contents increased with increase in potassium fertilizer rates up to a certain level and then decreased. The maximum dry matter contents (19.80 %) were observed in the treatment with K fertilizer rate of 183 kg $ha^{-1}$ followed by18.14 % in the treatment with 273 kg K $ha^{-1}$ (Table 4). In the present study, the increasing potassium levels increased the starch contents in tubers up to 81.20 % in the treatment $T_5$ with 228 kg $ha^{-1}$ of potassium (Table 5). This increase was of 9.21 % when compared with control plots. A decrease in starch contents was observed with further increase in potassium rate.

**Effect of K fertilizer on quality parameters**

The results showed that protein contents of potato tubers increased with increase in potassium fertilizer rates (Table 5). This increase ranged from 11.67 % in $T_0$ to the maximum values of 13.28% in the treatment when K fertilizer was applied at 139 kg $ha^{-1}$ i.e. 15 mg K $L^{-1}$ of soil solution. In the present study, the reducing sugars concentration was maximum (42 mg per 100g fresh weight) in control treatment that decreased with increase in K application rate. The minimum concentration (26 mg per 100g fresh weight) was observed in the treatment where 273 kg K $ha^{-1}$ was applied (Table 5).

The data showed that vitamin C contents in potato tubers increased with increasing level of K fertilizer and maximum (18.64 mg per 100g fresh weight) vitamin C contents were observed in the treatment with 228 kg K $ha^{-1}$ (Table 5). This increase was statistically significant when compared with control treatment.

**Evaluation of yield response models to predict economic optimum K rate for maximum tuber yield**

The optimum potassium fertilizer rates for potato tuber yield predicted by the models tested i.e. square root, linear plus plateau, quadratic plus plateau, exponential and quadratic model were 67, 100, 100, 32 and 179 kg K $ha^{-1}$, respectively (Table 6). The five tested models showed minor difference in $R^2$ values however, they showed large variation in calculated optimum K fertilizer rates with similar $R^2$ values. This is illustrated by present study which exhibited variations in K rates between 32 and 179 kg K $ha^{-1}$ (Table 6). The linear plus plateau models fitted the data with less bias on the basis of $R^2$ and S.E than the other models (Table 6).

**Evaluation of yield response models to predict economic optimum K rate for quality parameters**

The optimum potassium fertilizer rates for dry matter predicted by the models tested i.e. square root, linear plus plateau, quadratic plus plateau, exponential and quadratic model were 74, 103, 103, 428 and 144 kg K $ha^{-1}$, respectively (Table 7). In the present study, the optimum K fertilizer rate for potato dry matter predicted by quadratic model was lower than exponential model. It has been observed that the optimum K fertilizer rate predicted by exponential model was very high i.e. 428 kg K $ha^{-1}$ than the applied K rates.

The five tested models showed little difference in $R^2$ values however, they showed large variation in calculated optimum K fertilizer rates with similar $R^2$ values. The optimum K rate calculated by both quadratic plus plateau and linear plus plateau was (103 kg K $ha^{-1}$). Optimum K rate for protein predicted by square root, quadratic plus plateau, linear plus plateau, exponential and quadratic models were 74, 140, 180, 179 and 118 kg K $ha^{-1}$, respectively (Table 8). Optimum K rates for starch predicted by square root, quadratic plus plateau, linear plus plateau, exponential and quadratic models were 77, 222, 203, 561 and 150 kg K $ha^{-1}$, respectively (Table 9). Optimum K rates for vitamin C predicted by square root, quadratic plus plateau, linear plus plateau, exponential and quadratic models were 74, 103, 200, 430 and 84 kg K $ha^{-1}$, respectively (Table 10).

The optimum K rate for protein, starch and vitamin C calculated by the quadratic model was very less (0.283, 0.00142 and 0.00142 kg K $ha^{-1}$ respectively) than the applied K rates. The standard error of the estimate also varied greatly among models. The linear plus plateau models fitted the data with less bias on the basis of $R^2$ and S.E than the other models.

**External K requirements of potato crop**

Optimum K requirement for maximum potato tuber yield was 100 kg K $ha^{-1}$ (Table11). Regarding potato quality parameters K fertilizer was 180, 230 and 200 kg $ha^{-1}$ for protein, starch and vitamin C content respectively.

**Table 1: Physical and chemical properties of the soil used for experiment**

| **Determinant** | **Unit** | **Value** |
|---|---|---|
| $EC_e$ | dS $m^{-1}$ | 0.75 |
| $pH_s$ | - | 7.81 |
| Organic matter | % | 0.72 |
| Total N | % | 0.04 |
| Available P | mg $kg^{-1}$ | 5.80 |
| Available K | mg $kg^{-1}$ | 71.00 |
| Cation exchange capacity | $cmol^c$ $kg^{-1}$ | 9.26 |
| $CaCO_3$ | % | 8.71 |
| Sand | % | 43.00 |
| Silt | % | 35.00 |
| Clay | % | 22.00 |
| Textural Class | - | Loam (Typic Ustochrept) |

**Table 2: Freundlich model based K rate applied to potato crop in the field experiment**

| **Treatment** | $T_0$ | $T_1$ | $T_2$ | $T_3$ | $T_4$ | $T_5$ | $T_6$ |
|---|---|---|---|---|---|---|---|
| **Adjusted soil solution K levels (mg $L^{-1}$)** | 0 | 5 | 10 | 15 | 20 | 25 | 30 |
| **K rate (kg $ha^{-1}$)** | 0 | 49 | 94 | 139 | 183 | 228 | 273 |

**Table 3:Effect of potassium on plant height, leaf area and chlorophyll contents of potato crop under field conditions** (Average of three replicates)

| Treatment | Adjusted soil solution K levels (mg $L^{-1}$) | K rate (kg $ha^{-1}$) | Plant height (cm ) | Leaf area ($cm^{-2}$ $plant^{-1}$) | Chlorophyll (%) |
|---|---|---|---|---|---|
| $T_0$ | 0 | 0 | 40.67c | 2081d | 35.00d |
| $T_1$ | 5 | 49 | 44.00c | 2358c | 36.00cd |
| T2 | 10 | 94 | 48.67b | 2514c | 39.33bc |
| T3 | 15 | 139 | 56.33a | 2924b | 39.00bc |
| T4 | 20 | 183 | 56.00a | 3245a | 41.00ab |
| T5 | 25 | 228 | 55.33a | 3438a | 42.67ab |
| T6 | 30 | 273 | 55.67a | 3438a | 43.67a |

The means sharing same letters are not statistically different at 5% level of probability

**Table 4: Effect of potassium on potato tuber yield, dry matter percentage and starch contents of potato crop under field conditions** (Average of three replicates)

| Treatment | Adjusted soil solution K levels (mg $L^{-1}$) | K rate (kg $ha^{-1}$) | Potato yield (T $ha^{-1}$) | Dry matter (%) | Starch content (%) |
|---|---|---|---|---|---|
| $T_0$ | 0 | 0 | 20.04c | 12.62f | 73.56c |
| $T_1$ | 5 | 49 | 25.97b | 16.23e | 74.44c |
| T2 | 10 | 94 | 31.30a | 17.75d | 75.80bc |
| T3 | 15 | 139 | 34.05a | 18.80b | 75.76bc |
| T4 | 20 | 183 | 30.84ab | 19.80a | 78.20ab |
| T5 | 25 | 228 | 33.20a | 18.39c | 81.20a |
| T6 | 30 | 273 | 30.09ab | 18.14c | 79.56a |

The means sharing same letters are not statistically different at 5% level of probability

**Table 5: Effect of potassium on protein, reducing sugar, vitamin C contents of potato crop under field conditions** (Average of three replicates)

| Treatment | Adjusted soil solution K levels (mg $L^{-1}$) | K rate (kg $ha^{-1}$) | Protein (%) | Reducing sugar mg $(100g\ FW)^{-1}$ | Vitamin C mg $(100g\ FW)^{-1}$ |
|---|---|---|---|---|---|
| $T_0$ | 0 | 0 | 11.67c | 42.00a | 14.98c |
| $T_1$ | 5 | 49 | 11.80bc | 40.33ab | 15.40c |
| T2 | 10 | 94 | 12.11b | 41.33ab | 15.94bc |
| T3 | 15 | 139 | 13.28a | 37.33ab | 17.17bc |
| T4 | 20 | 183 | 13.16a | 36.00bc | 18.49ab |
| T5 | 25 | 228 | 13.27a | 31.00cd | 18.64a |
| T6 | 30 | 273 | 13.27a | 26.00d | 18.35a |

The means sharing same letters are not statistically different at 5% level of probability

**Table 6: Optimum rates of K fertilization predicted for potato yield by each model along with their coefficients of determination ($R^2$) and standard error of estimate values**

| Model | Optimum K rate (kg ha$^{-1}$) | Coefficien t of determination ($R_2$) | Coefficients of equations | | |
|---|---|---|---|---|---|
| | | | A | b | C |
| Square root | 67 | 0.86** | 2.16(19.46) * | 0.5375(1.775) | 0.03108(0.06043) |
| Quadratic plus plateau | 100 | 0.98** | 11.95(0.821) | 0.091(0.0389) | 0.00062(0.00037) Plateau value=32.70 (0.394) |
| Linear plus plateau | 100 | 0.99** | 10.83(0.776) | 0.1639(0.01012) | Plateau value=32.71 (0.545) |
| Exponential / Mitscherlich | 32 | 0.83** | - | 25.49(9.468) | 0.003737(0.00197) |
| Quadratic | 179 | 0.92** | 1.395(20.19) | 0.02314(0.1470) | 8.288(0.0004086) |

*Values in parenthesis are standard error of estimate ** = (p = 0.05)

**Table 7: Optimum rates of K fertilization predicted for dry matter by each model along with their coefficients of determination ($R^2$) and standard error of estimate values**

| Model | Optimum K rate (kg ha$^{-1}$) | Coefficient of determination ($R_2$) | Coefficients of equations | | |
|---|---|---|---|---|---|
| | | | A | b | c |
| Square root | 74 | 0.92** | 0.7592(12.42) | 0.1871(0.844) | 0.0585(0.02883) |
| Quadratic plus plateau | 103 | 0.98** | 0.7309(12.02) | 0.03946(0.06944) | 0.00041(0.000424) Plateau value=32.70 (0.394) |
| Linear plus plateau | 103 | 0.99** | 0.01112(0.054 86) | 0.6805(12.92) | 0.3696(18.78) |
| Exponential / Mitscherlich | 428** | 0.88** | - | 23.40(81.44) | 0.002725(0.01270) |
| Quadratic | 144 | 0.92** | 1.4113(12.83) | 0.006962(0.07125) | 2.44e-005(0.0001936) |

*Values in parenthesis are standard error of estimate ** = (p = 0.05)
**optimum rate is very high than applied rates

**Table 8: Optimum rates of K fertilization predicted for protein content by each model along with their coefficients of determination ($R^2$) and standard error of estimate values**

| Model | Optimum K rate (kg ha$^{-1}$) | Coefficient of determination ($R_2$) | Coefficients of equations | | |
|---|---|---|---|---|---|
| | | | A | b | c |
| Square root | 74 | 0.82** | 0.3849 (11.46) | 0.09484 (0.04050) | 0.05484 (0.004711) |
| Quadratic plus plateau | 140 | 0.98** | 0.05686 (11.67) | 0.003070 (0.0004450) | 3.159e-005(4.50e-005) |
| Linear plus plateau | 180 | 0.99** | 0.06247(0.01 14) | 0.2401 (11.44) | Plateau value=13.23 (0.1639) |
| Exponential / Mitscherlich | 179 | 0.80** | - | 0.003352 (0.01008) | 77.04 (196.8) |
| Quadratic | 118 | 0.81** | 0.2969 (11.45) | 0.005027(0.01315) | 1.76e-005(2026e-005) |

*Values in parenthesis are standard error of estimate ** = (p= 0.05)

**Table 9: Optimum rates of K fertilization predicted starch content by each model along with their coefficients of determination ($R^2$) and standard error of estimate values**

| Model | Optimum K rate (kg ha$^{-1}$) | Coefficient of determination ($R_2$) | Coefficients of equations | | |
|---|---|---|---|---|---|
| | | | A | b | c |
| Square root | 77 | 0.90** | 1.053 (73.59) | 0.2595(0.08634) | 0.015(0.03190) |
| Quadratic plus plateau | 222 | 0.97** | 0.6680 (73.68) | 0.1707(0.1220) | 8.942e-005(6.979e-005) |
| Linear plus plateau | 230 | 0.99** | 0.00437(0.03144) | 0.6087(72.98) | 0.8304(79.69) |
| Exponential / Mitscherlich | 561** | 0.83** | - | 0.001444 (0.005986) | 106.9 (38.14) |
| Quadratic | 150 | 0.89** | 0.9575 (73.30) | 0.001612(0.0276) | 5.69e-005(2.51e-006) |

*Values in parenthesis are standard error of estimate ** = (P= 0.05)
**optimum rate is very high than applied rates

**Table 10: Optimum rates of K fertilization predicted for vitamin C by each model along with their coefficients of determination ($R^2$) and standard error of estimate values**

| Model | Optimum K rate (kg ha$^{-1}$) | Coefficient of determination ($R_2$) | Coefficients of equations | | |
|---|---|---|---|---|---|
| | | | A | b | c |
| Square root | 74 | 0.89** | 0.5945(14.86) | 0.1465(0.01918) | 0.08469(0.01405) |
| Quadratic plus plateau | 180 | 0.96** | 0.1360(15.01) | 0.004653(0.001454) | 3.215e-005(18.46) Plateau value=32.70 (0.394) |
| Linear plus plateau | 200 | 0.99** | 0.002625 (0.01919) | 0.2971(14.61) | 0.2677(18.50) |
| Exponential / Mitscherlich | 430** | 0.65** | - | 0.008223(0.002060) | 183.6(58.24) |
| Quadratic | 84 | 0.54** | 0.4814(14.62) | 0.00815(0.022631) | 2.860e-005(2.749e-005) |

*Values in parenthesis are standard error of estimate ** = (P= 0.05)
**optimum rate is very high than applied rates

**Table 11: Economic optimum K rate predicted by Linear plus plateau model**

| Variable | Optimum K rate (kg ha$^{-1}$) |
|---|---|
| Potato tuber yield (Mg ha$^{-1}$) | 100 |
| Dry matter (%) | 103 |
| Protein (%) | 180 |
| Starch (%) | 230 |
| Vitamin C mg (100g FW)$^{-1}$ | 200 |

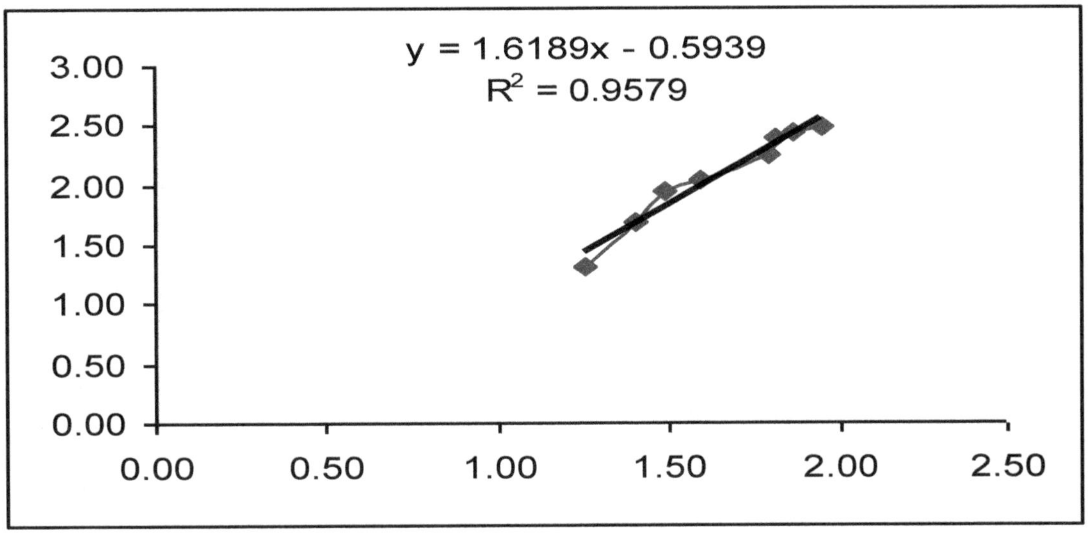


**Figure 1. Freundlich adsorption isotherm for the selected soil**

## Discussion

Field studies were conducted to see the effect of model based potassium fertilizer rate using adsorption isotherm technique for potato crop. The fate of nutrients added in the soil depends upon the initial nutrient level in the soil while the soil solution concentration depends upon the rate of nutrient removal by the plants. It also depends desorption rate of nutrient from solid phase. The adsorption based K equilibrium solution level serves as an index of K availability. So, it has been reported that equilibrium K concentration using adsorption isotherm technique provides a better index of fertility of the soil (Singh and Jones, 1975). In the present study, Freundlich isotherm of equilibrium K concentration against adsorbed K gave a highly significant linear relationship (Fig. 1). It might be due assumption of unlimited sorption sites for heterogeneous medium in Freundlich model which in turn gave better correlation in soil with illite as the dominant clay mineral. The K fertilizer doses were calculated against these specific soil solution levels to see their effect on growth, yield and quality of potato crop. All potassium fertilizer rates were applied as basal dose during potato planting.

The results of the present study showed that plant height increased significantly with the application of K fertilizer. Although plant height is a genetic factor but it can be used as an indicator of crop performance and can be improved through balanced nutrition. Different growth, physiological and yield parameters contribute to the yield of the cop. For potato crop, the plant height, leaf area and chlorophyll contents are considered as important determinations used to describe crop performance (Taiz and Zeiger, 2006). These indices indicate the carbon assimilation rate of the potato plant system and its ultimate conversion in to sink (tubers).

The results of our study showed that potassium fertilization improved leaf area of potato crop. Leaf area is used an indicator for photosynthetic efficiency of the plants as it captures light thus an increase in leaf area results an improvement in photosynthetic rate. It has been observed that leaf area has critical role in studies of plant competition, plant nutrition, plant protection measures, plant soil-water relations and crop ecosystems (Mohsenin, 1986). Further increase in leaf area with increasing K rate was statistically non-significant. Similarly, Al-Moshileh et al. (2005) reported significant the improvement in growth parameters such as plant height, and leaf area of potato crop with increasing level of potassium fertilizer.

Chlorophyll is the real plant factory for manufacturing food. Adequate supply of K improved chlorophyll contents of potato plant significantly. The improvement in plant height, leaf area and chlorophyll contents (%) might be due to increase in photosynthetic rate which in turn related to improved stomatal conductance and higher ribulase bisphosphate carboxylase activity resulting in rapid rate of $CO_2$ fixation (Cakmak and Engels, 1999).

Results of our study showed that tuber yield was increased with increasing rate of potassium fertilizer. The response of the crop to K application might be due to low available status of K in soil before planting (Table 1). The farmers, in general depend on the native sources of K in irrigation water and soil minerals. The K requirement of potato crop is more that is required for carbohydrate metabolism and other physiological functions (Singh and Trehan, 1998; Incrocci et al., 2017) along with conversion of N and P in plants (Mengel and Kirkby, 1987). Potassium is known to facilitate the efficient translocation of photosynthates to the developing tubers (Beringer, 1978; Kavvadias et al., 2012). This fact is evident from the present study, that there was a progressive increase in tuber yield with each incremental level of added K up to 155 kg $ha^{-1}$. These results are in line with those of Hannan et al. (2011) who reported an increase in potato tuber yield with increasing K rate using isotherm technique. These might be due to translocation of more photosynthates to the tubers. In fact, potassium has critical role in the translocation photosynthates to storage organs (Romheld and Kirkby, 2010).

The results showed that tuber dry matter contents increased with increase in potassium fertilizer rates up to a certain level and then decreased. The decrease in dry matter contents with higher levels of potassium might be due to increase in water contents to maintain cell turgor pressure (Hannan et al., 2011). The results are in line with the previous studies conducted by Kumar et al. (2004) who have reported a decrease in dry matter contents with increasing K fertilizer rates. The potassium affects the water contents of the plasma thus increases the water contents of tubers. The reduction in dry matter contents with increase in potassium rate might also be due to translocation of more photosynthates and water to the tubers which in turn decreased the dry matter contents (Kavvadias et al., 2012).

Potassium stimulates the activity of different important enzymes in the plants such as starch synthatase that catalyzes simple sugars into complex polysaccharide i.e. starch (Mengel and Kirkby, 1987). In the present study, the increasing potassium levels increased the starch contents in tubers. A decrease in starch contents was observed in the present study that might be due to increase in water uptake through potassium application that restricted the number of amyloplasts in cytoplasm (Perrenoud, 1993). The same results have also been reported in previous work conducted by Hannan et al. (2011).

The results showed that protein contents of potato tubers increased with increase in potassium fertilizer rates in our study. These results are supported with the work of Khan et al. (2012 who reported that protein contents were improved with the application of potassium and these increased with increasing rates. This improvement in protein contents might be due to more conversion of photosynthates to protein and other storage components such as starch.

The concentration of reducing sugars in tubers is an important quality indicator for potato processing industries. In the present study, the reducing sugars concentration decreased with increase in K application rate. It appeared that still higher K levels need to be tested for finding out optimum K fertilizer rate for minimum reducing sugars in the potato tubers. Similarly, Hannan et al. (2011) reported a decrease in reducing sugars contents with increasing K fertilizer levels.

The data showed that vitamin C contents in potato tubers increased with increasing level of K fertilizer. This increase was statistically significant when compared with control treatment. It has been reported that potassium triggers the photosynthates production their translocation to tubers thus improves the conversion of photosynthates in to secondary metabolites such as vitamins (Mengel and Kirkby, 1987: Khan et al., 2012). Potassium has been reported to be engaged in the conversion of radiant energy into chemical energy. This metabolic energy is required for plant metabolism that results in increased production of proteins and starch, and decreases the reducing sugars contents thus improving the quality of potato tubers (Incrocci et al., 2017). The higher energy status resulting with optimum K supply promotes the synthesis of Vitamin C and other secondary metabolites.

In the present study, the optimum K fertilizer rate for tuber yield predicted by quadratic model was higher than exponential model while the optimum K fertilizer rate for potato dry matter predicted by quadratic model was lower than exponential model. These results are supported by the work of Hannan et al. (2011) where they reported the large variation in optimum K rate due to inappropriate model selection. They reported that Linear plus plateau was the most suitable model and best option for optimum K fertilizer recommendations as observed in these studies. The optimum K fertilizer rates predicted by different models varied greatly (Neeteson and Wadman, 1987; Incrocci et al., 2017)).

It has been observed that coefficient of determination is a poor criterion to select a model for optimization of K fertilizer rate. The same results have been reported in previous studies (Colwell, 1994; Hannan et al., 2011). The standard error of the estimate also varied greatly among models. So, it can be concluded from this study that the linear plus plateau model is best suited to describe the yield response of potato to K fertilizer and to predict the economic optimum K rates.

## Conclusion

The results showed that growth parameters like plant height, leaf area and chlorophyll contents contribute significantly to potato tuber yield. The quality parameters like protein, starch and vitamin C contents were significantly affected by increasing K rates. Out of the tested models, linear plus plateau and Quadratic plus plateau predicted (equally well) the optimum fertilizer K rate for both for yield and quality attributes of potato crop. It was observed that the optimum fertilizer K rates (at 95 % relative yield) for potato tuber yield, protein, starch and vitamin C contents were 100, 180, 230 and 200 kg K $ha^{-1}$, respectively. So, it is suggested that adsorption isotherm technique should be used to calculate site specific and crop specific fertilizer requirements of crops. It can overcome the problem of sporadic responses to fertilizer as well. The K fertilizer application would increase the quality of crop that would support the quality based marketing system in Pakistan leading to improved profitability for the farmers.

## Acknowledgement

The authors acknowledge the support of technical staff during the laboratory analysis. The first author also acknowledges the guidance from Dr. Abdul Hannan during the conduct and write-up of the work.

## References

Al-Moshileh A, Errebhi MMA and Motawei MI, 2005. Effect of various potassium and nitrogen rates and splitting methods on potato under sandy soil and arid environmental conditions. Emir. J. Agric. Sci. 17:01-09.

AOAC, 2005. Official Methods of Analysis. 18th edn. Association of Official Analytical Chemists, Arlington, VA, USA.

Beringer H, 1978. Functions of potassium in plant metabolism with particular reference to yield, pp. 185-202. in Sekhon G.S (ed.), Potassium in soils and crops. Gurgaon, Haryana, India: Potash Research Institute of India.

Blankensh SM, Ellsuworth DD and Powell RL, 1993. A ripening index for banana fruit based on starch content. J. Hort. Technol. 3:338-339.

Cakmak I and Engels C, 1999. Role of mineral nutrition in photosynthesis and yield formation, pp. 141-168.in Rengel Z (ed.), Mineral nutrition of crops: mechanisms and implications. The Haworth Ptress, New York, USA.

Chapman HD and Parker F, 1961. Determination of NPK, pp. 150-179. In methods of analysis for soil, plant and water. Division of Agriculture, University of California, USA.

Colwell JD, 1994. Estimating fertilizer requirements: A quantitative approach. CAB International, Wallingford, UK.

FAO, 2016. Fertilizer use by crop in Pakistan. Natural Resources Management and Environment Department, Food and Agriculture Organization of the United Nations, USA. Available at: http://www.fao.org/docrep/007/y5460e/y5460e0.htm. Accessed on 22-08-2016.

Hannan A, Arif M, Ranjha AM, Abid A, Fan XH and Li YC, 2011. Response models to determine potassium fertilizer rates for potato crop on a calcareous soil in Pakistan. Comm. Soil Sci. Plant Anal. 42(6): 645-655.

Hortwitz W, 1960. Official methods of analysis. Association of Official Agric. Chemists, Washington, DC.9:314-320.

Incrocci L, Massa D and Pardossi A, 2017. New trends in the fertigation management of irrigated vegetable crops. Horticulturae 3, 37; doi:10.3390/horticulturae3020037.

Kavvadias V, Paschalidis C, Akrivos G and Petropoulos D, 2012. Nitrogen and potassium fertilization responses of potato (*Solanum tuberosum*) cv. Spunta. Comm. Soil Sci. Plant Anal. 43(1-2): 176-189.

Kenyanya O, Mbuvi HM, Muthengia JM and Omayo EK, 2014. Use of adsorption isotherm models to determine potassium fertilizer acreage doses for optimum maize growth and yields in Nyamira County, Kenya. Intl. J. Agri. Crop Sci. 8 (15): 1525-1536.

Khan MZ, Akhtar ME, Mahmood-ul-Hassan M, Mahmood MM and Safdar MN, 2012. Potato tuber yield and quality as affected by rates and sources of potassium fertilizer. J. Plant Nutr. 35(5): 664-677.

Kumar D, Singh BP and Kumar P, 2004. An overview of the factors affecting sugar content of potatoes. Ann. Appl. Biol. 145: 247-256.

Louison L, Omrane A, Ozier-Lafontaine H and Picart D, 2015. Modeling plant nutrient uptake: Mathematical analysis and optimal control. Lect. Notes Pure Appl. 4: 193–203.

Mengel K and Kirkby EA, 1987. Principles of plant nutrition, 4th Ed. International Potash Institute, Worblaufen-Bern, Switzerland.

Mengel K, Rahmatullah and Dou H, 1998. Release of potassium from the silt and sand fraction of loss-derived soils. Soil Sci. 163: 805-813.

Mohsenin N, 1986. Physical properties of plant and animal materials, Gordon and Breach Science Publishers, New York, USA.

Neeteson JJ and Wadman WP, 1987. Assessment of economically optimum application rates of fertilizer N on the basis of response curves. Fert. Res. 12: 37-52.

Perrenoud S, 1993. Fertilizing for higher yield potato, IPI Bull. 8, 2nd Ed. International Potash Institute, Worblaufen-Bern, Switzerland.

Romheld V and Kirkby EA, 2010. Research on potassium in agriculture: Needs and prospects. Plant Soil 335: 155-180.

Ruck JA, 1961. Chemical methods for analysis of fruits and vegetable products. Res. St. Summer Land. Res. Branch Canada. Deptt. of Agric. 1154.

Ryan J, Estefan G and Rashid A, 2001. Soil and plant analysis laboratory manual, 2nd edition, International Center for Agriculture in Dry Areas (ICARDA), Syria, 172p.

Samadi A, 2003. Predicting phosphate requirement using sorption isotherms in selected calcareous soils of western Azarbaijan province, Iran. Comm. Soil Sci. Plant Anal. 34(19-20): 2885-2899.

Samadi A, 2006. Potassium exchange isotherms as a plant availability index in selected calcareous soils of Western Azarbaijan Province, Iran.Turk. J. Agric. 30: 213-222.

Singh BB and Jones JP, 1975. Use of sorption-isotherms for evaluating potassium requirement of plants. Soil Sci. Soc. Am. J. 39:881-886.

Singh JP and Trehan SP, 1998. Balanced fertilization to increase the yield of potato. in Barar MS and Bansal SK (eds.), Proceedings of the IPI-PRII-PAU Workshop on Balanced Fertilization in Punjab Agriculture, International Potash Institute, Worblaufen-Bern, Switzerland.

Steel RGD, Torrie JH and Dicky DA, 1997. Principles and procedures of statistics a biometrical approach. (3rd Ed.). pp. 204-227. McGraw Hill Book International Co., Singapore.

Taiz L and Zeiger E, 2006. Plant defenses: Surface protectants and secondary metabolites, pp. 283-308. In Taiz L and Zeiger E (ed.), Plant Physiology, Sinauer Associates, Sunderland, MA.

Thompson RB, Incrocci L, Voogt W Pardossi A and Magán JJ, 2017. Sustainable irrigation and nitrogen management of fertigated vegetable crops. Acta Hortic. 1150: 363–378.

Wang C, Boithias L, Ning Z, Han Y, Sauvage S, Sánchez-Pérez JM, Kuramochi K and Hatano R, 2016. Comparison of Langmuir and Freundlich adsorption equations within the SWAT-K model for assessing potassium environmental losses at basin. Agric. Water Manage. doi.org/10.1016/j.agwat.2016.08.001.

# 24

# Combined application of sorghum and mulberry water extracts is effective and economical way for weed management in wheat

**Shahbaz Khan[1], Sohail Irshad[2]*, Faisal Mehmood[1], Muhammad Nawaz[3]**
[1]Department of Agronomy, University of Agriculture, Faisalabad, Pakistan.
[2]In-Service Agricultural Training Institute, Rahim Yar Khan, Pakistan.
[3]Department of Agronomy, College of Agriculture, BZU, Bahadur Campus Layyah, Pakistan.

## Abstract
Wheat (*Triticum aestivum* L.) is of prime importance being staple food of the masses of Pakistan. Weed infestation in wheat not only reduce the yield but also affects the efficiency of other production factors. Allelopathy is sustainable and ecofriendly method for the management of weeds and diseases. A field trail was carried out to assess the allelopathic potential of mulberry and sorghum water extracts against invasive winter weeds like *Phalaris minor*. Retz, *Chenopodium album* L., *Avena fatua* L. and *Convolvulus arvensis* L. in wheat at Agronomic Research Farm, University of Agriculture, Faisalabad. Experiment was laid out in randomized complete block design (RCBD) with four replications. All the treatment application showed better weed management and enhanced final yield of the crop. Application of sorghum water extract (SWE) at the rate of 18 L $ha^{-1}$ and mulberry water extract (MWE) at the rate of 18 L $ha^{-1}$ indicated better weed management (51-55%) and improved grain yield (28%) as compared to control. SWE at the rate of 27 L $ha^{-1}$ and MWE at the rate of 9 L $ha^{-1}$ in combined application also showed better weed management and improved the grain yield while Atlatis (Standard herbicide) had 66-68% weed control and 32% increase in final yield but it was not found economical. Combined application of SWE at the rate of 18 L $ha^{-1}$ and MWE at the rate of 18 L $ha^{-1}$ was found best combination with maximum net return while hand weeding and herbicide application were extravagant because of high cost and low net return.

*Corresponding author email: sohail_99uaf@yahoo.com

**Keywords**: Allelopathy, Wheat, Water Extracts, Weeds

## Introduction

Wheat is of prime importance being the main source of food due to its nutritional value since long. Most of the nutritional requirements of human body are met through this crop. It has main role in national economy. The final yield of the crop is not enough to meet dietary requirements of ever increasing population. Multiple factors including weed infestation, late sowing, improper nutrition and water use efficiency are responsible for limiting the yield (Jabran et al., 2011). A number of constraints hinder wheat production like poor quality seed, poor storage conditions, high prices of fertilizers and inadequate measures of plant protection principally weed infestation. Noxious weeds of wheat like *Phalaris minor* and *Avena fatua* reduce the yield up to 30%. Reduction in the wheat yield up to 20-30% was observed in case of high infestation of field with *Avena fatua* L. (Hassan and Khan, 2007).

Most of the weeds have established resistance against herbicides (Heap, 2008) and use of these agrochemicals have many health concerns (Kudsk and Streibig, 2003) with the problems like their residual effects and degradation in soil. Application of Mesotrione (Selective herbicide) to soil affects the population of microbes (Crouzet et al., 2010). Allelopathy is a novel method for the management of weeds and diseases. It refers to positive or negative impact of one plant to other including both weed and crop species by the release of some allelochemicals from plant parts through leaching, residue decomposition, root exudates and other ways in agro ecosystem (Ferguson and Rathinasabapathi, 2003). It has stimulatory along with inhibitory effect on plants as well as on microorganism. Number of studies depicted the effectiveness of allelochemicals against crop pests while working on the effect of sorghum, sunflower and brassica water extracts along with reduced dose of glyphosate for the management of purple nutsedge in cotton, Iqbal et al. (2009) investigated a reduced density and biomass of purple nutsedge.

Allelochemicals can interrupt the normal functioning of plant and can affect the physiological functions like respiration, photosynthesis and cell elongation (Field et al., 2006). There is mounting problem of pesticide and herbicide resistance in modern agriculture. Owing to pesticide resistance number of health related problems are increasing, unselective use distressed the ecosystem with soil and environment concerns, in this scenario allelopathy can play its role to cope these problems (Zhu and Li, 2002; Roseveld and Bretveld, 2008; Dayan et al., 2009). It is common observation of reduced vegetation under mulberry tree and its leaf defoliation in autumn suppress the growth of neighboring weeds. Use of mulberry water extracts reduced germination, root and shoot length with reduced dry weight of raddish (Hong et al., 2003). Wheat yield is affected by different broad and narrow leaf weeds including broad leaf dock, sweet clover, swine cress, field bind weed, fumitory, wild medic, little seed canary grass and wild oat (Shamsi and Ahmed, 1984).

Hong et al. (2004) reported decrease in weed density (65%) and dry weights (70%) when mulberry leaves were incorporated in soil at the rate of 2 t ha$^{-1}$ in rice crop which ultimately increased the yield of the paddy by 23%. Some of the studies report the stimulatory effect of mulberry e.g. mulberry water extract of 50% concentration promoted the growth and development in broad bean and pea while in case of lentil, germination and growth showed positive impact by the single application of 25% concentration of mulberry water extract (Mughal, 2000).

Present study was conducted to assess the combined and alone application of mulberry and sorghum water extracts for the management of weeds in wheat and in particular to endorse an ecofriendly approach to reduce environmental and human health hazards by indiscriminate use of agrochemicals.

## Material and Methods

A field trail was conceded to assess the allelopathic impact of sorghum and mulberry water extracts on weeds of wheat during the season 2014-2015 at Agronomic Research Area, University of Agriculture, Faisalabad, using randomized complete block design (RCBD) with four replications. Wheat cultivar AARI-2011 was sown by hand drill with R*R of 22 cm with gross plot size of 7 m × 2.2 m and 10 rows in each plot. Sowing was done in well pulverized soil (soil type: Lyallpur soil series (Aridsol-fine silty, mined, hyperthermic, ustalfic, Haplagrid) in USDA classification). Fertilizers were applied at the rate of 100-90-75 kg ha$^{-1}$. Irrigations were given according to the crop requirement. Treatments combination plan has been shown in table 1.

### Preparation of extracts and herbicidal solution

Water extracts of sorghum and mulberry were prepared according to Cheema and Khaliq (2000a). Briefly, mulberry leaves were collected from the Agronomic Research Area, University of Agriculture, Faisalabad, Pakistan. The leaves were washed three times under running tap water and were dried in an oven at 70°C for 48 h till a constant weight was attained. The dried leaves were soaked in tap water at 10% (w/v) at room temperature (25 _ 3°C) for 24 h. After that, the material was filtered to obtain a 10% extract. Then, the extract was boiled to concentrate it by 95%. This concentrated extract was named as 100% (stock solution). Mature sorghum crop was used for all these treatment preparations. It was placed in the shade for couple of days to avoid leaching by rain. After drying in the shade , it was chopped by ordinary fodder cutter ,thereafter, then this material was soaked in water in 1:10 ratio for 24 hours and concentrated to reduce its volume by 95% via boiling at100◦C (Cheema and khaliq, 2000a). The volume of the spray was 330 L ha $^{-1}$. Recommended doses of iodosulforun + mesosulforun at the rate of 14.4 g ha$^{-1}$ a.i. (Atlantis 3.6 WG); product of Bayer crop sciences, was used.

**Application of treatments**

There was no application of treatment in weedy check plot. It was left untreated throughout the crop growth period. Hand weeding was done twice at 30 and 45 days after sowing the crop. Mixture of Iodosulfuran and Mesosulfuron (Atlantis 3.6WG) at the rate of 14.4 g a.i. ha$^{-1}$ was sprayed after 30 days of sowing of crop. Similarly all other combinations of mulberry and sorghum water extracts were applied once in crop growth period after 30 days of sowing the wheat crop.

**Weed Parameters**

Weed density was calculated by counting all weeds present in 1 m$^2$. Fresh weight of all weeds was recorded by pulling out all of weeds plants and weighing them by ordinary scientific balance. After recording the fresh weight, weed samples were first sun dried, thereafter placed in an oven at 70°C for 72 hours till then all of the moisture was removed. These samples were weighed by weighing balance to get dry weight.

**Yield and yield components**

Total number of tillers was calculated by counting the number of tiller from an area of 1 m$^2$. Regarding the data of number of spikelets/spike, five spikes were selected at random from each plot and number of spikelets was calculated and means of the calculated values was determined. For the estimation of final grain yield, an area of 1 m$^2$ of fully mature crop was harvested from each plot. After threshing, the weight of grains was taken.

**Economic analysis**

Economic analysis was carried to determine which treatment was much cost effective. Economic analysis was done through method prescribed by Byerlee et al. (1986).

**Statistical analysis**

Data were analyzed statistically by using Statistix 8.1 version (a computer software package for statistical analysis) and the difference between treatments' means was compared by employing LSD test at 5% level of probability (Steel et al., 1997).

## Results

**Weed Parameters**

**Total weed density/m$^{-2}$**

Significant impact was observed on growth parameters of weed when applied with allelopathic water extracts. A number of weed growth parameters like fresh and dry biomass of weeds and weed density provides the information of weed infestation in the field. In the present study, weed density was calculated after 45 and 60 after sowing of the crop. Different broad and narrow leaf weeds like *Phalaris minor, Avena fatua, chenopodium album* and *convolvulus arvensis* infestation in the field was observed. Most of treatments applied showed reduced weed population (Table 3) but combined application of SWE at the rate of 18 L ha$^{-1}$ and MWE at the rate of 18 L ha$^{-1}$ reduced extreme weed density by 55% at 45 days after sowing (DAS) and 60% at 60 DAS while combined application of SWE at the rate of 27 L ha$^{-1}$ and MWE at the rate of 09 L ha$^{-1}$ depressed weed density by 41% at 45 DAS and 54% at 60 DAS. Total fresh and dry weed biomass was also affect significantly by the phytotoxic effect of allelochemicals of extracts when compared to control. Fresh biomass of weeds was also reduced by the application of mixture of Iodosulfuran + Mesosulfuron (Atlantis 3.6 WG) at the rate of 14.4 g a.i. ha$^{-1}$ up to 65-75% in comparison to other treatments this trend was followed by combined application of water extracts of both sorghum and mulberry at the rate of 18 L ha$^{-1}$ each by 51-56% while alone application of both the extracts found less effective in reducing the fresh biomass of weeds. In case of dry weights of the weeds similar observations were recorded and it was reduced up to 66-68% in case of application of (Atlantis 3.6WG) at the rate of 14.4 g a.i. ha$^{-1}$) this trend was followed by combined application of both the water extracts at the rate of 18 L ha$^{-1}$ of each with the reduction in dry weight up to 51-55%. It was observed that the combine application of both water extracts reduced the dry weight of weeds as compared to sole application but alone application of MWE was found less effective as compared to SWE in reducing dry weight of weeds.

**Yield parameters**

Total tillers, number of spikelets/spike and grain yield were expressively affected by the application of definite treatments in comparison with control. In case of hand weeding, number of tillers and spikelets/spike were more and this trend was at par where herbicide application was done. In case of crop water extracts, combined application of both the extracts at the rate of 18 L ha$^{-1}$ of both produced maximum number of tillers which were statistically similar to application of herbicide (Table 2). Least number of tillers and spikelets/spike were observed in case of weedy check which was found statistically similar to alone application of MWE at the rate of 18 L ha$^{-1}$. Same was the case with the grain yield; maximum grain yield was recorded in hand weeding which was similar to herbicide application while minimum grain yield was observed in case of weedy check (Table 2).

**Economic analysis**

Effectiveness of the production scheme can be assessed for its economic fitness for the farmers. Present study was evaluated by conducting economic analysis along with marginal rate of return to determine best treatments applied. Results of the analysis depicted increase in income of Pakistani rupees 2570 to 25803 ha$^{-1}$ in different combination of treatments as compared to control treatment. Maximum net benefits (Pak Rs. 25803) (236.72$) were obtained with combined application of both crop water extracts at the rate of 18 L ha$^{-1}$ each while it was followed by application of herbicide (Pak Rs. 25463) (233.60$) and hand weeding with Pakistani rupees of 24850.5. Minimum net return (Pak Rs. 2570) (23.57$) was obtained in case of sole application of MWE at the rate of 18 L ha$^{-1}$.

**Table 1: Treatment Combinations**

| | Description |
|---|---|
| $T_1$ | Weedy check (Weed was not removed throughout growth period) |
| $T_2$ | Hand weeding (30 and 45 DAS; days after sowing) |
| $T_3$ | Iodosulfuran + Mesosulfuron (Atlantis 3.6WG) at the rate of 14.4 g a.i. ha$^{-1}$ (30 DAS) |
| $T_4$ | Sorghum water extract18 L ha$^{-1}$ (30 DAS) |
| $T_5$ | Sorghum water extract 27 L ha$^{-1}$ (30 DAS) |
| $T_6$ | Sorghum water extract 36 L ha$^{-1}$ (30 DAS) |
| $T_7$ | Mulberry water extract 18 L ha$^{-1}$ (30 DAS) |
| $T_8$ | Mulberry water extract 27 L ha$^{-1}$ (30 DAS) |
| $T_9$ | Mulberry water extract 36 L ha$^{-1}$ (30 DAS) |
| $T_{10}$ | Sorghum water extract 27 L ha$^{-1}$ + Mulberry water extract 09 L ha$^{-1}$ (30 DAS) |
| $T_{11}$ | Sorghum water extract 18 L ha$^{-1}$ + Mulberry water extract 18 L ha$^{-1}$ (30 DAS) |
| $T_{12}$ | Sorghum water extract 09 L ha$^{-1}$ + Mulberry water extract 27 L ha$^{-1}$ (30 DAS) |

**Table 2: Effect of sorghum and mulberry water extracts on yield parameters of wheat**

| **Treatments** | **Number of tillers m$^{-2}$** | **Number of spikelets per spike** | **Grain yield (t ha$^{-1}$)** |
|---|---|---|---|
| Weedy check | 331 g | 11 e | 3.88 f |
| Hand weeding | 382 a | 18 a | 5.31 a |
| Iodosulfuran + Mesosulfuron | 374 ab | 17 ab | 5.17 ab |
| SWE at the rate of 18 L ha$^{-1}$ | 347 ef | 13 de | 4.08 de |
| SWE at the rate of 27 L ha$^{-1}$ | 350 de | 13 de | 4.10 de |
| SWE at the rate of 36 L ha$^{-1}$ | 357 cd | 14 cd | 4.44 d |
| MWE at the rate of 18 L ha$^{-1}$ | 337 fg | 13 de | 4.02 e |
| MWE at the rate of 27 L ha$^{-1}$ | 343 ef | 13 de | 4.08 de |
| MWE at the rate of 36 L ha$^{-1}$ | 352 de | 14 cd | 4.27 d |
| SWE at the rate of 27 L ha$^{-1}$ + MWE at the rate of 09 L ha$^{-1}$ | 358 cd | 16 bc | 4.70 c |
| SWE at the rate of 18 L ha$^{-1}$ + MWE at the rate of 18 L ha$^{-1}$ | 368 bc | 16 abc | 5.00 b |
| SWE at the rate of 09 L ha$^{-1}$ + MWE at the rate of 27 L ha$^{-1}$ | 361 cd | 15 bc | 4.50 c |
| LSD at 5% probability level | 11.695 | 2.2781 | 0.2166 |

Different lettering shows the statistically significant difference among the performance of treatments ($P < 0.05$) DAS = Days after sowing; SWE = Sorghum water extract; MWE = Mulberry water extract; LSD = Least significant difference.

**Table 3: Effect of sorghum and mulberry water extracts on total weed density ($m^{-2}$)**

| **Treatments** | **Weed density at 45 DAS ($m^{-2}$)** | **Weed density at 60 DAS ($m^{-2}$)** | **Fresh weight at 45 DAS (g $m^{-2}$)** | **Fresh weight at 60 DAS (g $m^{-2}$)** | **Dry weight at 45 DAS (g $m^{-2}$)** | **Dry weight at 60 DAS (g $m^{-2}$)** |
|---|---|---|---|---|---|---|
| Weedy check | 35.25 a | 38.25 a | 47.25 a | 56.07 a | 5.97 a | 7.09 a |
| Hand weeding | 0.00 i | 0.00 i | 0.00 h | 0.00 h | 0.00 i | 0.00 h |
| Iodosulfuran + Mesosulfuron | 8.25 h | 7.250 h | 15.00 g | 14.00 g | 1.89 h | 2.37 g |
| SWE at the rate of 18 L $ha^{-1}$ | 27.00 bc | 23.75 bc | 40.25 bc | 48.18 bc | 5.09 bc | 6.09 bc |
| SWE at the rate of 27 L $ha^{-1}$ | 25.25 d | 24.25 bc | 39.00 bc | 45.01 bcd | 4.93 cd | 5.94 bc |
| SWE at the rate of 36 L $ha^{-1}$ | 23.75 e | 21.75 de | 33.50 d | 38.23 d | 4.23 e | 5.09 d |
| MWE at the rate of 18 L $ha^{-1}$ | 28.25 b | 25.50 b | 44.00 ab | 50.37 ab | 5.48 b | 6.62 ab |
| MWE at the rate of 27 L $ha^{-1}$ | 26.75 c | 23.50 cd | 42.00 b | 49.65 ab | 5.34 b | 6.53 ab |
| MWE at the rate of 36 L $ha^{-1}$ | 26.50 cd | 23.00 cd | 35.75 cd | 40.87 cd | 4.52 de | 5.42 cd |
| SWE at the rate of 27 L $ha^{-1}$ + MWE at the rate of 09 L $ha^{-1}$ | 20.75 f | 17.50 f | 24.50 ef | 31.42 ef | 3.09 f | 3.92 ef |
| SWE at the rate of 18 L $ha^{-1}$ + MWE at the rate of 18 L $ha^{-1}$ | 15.75 g | 15.00 g | 20.75 f | 26.90 f | 2.63 g | 3.44 f |
| SWE at the rate of 09 L $ha^{-1}$ + MWE at the rate of 27 L $ha^{-1}$ | 22.00 f | 20.75 e | 27.50 e | 32.97 e | 3.42 f | 4.17 e |
| **LSD at 5% probability level** | 1.3849 | 1.8073 | 5.0765 | 6.9199 | 0.4145 | 0.7176 |

Different lettering shows the statistically significant difference among the performance of treatments ($P < 0.05$) DAS = Days after sowing; SWE = Sorghum water extract; MWE = Mulberry water extract; LSD = Least significant difference

## Discussion

Weed density is the key parameter that gives index of weed infestation in the field. In present study, weed density was recorded after 45 DAS and 60 DAS. Infestation of different braod and narrow leaf weeds like *convolvulus arvensis*, *Avena fatua*, *chenopodium album* and *Phalaris minor* was observed during the study in the field. Combined application of SWE and MWE at the rate of 18 L $ha^{-1}$ of both showed maximum weed control as compared to other treatments as depicted in Table 3. There was substantial variation in weed control between 45 and 60 DAS; this might be due to natural variation of weed seed bank in soil. Findings of our study depicted that sole or combine application of both crop water extracts had more or less effect on weed density in field. These results are in line with the findings of Cheema and Khaliq (2000) who reported less population of weeds when applied with sorghum water extract. Our findings are quite similar to the findings of Cheema et al. 2000 who reported weed control up to 16-55% by spraying sorghum water extract. Hong et al. (2004) reported 65% reduction in weed density in paddy field when applied with mulberry water extract. Parameters like fresh and dry weight are also of prime importance as it regulates the uptake of different nutrients from soil profile by altering it into micro and macro molecules. Allelopathic extracts when applied to the crop, affected the fresh as well as dry weight (Table 3).

Alone application of MWE was less effective in suppressing weed fresh weight as compared to SWE. It might be possible due to the presence of allelochemicals present in water extracts that possessed inhibitory effect rather than killing weeds. Our results are in line with the findings of Putnam and Defrank (1979) who reported that crop plants release some of the chemicals that have inhibitory effect which could be used for the management of weeds. Our findings also confirm the study of Purvis et al. (1985) who stated inhibitory effect of sorghum that suppressed the dry biomass of weeds due to these allelochemicals.

Improvement in the yield parameters of wheat was observed due to less wheat density by the application of water extracts. Present study revealed that presence of allelochemicals in water extracts have promoting effect on wheat while negative impact on weed growth and development. Application of water extracts significantly affected yield parameters like number of spikelets/spike, total tillers and grain yield. Maximum number of tillers, spikelets/spike, total tillers and grain yield were observed in case of hand weeding which was statistically at par with herbicide application. In case of water extracts, combined application of water extracts of SWE and MWE at the rate of 18 L $ha^{-1}$ produced maximum number of tillers and other yield parameters (Table 2). Improvement in spikelets/spike may be due to better weed management. Our results are supported the findings of Einhelling and Rasmussen (1989) who reported inhibitory effect of sorghum on weeds resulted in improvement in the yield of wheat. Likewise, findings of Hejl and Koster (2004) and Jabran et al. (2008) also supports our findings who reported inhibitory effects of sorghum and mulberry on weeds and their growth. Increase in grain yield might be due to suppressing ability of water extracts to weeds which ultimately caused in less competition of nutrients between crop and weeds resulted in more number of tillers, more spikelets/spike, increase in spike length, more number of grains/spike and grain weight. These results are in accordance with Haq et al. (2010) who reported better weed management in wheat by the application of MWE resulted in increase in grain yield.

## Conclusion

It is concluded from present study that combined application of SWE at the rate of 18 L/ ha and MWE at the rate of 18 L/ha at 45 and 60 days after sowing of wheat helped to achieve maximum weed control. It also produced highest grain yield by improvement in yield contributing parameters with maximum net benefit returns. Sorghum and mulberry water extracts can be used effectively to control weed in wheat crop. It is recommended to use both these water extracts as tool for better weed management approach in wheat as these are safe and ecofriendly.

## References

Cheema ZA and Khaliq A, 2000. Use of sorghum allelopathic properties to control weeds in irrigated wheat in a semi-arid region of Punjab. Agric. Eco. Envt. 79: 105-112.

Cheema ZA, Sadiq HMI and Khaliq A, 2000. Efficacy of Sorgaab (Sorghum water extract) as a natural weed inhibitor in wheat. Int. J. Agri. Biol. 2: 144-146.

Crouzet O, Batisson I, Hoggan PB, Bonnemoy FEE, Bardot C, Poly F, Bohatier J and Mallet C, 2010. Response of soil microbial communities to the herbicide mesotrione: A dose-effect microcosm approach. Soil Biol. Biochem. 42: 193-202.

Dayan FE, Cantrell CL and Duke SO, 2009. Natural products in crop protection. Bioorg. Med. Chem. 17: 4022-4034.

Einhelling FA and Rasmussen JA, 1989. Prior cropping with grain sorghum inhibits weeds. J. Chem. Ecol. 15: 951-960.

Ferguson JJ and Rathinasabapathi B, 2003. Allelopathy: How plants suppress other plants. online. Internet. Available from URL: http://edis.ifas.ufl.edu./ Accessed October 2011.

Field B, Jordan F and Osbourn A, 2006. First encounters – deployment of defence-related natural products by plants. New Phytol. 172: 193-207.

Haq RA, Hussain M, Cheema ZA, Mushtaq MN and Farooq M, 2010. Mulberry leaf water extract inhibits bermuda grass and promotes wheat growth. Weed Biol. Manage. 10: 234-240.

Hassan G and Khan AI, 2007. Yield and yield components of wheat affected by wild oat *Avena fatua* L densities under irrigated conditions in: Afri. Crop Sci. Proceed. 8: 33-36.

Heap I, 2008. The international survey of herbicide resistant weeds. Online. Internet. Available from URL: http://www. Weedscience.com/. Accessed August 2011.

Hejl AM and Koster KL, 2004. The allelochemical sorgoleone inhibits root H+-ATPase and water uptake. J. Chem. Ecol. 3: 2181-2191.

Hong NH, Xuan TD, Tsuzuki E and Khanh TD, 2004. Paddy weed control by higher plants from Southeast Asia. Crop Prot. 23: 255-261.

Hong NH, Xuan TD, Tsuzuki E, Hiroyuki T, Mitsuhiro M and Khanh TD, 2003. Screening for allelopathic potential of higher plants from Southeast Asia. Crop Prot. 22: 829-836.

Iqbal J, Cheema ZA and Mushtaq MN, 2009. Allelopathic crop water extracts reduce the herbicide dose for weed control in cotton (*Gossypium hirsutum*). Int. J. of Agri. Biol. 11: 360-366.

Jabran K, Cheema ZA, Farooq M and Khan MB, 2011. Fertigation and foliar application of fertilizers alone and in combination with canola extracts enhances yield in wheat crop. Crop Environ. 2: 42-45.

Jabran K, Cheema ZA, Farooq M, Basra SMA, Hussain M and Rehman H, 2008. Tank mixing of allelopathic crop water extracts with pendimethalin helps in the management of weeds in canola (*Brassica napus*) field. Int. J. Agric. Bio. 10: 293-296.

Kudsk P and Streibig JC, 2003. Herbicides – a two-edged sword. Weed Res. 43: 90-102.

Mughal AH, 2000. Allelopathic effect of leaf extract of *Morus alba* L. on germination and seedling growth of some pulses. Range Manage. Agroforest. 21: 164-169.

Purvis CE, Jessop RS and Lovett JV, 1985. Selective regulation of germination and growth of annual weeds by crop residues. Weed Res. 25: 415-421.

Putnam AR and Duke WO, 1979. Allelopathy in agro ecosystem Ann. Rev. Phytopathol. 16, 431-451 in Rice, E.L. Allelopathy 2nd ed. pp. 72.

Roseveld N and Bretveld R, 2008. The impact of pesticides on male fertility. Curr. Opin. Obstet. Gyn. 20: 229-233.

Shamsi SRA and Ahmed B, 1984. Ecophysiological studies on some important weeds of wheat. Final technical report, Pakistan Science foundation Research. P-PU/agriculture (64). Dept. Bot. Univ. Punjab, Lahore, Pakistan.

Steel RGD, Torrie JH and Dickey D, 1997. Principles and Procedures of Statistics: A Biometrical Approach. 3rd Ed. McGraw Hill Book Co. Inc. New York, USA. pp: 172-177.

Zhu Y and Li QX, 2002. Movement of bromacil and hexazinone in soils of Hawaiian pineapple fields. Chemosphere. 49: 669-674.

# 25

# Quality response of maize fodder cultivars to harvest time

**Abdul Rehman[1], Aurangzeb[1], Rafi Qamar[1*], Atique-ur-Rehman[2], Muhammad Shoaib[3], Jamshaid Qamar[1] and Farwa Hassan[1]**

[1]Department of Agronomy, University College of Agriculture, University of Sargodha, Pakistan
[2]Department of Agronomy, Bahauddin Zakariya University, Multan, Pakistan
[3]Maize and Millets Research Institute, Yousufwala, Sahiwal, Pakistan

*Corresponding author email: drrafi1573@gmail.com

## Abstract

The study was conducted to investigate the fodder quality of four maize cultivars; DK919, 30R50, 31R88 and 6621 as influenced by harvest time at Agronomic Research Area of University of Sargodha. Maize cultivars were harvested at three different times viz. 80, 90 and 100 days after sowing (DAS). Significant differences were recorded among the cultivars for plant height, acid and neutral detergent fiber contents, lignin and crude protein. Maximum acid detergent fiber content, neutral detergent fiber content and lignin were observed at 100 DAS while crude protein was maximum at 80 DAS. However, plant height was remained unaffected with respect to harvest times. Moreover, maize cultivars had distinct differences in plant height and fodder quality parameters. Maximum plant height and crude protein were recorded in cultivar 31R88. The cultivar DK919 showed maximum values of acid detergent fiber content and neutral detergent fiber while lignin content was higher in V6621. Fodder quality parameters of cultivars 31R88, DK919 and V6621 were superior than 31R88 under the present climatic conditions of Sargodha.

**Keywords:** Harvest Time, Cultivars, Fodder Yield, Maize, Fodder Quality

## Introduction

Maize (*Zea mays* L.) is a dual-purpose crop universally grown for grain and forage. It produces ample quantity of green herbage with high nutritional and appetizing value (Akdeniz et al., 2004; Erdal et al., 2009). It has a distinct position in the national economic system of Pakistan and have 6.4% share in the total grain production. Moreover, it is a quality source of food, feed and fodder (Abdullah et al., 2007). In Pakistan, at least two crops of maize can be harvested in a year i.e., in spring and autumn seasons. According to the Economic Survey of Pakistan, the maize cultivation during the year (2014-15) was 1.13 million hectares with grain production of 4.69 million tons with average production was 4155 kg $ha^{-1}$ (GOP, 2014-15). Maize production in Pakistan is still low compared to other countries in spite of favourable environmental conditions and high yielding varieties. There are various limiting factors like water shortage, unpredictable rainfall, unavailability and high cost of fertilizers, less significance of fodder production and human population pressure adversely affecting fodder production (Rashid et al., 2007).

Quality of grain and fodder is considered as most important in maize production. Environment, planting time, stage at harvest, type of hybrids, agronomic management, hygienic quality, digestibility and consumption by animal are the most important grain, fodder and silage quality determining and limiting factors (Bal et al., 2000; Widdicombe and Thelen, 2002; Geren, 2000; Yilmaz et al., 2003). Maize fodder and feed products provides all forms of elementary nutrients and source of energetic nutrients with relatively low content of crude protein (Mlynár et al., 2004). Fodder harvesting at appropriate time is a main aspect for a successful forage production. Fodder cutting at maturity resulted in higher lignin content while lower concentration of fodder quality traits like plant protein, neutral detergent fiber (NDF), acid detergent fiber (ADF) and leaf proportion (Atis et al., 2012). Forage quality is directly influenced by the stage of maturity, which decreases as plant advances towards maturity and results in lower forage digestibility and consumption by animals (Ball et al., 2001). The neutral detergent fiber and acid detergent fiber are the most important fodder quality constituent, which are used as standard forage testing techniques. Moreover, these quality parameters are used to calculate fodder digestibility and intake potential (Ball et al., 2001).

Fodder quality characters may vary among different maize cultivars. Similarly, the time of fodder harvesting is affecting the fodder quality of each cultivar. The present study, therefore, was undertaken to evaluate the quality of four maize cultivars harvested at three different harvest time of autumn sown maize under semi-arid condition of Pakistan.

## Materials and Methods

### Study site

The current study referred to know the fodder quality response of four autumn maize cultivars to three harvest times was conducted at the research area of University College of Agriculture, University of Sargodha, Sargodha during the year 2012. Sargodha lies at 32.08° N and 72.67° E. General elevation of land from sea level is 193 m.

### Soil collection and analysis

The soil on which experiment was conducted was clay loam. Before starting the experiment, the samples of soil were collected to a depth of 30 cm and were analyzed for various physical and chemical properties (Jackson, 1962; Moodie et al., 1959; Watanable and Olsen, 1965). Soil characteristic recorded are presented in Table-1.

### Experimental design and crop husbandry

Experiment was conducted applying randomized complete block design in a split plot arrangement having three replications with net plot size assigned to single treatment of 4 m x 6 m. Three harvest times 80, 90 and 100 days were allocated to main plots while four cultivars (DK919, 30R50, 31R88 and 6621) were assigned to subplots.

**Table – 1: Pre-sowing analysis of experimental soil during 2012**

| Characteristic | Unit | Values | |
|---|---|---|---|
| | | 0-15 cm | 15-30 cm |
| **Chemical analysis** | | | |
| Saturation | % | 37 | 36 |
| pH | ---- | 7.8 | 7.9 |
| EC | dS $m^{-1}$ | 1.64 | 1.65 |
| Organic matter | % | 0.96 | 0.84 |
| Total nitrogen | % | 0.047 | 0.043 |
| Available phosphorous | Ppm | 12.5 | 8.5 |
| **Mechanical analysis** | | | |
| Sand | % | 29 | 25 |
| Silt | % | 47 | 46 |
| Clay | % | 22 | 33 |
| Soil texture | ---- | Silt loam | Loam |

Field was prepared by using tractor mounted cultivator, cultivating thrice which resulted in well pulverized. There were 36 plots prepared and each plot had 5 ridges in East-West direction. Sowing was done manually on July 26, 2012. Maize hybrids seed was used at the rate of 10 kg $acre^{-1}$. Nitrogen fertilizer was applied in three splits in the form of Urea i.e. $1/3^{rd}$ at the time of sowing, $1/3^{rd}$ when crop was at knee height and $1/3^{rd}$ at tasseling stage. Whole of the phosphorus and potash fertilizers, in the form of Di-ammonium phosphate and murate of potash were applied as basal dose. On the basis of need of the crop, seven irrigations were applied in addition to rainfall during the whole growth period. First irrigation was applied after 05 days of sowing and later irrigations were applied as when needed. To maintain inter plant spacing thinning was done when crop reached at the height of 15 cm. Plant protection measures like application of weedicide and insecticide etc. were kept normal for all treatments. Harvesting of different samples was done 80, 90 and 100 days after sowing respectively.

### Fodder quality parameters of maize

Ten plants were selected at random from each plot to record individual plant observation like plant height (cm) by using standard procedure. On each harvest date one plot of each genotype was harvested. Dried samples were ground using a hammer mill to pass a 1 mm screen. Whole plants samples were analyzed for NDF, ADF and CP content. A 0.5 g sample was used for sequential detergent analysis to determine NDF and ADF contents (Soest et al., 1991). Total N was determined by the Kjeldahl procedure and CP content was calculated by multiplying total N by 6.25. All compositional data were calculated on a dry matter basis. The concentration of lignin was measured by using the simplified method adopted by Goering and Van Soest, (1970).

**Statistical analysis**

Data were analyzed statistically using SAS (SAS Institute, 2008). The effects of harvest time and cultivars and their interaction were evaluated by the Duncan's Multiple Range test (DMRT) at $p \leq 0.05$ unless otherwise mentioned. The computer package MS-Excel was used to prepare the graphs.

## Results

Different maize cultivars were affected by harvest time, cultivars and their interactive effect were significantly different ($p > 0.05$) on plant height (cm), acid and neutral detergent fiber (%), lignin (%) and crude protein (%). Plant height is an important yield-contributing as well as fodder production factor. It is apparent from Figure-1 that there is significant difference in plant height of maize cultivars when harvested at various times. Statistically taller (212 cm) plant height was noted at 100 DAS. Cultivars 31R88 showed taller (252.8 cm) plant height among other cultivars. The interaction between harvest time and cultivars showed that cultivar 31R88 recorded (262.4 cm and 259.5 cm) taller plants ($p > 0.05$) at harvest time of 100 and 80 DAS, respectively. However, shorter (174.2 cm) plant height was recorded in V6621 at 100 DAS compared than other treatments. An overview of the acid detergent fiber data (Figure-2) revealed that maximum (32.7 and 32.3 %) ADF was attained in V6621 and DK919 at 100 DAS respectively. Neutral detergent fiber is the percent of cell wall material of the forage cellulose hemicelluloses and unavailable protein. It is the evident from the results (Figure 3) that harvest time and varieties significantly affected the neutral detergent fiber. Moreover, significantly maximum (50.7, 48.3 and 48.3 %) neutral detergent fiber was recorded in DK919, 31R88 and V6621 at 100 DAS respectively. Lignin is an indigestible plant structural constituent which binds the other plant molecules and makes them indigestible. Statistically maximum lignin contents (3.6 and 3.4 %) were attained in variety V6621 and DK919 at 100 DAS while statistically minimum lignin (2.4 %) was observed in 31R88 at 80 DAS (Figure 4). Crude protein affects the nutritional value and palatability of the forage crop. It is apparent from Figure 5 that harvest time and cultivars had a significant effect on crude protein of maize. The interaction between harvest time and cultivars showed maximum crude protein (8.8, 8.7, 8.3 and 8.1 %) in 31R88, V6621, 30R50 and DK919 at 80 DAS respectively.

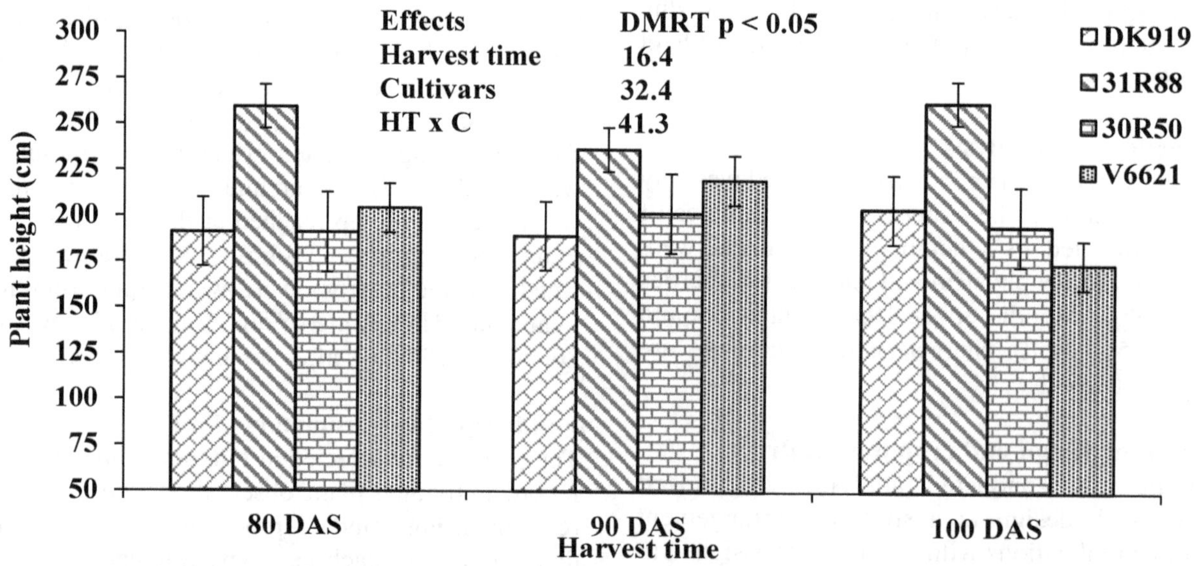


**Figure – 1: Effect of harvest time and cultivars on plant height (cm) of autumn maize.**

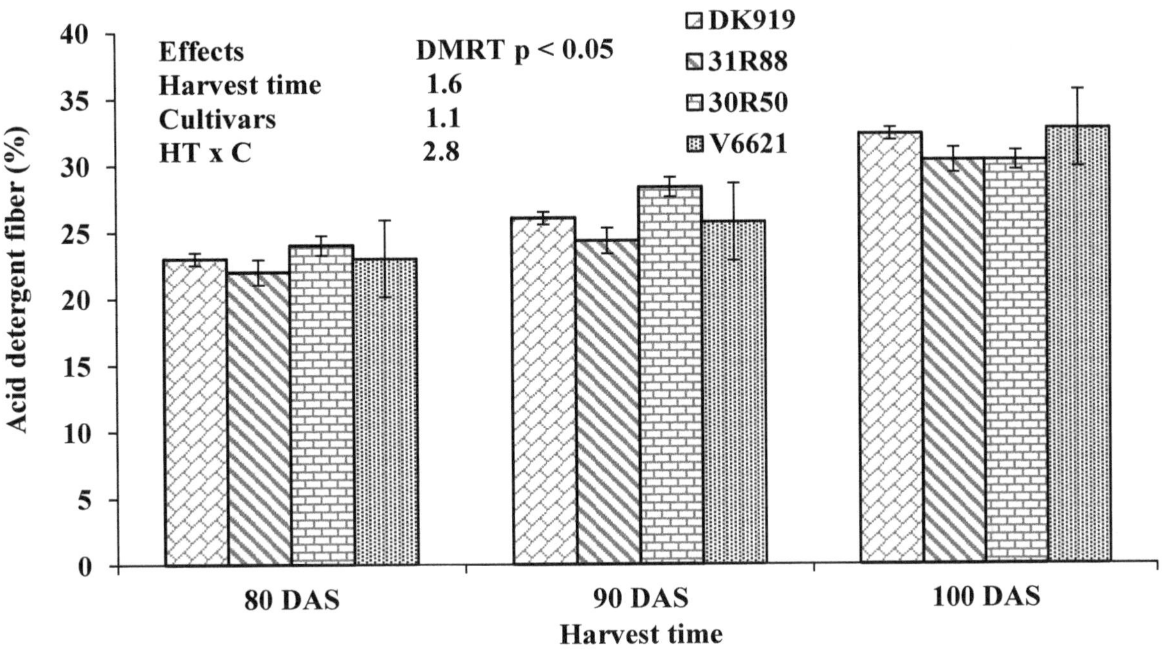


**Figure - 2: Effect of harvest time and cultivars on acid detergent fiber (%) of autumn maize.**

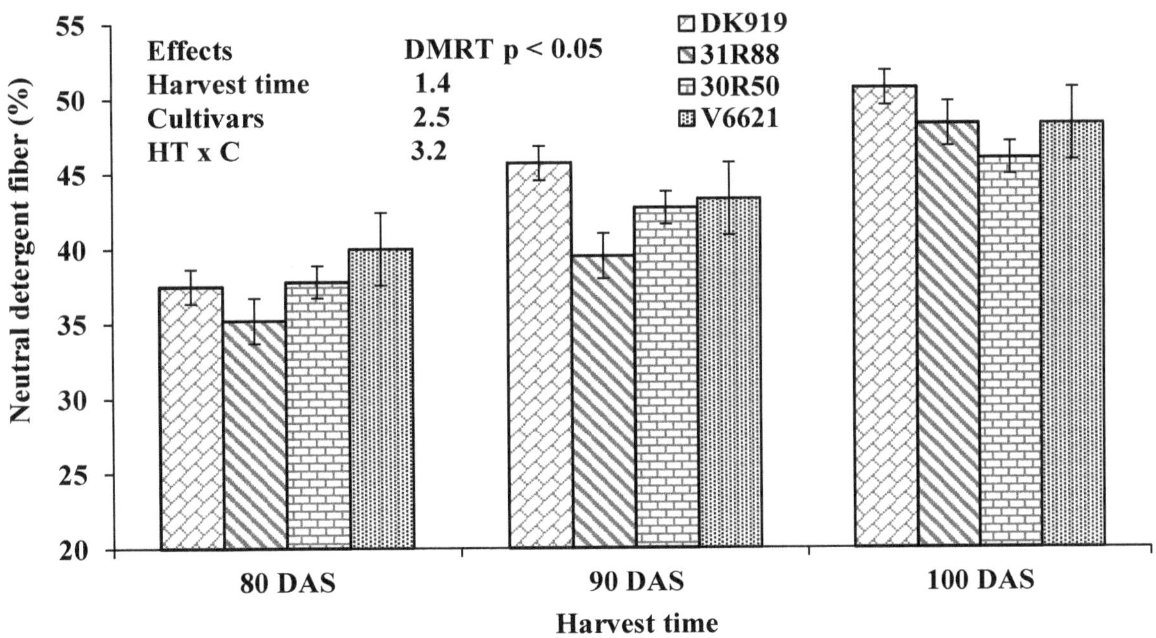


**Figure - 3: Effect of harvest time and varieties on neutral detergent fiber (%) of autumn maize.**

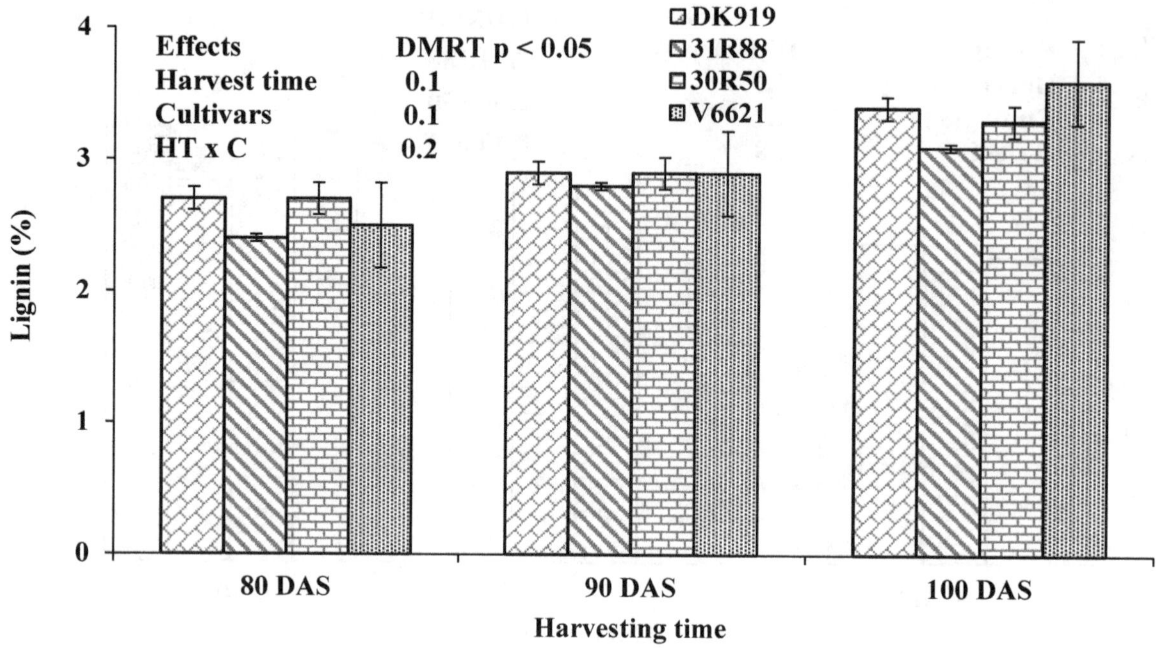


**Figure – 4: Effect of harvest time and cultivars on lignin (%) of autumn maize.**

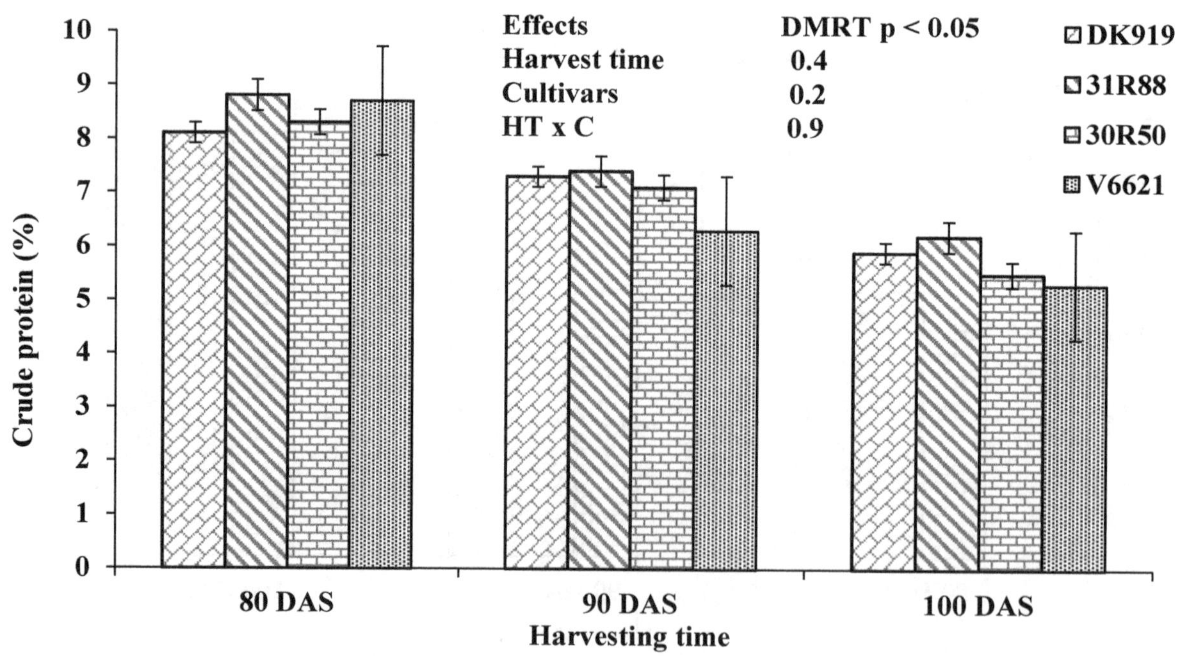


**Figure – 5: Effect of harvest time and cultivars on crude protein (%) of autumn maize.**

## Discussion

Maize forage yield and quality were influenced by harvest time and cultivars. During plant maturity the trend in plant height was increased from early to late harvest due to prolonged exist in field (Ayub et al., 2002; Xie et al., 2012). On the contrary, Carmi et al. (2006) testified that plant height at early harvest was not changed from late harvest. The inconsistent findings might be due to genetic variation in plants traits while in the current study shorter plant height was recorded. The variation among cultivars might be due to environment, soil fertility, harvesting stage and cultivars genome. Moreover, under similar agronomic and climatic conditions genetic character dominated and showed significant variation in cultivars tallness among various maize cultivars (Hussain et al., 2010; Awan et al., 2001). Forage quality and yield must be optimized to determine the best time for harvest. If forage is harvested too early, excessive loss of nutrients, from soil run off, occurs due to poor starch development in the kernel and low energy concentration. Fodder harvested too late above 100 DAS has decreased nutritive value due to poor starch and fiber digestion of silages (Neylon and Kung, 2003). The effect of harvest time and cultivars were significant for quality parameters. Fraction of leaves was constantly reduced as improvement in maturity (Carmi et al., 2005 and 2006). Harvest time significantly influenced the fodder quality parameters like acid detergent fiber, neutral detergent fiber and lignin which were maximum at late harvest stage (Butler and Muir, 2003; Carmi et al., 2005). Lignin accumulation and synthesis occur at the stage of secondary cell wall development (Carmi et al., 2006). The reason might be that at early harvest, the moisture contents in the plant were high and the concentration of dry matter was less as the plant matures dry matter accumulation increased in the plant, which resulted in maximum content of ADF, NDF and lignin (Carmi et al., 2006). However, the crude protein content was minimum observed in our study due to late harvest stage. Our results supported the findings of Huang et al. (2012) they reported that delayed harvest stage produced lower crude protein concentration which might be due to the higher dry matter yield per land area. Differences in crude protein content among genotypes were also reported by Carmi et al. (2005); Miron et al. (2005); Miron et al. (2006); Yosef et al. (2009). They reported that crude protein concentration was maximum at the first harvest stage and declined with maturity of plant due to increase in concentration of acid and neutral detergent fiber and lignin. They have a reverse trend and reached at maturity. Additionally, cultivars may affect rates of nutrient translocation (nutrients moving from the stalk and leaves to the ear) and rates of maturation (Lewis et al., 2004; Owens, 2005).

## Conclusion

The results of the research showed that investigated fodder quality parameters of cultivars were influenced by harvest time. Fodder growers who are interested in increased concentration of ADF, NDF and lignin content than maize fodder was harvested late. Keeping in view the yield and silage quality the cultivars 30R50 and 31R88 should be preferred respectively over other cultivars under the present climatic conditions of Sargodha.

## References

Abdullah G, Hassan I Khan A and Munir M, 2007. Effect of planting methods and herbicides on yield and yield components of maize. Pak. J. Weed Sci. Res. 13: 39-48.

Akdeniz H, Yilmaz I Andic N and Zorer S, 2004. A Study on Yield and Forage Values of Some Corn Cultivars. Univ. of Yuzuncuyil. J. Agric. Sci. 14: 47-51.

Atis I, Konuskan O Duru M Gozubenli H and Yilmaz S, 2012. Effect of harvesting time on yield, composition and forage quality of some forage sorghum cultivars. Int. J. Agric. Biol. 14: 879-886.

Awan TH, Mahmood MT Maqsood M Usman M and Hussain MI, 2001. Studies on hybrid and synthetic cultivars of maize for forage yield and quality. Pak. J. Agri. Sci. 38: 50-52.

Ayub M, Nadeem MA Tanveer A and Husnain A, 2002. Effect of different levels of nitrogen and harvesting times on the growth, yield and quality of sorghum fodder. Asian J. Plant Sci. 1: 304-307.

Bal MA, Shaver RD Shinners KJ Coors JG Lauer JG Straub RJ and Koegel RG, 2000. Stage of maturity, processing and hybrid effects on ruminal in situ disappearance of whole-plant corn silage. Anim. Feed Sci. Technol. 86: 83-94.

Ball DM, Collins M Lacefield GD Martin NP Mertens DA Olson KE Putnam DH Undersander DJ and Wolf MW, 2001. Understanding Forage Quality.

American Farm Bureau Federation Publication 1-01, Park Ridge, Illinois, USA.

Butler TJ and Muir JP, 2003. Row Spacing and Maturity of Forage Sorghum Silage in North Central Texas. Forage Research in Texas, http://forageresearch.tamu.edu/2003/Forage Sorghum.

Carmi A, Umiel N Hagiladi A Yosef E Ben-Ghedalia D and Miron J, 2005. Field performance and nutritive value of a new forage sorghum variety 'Pnina' recently developed in Israel. J. Sci. Food Agric. 85: 2567–2573.

Carmi A, Aharoni Y Edelstein M Umiel N Hagiladi A Yosef E Nikbachat M Zenou A and Miron J, 2006. Effects of irrigation and plant density on yield, composition and *in vitro* digestibility of a new forage sorghum variety, Tal, at two maturity stages. Anim. Feed. Sci. Tech. 131: 120-132.

Erdal S, Pamukcu M Ekiz H Soysal M Savur O and Toros A, 2009. The determination of yield and quality traits of some candidate silage maize hybrids. Univ. of Akdeniz. J. Agric. Sci. 22: 75-81.

Geren H, 2000. Investigations on the effect of sowing dates on the forage yields and agronomical characteristics related to silage of different maize (*Zea mays* L.) cultivars grown as main and second crops. Univ. of Ege, Graduate School of Natural and Applied Sciences, (Unpublished Ph.D. Thesis), Izmir Turkey, pp: 36-43.

Goering HK and Van Soest PJ, 1970. Forage fiber analysis. In: Agricultural Handbook No 379. Agricultural Research Service, United States Department of Agriculture, Washington, D.C. pp. I-20.

Govt. of Pakistan, (2014-15). Economic Survey of Pakistan. Finance Division Economic Advisor's Wing, Islamabad.

Huang H, Faulkner DB, Singh V Danao MC and Eckhoff SR, 2012. Effect of harvest date on yield, composition and nutritive value of corn stover and DDGS. Transactions of the ASABE. 55: 1859-1864.

Hussain N, Zaman Q Nadeem MA and Aziz A, 2010. Response of maize varieties under Agro-ecological conditions of Dera Ismail Khan. J. Agri. Res. 48: 59-63.

Jackson ML, 1962. Soil Chemical Analysis. Printice Hall. Inc., Englewood Cliffs, New Jerrsy. USA.

Lewis AL and Cherney WJH, 2004. Hybrid, maturity and cutting height interactions on corn forage yield and quality. Agron. J. 96: 267-274.

Miron J, Zuckerman E Sadeh D Adin G Nikbachat M Yosef E Ben-Ghedalia D Carmi A Kipnis T and Solomon R, 2005. Yield, composition and *in vitro* digestibility of new forage sorghum varieties and their ensilage characteristics. Anim. Feed Sci. Tech. 120: 17-32.

Miron J, Solomon R Adin G Nir U Nikbachat M Yosef E Carmi A Weinberg ZG Kipnis T Zuckerman E and Ben-Ghedalia D, 2006. Effects of harvest stage and re-growth on yield, composition, ensilage and *in vitro* digestibility of new forage sorghum varieties. J. Sci. Food Agric. 86: 140-147.

Mlynár R, Rajčáková Ľ and Gallo M, 2004. The effect of Lactobacillus buchneri on fermentation process and aerobic stability of maize silage. Risk factors of food chain, Nitra, SPU. 170-172.

Moodie CD, Smith HW and McCreery RA, 1959. Laboratory Manual for Soil Fertility. Washington State College, Mimeograph, USA.

Neylon JM and Jr. Kung L, 2003. Effects of Cutting Height and Maturity on the Nutritive Value of Corn silage for Lactating Cows. J. Dairy Sci. 86: 2163-2169.

Owens F, 2005. Corn genetics and animal feeding value. 66th Minnesota Nutrition Conference. St. Paul, MN. 20-21.

Rashid M, Ranjha AM and Rehim A, 2007. Model based P fertilization to improve yield and quality of sorghum (*Sorghum bicolor* L.) fodder on an ustochrept soil. Pak. J. Agri. Sci. 31: 27-29.

SAS Institute, 2008. SAS online doc 9.13. SAS Institute, Inc., Cary, NC.

Soest Van PJ, Robertson JB and Lewis BA, 1991. Methods for dietary fiber, neutral detergent fiber and non-starch polysaccharides in relation to animal nutrition. J. Dairy Sci. 74: 3583–3597.

Watanabe FS and Olsen SR, 1965. Test of ascorbic acid method for determining phosphorous in water and $NaHCO_3$ extract from soils. Soil Sci. Soc. Am. Pro. 29: 677-678.

Widdicombe WD and Thelen KD, 2002. Row width and plant density effect on corn forage hybrids. Agron. J. 94: 326-330.

Xie T, Su P Shan L and Ma J, 2012. Yield, quality and irrigation water use efficiency of sweet sorghum [*Sorghum bicolor* (Linn.) Moench] under different land types in arid regions. Austra. J. Crop Sci. 6: 10-16.

Yilmaz S, Gozubenli H Can E and Atis I, 2003. Adaptation and silage yield of some maize (*Zea mays* L.) lines in a milk plain condition. Turkey 5th Field Crops Congress, 13-17 October, Diyarbakr-Turkey. Volume: I, ISBN: 975-7635-19-7. pp: 341-345.

Yosef E, Carmi A Nikbachat M Zenou A Umiel N and Miron J, 2009. Characteristics of tall *versus* short-type varieties of forage sorghum grown under two irrigation levels, for summer and subsequent fall harvests, and digestibility by sheep of their silages. Anim. Feed Sci. Tech. 152: 1-11.

# 26

# Economics of direct seeding methods of upland rice production in the Northern Guinea Savanna

**Nwokwu Gilbert Nwogboduhu**
Department of Crop Production and Landscape Management, Faculty of Agriculture and Natural Resources Management, Ebonyi State University, Abakaliki, Nigeria.

*Corresponding author email: g.nwokwu@yahoo.com

## Abstract

This study assessed the economic impacts of direct seeding of rice as an alternative crop establishment method for farmers in Samaru, Zaria in the Northern Guinea Savanna of Nigeria. Specifically, it examined the changes in farmers' inputs (labour and inputs) and level of productivity and incomes among direct-seeded methods such as broadcasting, drilling and dibbling and measured the economic returns on investment in direct seeding. Analyses included cost and return, and economic surplus framework. The economic analysis of upland rice production at both locations indicated that production of NERICA 8 and JAMILA by either broadcasting or drilling method at the seed rate of 80 kg ha$^{-1}$ gave the highest gross margin as well as return on investment. The result revealed that NERICA 8 and JAMILA sown by broadcasting method at 80 kg ha$^{-1}$ was the most profitable with gross margin of ₦246, 166.50 with return on investment of ₦6.72. This was followed by broadcasting JAMILA at 120 kg ha$^{-1}$ seed rate which gave a gross margin of ₦194, 583.50 and return on investment of ₦4.32. However, the least gross margin of ₦61, 249.85 was observed when NERICA 4 was dibbled at 120 kg ha$^{-1}$ which brought a loss of ₦16, 716.50 and ₦ 0.62k was lost per every naira invested.

**Keywords:** Economics, Direct Seeding, Upland Rice, Production

## Introduction

Rice is the world's most important crop and is a staple food for more than half of the world's population. Worldwide, rice is grown on 161 million hectares, with an annual production of about 678.7 million tons of paddy (FAO, 2009). About 90% of the world's rice is grown and produced (143 million ha of area with a production of 612 million tons of paddy) in Asia (FAO, 2009). Rice provides 30–75% of the total calories to more than 3 billion people (Khush, 1997; Von Braun and Bos, 2004). To meet the global rice demand, it is estimated that about 114 million tons of additional milled rice need to be produced by 2035, which is equivalent to an overall increase of 26% in the next 25 years (Yamano et al., 2016). The possibility of expanding the area under rice in the near future is limited. Therefore, this extra rice production needed has to come from a productivity gain. The major challenge to achieve this gain is less water, labour and chemicals (Hira, 2009)

A major reason for farmers' interest in direct seeding is the rising cost of cultivation and decreasing profits with conventional practice. Farmers likely prefer a technology that gives higher profit despite similar or slightly lower yield. Studies shows that various methods of direct seeding reduced the cost of production by US$9125 ha$^{-1}$ compared with conventional practice of transplanting (Inayat-Ali et al., 2012; Santhi et al., 1998). The largest reductions in cost occurred in practices in which zero tillage was combined with direct seeding. These cost reductions were largely due to either reduced labour cost or tillage cost or both under direct seeding method. In areas where wages are high the labour cost savings in rice establishment can reach US$50 $ha^{-1}$ (Kumar et al., 2009).

The conventional method of rice growing is not only water-guzzling but also cumbersome and laborious. Rice transplanting requires 200-250 man-hour $ha^{-1}$, which is 25% of the total requirement for the rice crop production (Anoop et al., 2007). The problem has further been intensified with the timely unavailability of labour. Delay in transplanting beyond optimum time due to labour scarcity is creating a reduction in rice yield. Further, reduced labour availability is increasing the cost of transplanting and squeezing the farmer's profit as the cost of transplanting is increasing continuously. Paddy transplanting by labour also results in low and non-uniform plant population due to which crop yields are reduced (Mahajan et al., 2009). The productivity and sustainability of rice-based systems are threatened because of the inefficient use of inputs; increasing scarcity of resources, especially water and labour; changing climate; the emerging energy crisis and rising fuel prices; the rising cost of cultivation and emerging socio-economic changes such as urbanization, migration of labour, preferences of non-agricultural work, concerns about farm-related pollution (Kumar and Ladha, 2011). Conventional tillage and crop establishment by transplanting is the most input intensive process in crop production and, therefore, more efficient alternatives are urgently needed. Potential solutions is a shift from manual transplanting to direct seeded rice. Direct seeded rice with zero tillage system performed as well as the conventional practice but with significant savings in water and labour use (Bhusan et al., 2007). Direct-seeding is cost-effective, can save water through earlier rice crop establishment (Ladha et al., 2003a; Singh et al.,, 2003). With alternate wetting and drying cycles in direct seeding, the crop is subjected to greater weed competition than transplanted rice because weeds emerge before or at the same time as the rice (Chuhan, 2012). Therefore, heavy weed infestation is a major problem in direct seeded rice and its success lies in effective weed control measures (Singh et al.,, 2003; Rao et al.,, 2007), as failure to eliminate weeds may result in low or no yield (Estorninos and Moody, 1988). Therefore, the study aimed at evaluating the economics of direct seeding methods of upland rice varieties with a goal for finding most suitable ones with a potential to cover large area with similar agro ecological conditions.

## Materials and Methods

The field trials were conducted on the experimental farm of the Institute for Agricultural Research, Ahmadu Bello University Samaru, Zaria and on the Research Farm of the Kaduna State Agricultural Development Programme, Maigana in 2011 and 2012 cropping seasons. Samaru is on the Latitude 11°11$^{1}$ N and Longitude 7 °38$^{1}$ E and is 686 m above sea level while Maigana is located on Latitude 11°11.06$^{1}$ N and Longitude 7.54° 7.58$^{1\ E}$ both in the Northern Guinea Savannah Agro ecological zone of Nigeria. Random samples of soils were taken at depth of 0-30cm from the experimental sites using an auger of 10 cm diameter before land preparation and were analysed for physical and chemical properties.

The treatments consisted of three seeding methods (broadcasting, drilling and dibbling), three seed rates (40, 80 and 120 kg $ha^{-1}$) and three upland rice varieties (NERICA 4, 8 and JAMILA). The treatments were laid out in a split plot design with the combination of sowing methods and seed rates in the main plots and three rice varieties in the sub plots measuring 3 m x 3 m with a net plot of 2 m x 2 m and were replicated three times. The main plots were separated by a distance of 1 m and the sub plots by 0.5 m. Pre-planting herbicides, glyphosate (round up) was applied to the experimental sites at the rate 2 kg active ingredients $ha^{-1}$ two weeks before land preparation in each year of the study in order to control the prevalent weeds on the field. Thereafter, the field was harrowed twice to ensure fine tilth of the soil and the soil levelled manually.

Seeds of each variety were treated with Apron star as seed dressing chemical at the rate of 1.0 g of metalaxy to 3.0 kg seed to prevent pest attack and the three rice varieties were planted on 13th July 2011 and 9th June 2012 at Samaru while at Maigana, the varieties were planted on 30th July 2011 and 12th June 2012 when the rains were fully stabilized using direct seeding such as broadcasting, drilling and dibbling methods. Also hand pulling methods of weed control was used to control the weeds that later emerged at four and eight weeks after sowing. Fertilizers were applied at the recommended rate of 100 kg N $ha^{-1}$, 50 kg $P_2O_5$ $ha^{-1}$ and 50 kg $K_2O$ $ha^{-1}$. The nitrogen fertilizer were applied as split, half of the nitrogen fertilizer together with 50 kg $P_2O_5$ and 50 kg $K_2O$ $ha^{-1}$ were applied once at two weeks after sowing using NPK (15:15:15) while the second half of the nitrogen were applied at panicle initiation stage using urea (46% N). Economics and energy analysis was done by taking pooled data of both the years. The economic analysis was measured using the partial budget procedure to determine the economic return. The data obtained was subjected to economic analysis where the returns and variable costs were calculated and gross margin (GM) and return on investment (ROI) was also determined. This depended on the prevailing market prices of inputs, labour and outputs. This was computed as:

GM=TR-TVC (Olukosi and Erhabor, 1988).

Where;
TR= Total revenue per hectare.
TVC=Total variable cost (sum of labour and material input cost) per hectare.

ROI =Net return /TVC X 100/1

Where; Net return = Total revenue (yield x output price) – TVC

## Results and Discussion

The physical and chemical properties of the soil used during the experimental periods at Samaru and Maigana are summarized in Table1. In 2011 at Samaru, the physico-chemical properties of the field on which the experiment was conducted showed that the soil was clay loam, slightly acidic with moderate total nitrogen and available phosphorus and low potassium and CEC and high organic carbon. In 2012 cropping season the soil was silt loam, slightly acidic with high total nitrogen; available phosphorus and potassium were moderate while organic carbon and CEC were low. At Maigana in 2011, the soil was silt clay loam, slightly acidic with high total nitrogen and organic carbon while available phosphorus, potassium and CEC were low. In 2012, the soil was sandy loam, slightly acidic with high total nitrogen while available P, K was moderate and organic carbon was high

The cost and return analysis per hectare on investment of growing upland rice varieties using different sowing methods and seed rates in 2011, 2012 wet seasons and the combined at Samaru and Maigana are presented in Tables 3 to 8. At Samaru in 2011, the result revealed that NERICA 8 sown by broadcasting method at 80 kg ha$^{-1}$ was the most profitable with gross margin of ₦129, 793.25 and Return on Investment of ₦3.02 which implied that for every ₦1.00 invested ₦2.02 was realized. Similarly, it was followed by NERICA 8 sown by broadcasting method at 120 kg ha$^{-1}$ seed rate which gave a gross margin of ₦95, 787.65 and Return on Investment of ₦2.23 which implied that for every ₦1.00 invested ₦1.23 was realized. However, the sowing of JAMILA with dibbling method at 120 kg ha$^{-1}$ brought a loss of ₦17, 932.50 and ₦ 0.34k was lost per naira invested.

In 2012, the use of broadcasting method to sow JAMILA at 80 kg ha$^{-1}$ seed rate was the most profitable with gross margin of ₦198, 750.00 and return on investment of of ₦4.42 indicating that for ₦1.00 invested ₦3.42 was realized. This was followed by broadcasting JAMILA at 120 kg ha$^{-1}$ seed rate which gave a gross margin of ₦194, 583.50 and return on investment of ₦4.32. There was no loss incurred during this year's production. However, the least gross margin of ₦61, 249.85 and return on investment of ₦1.11 was observed when NERICA 4 was dibbled at 80 kg ha$^{-1}$ seed rate.

The combined economic analysis indicated that broadcasting NERICA 8 at 80 kg ha$^{-1}$ seed rate was the most profitable with gross margin of ₦ 146, 215.00 and return on investment of ₦4.32 which indicated that for every ₦1.00 invested, ₦ 3.32 was realized. This was closely followed by either drilled NERICA 8 or JAMILA at 80 kg ha$^{-1}$ seed rate which gave a gross margin of ₦141, 194.30 and return on investment of ₦3.88.The least gross margin of ₦35, 467.65 and return on investment of ₦1.66 was observed when NERICA 4 was dibbled at 120 kg ha$^{-1}$ seed rate. Net returns in broadcasting was at par with drilling method and were higher than dibbling method. Kumar (2011) also observed similar findings and found higher B:C ratio in direct seeding as compared to manual puddled transplanted rice .

At Maigana in 2011, the result indicated that the highest gross margin of ₦119, 581.50 and return on investment of ₦3.85 was achieved by broadcasting NERICA 8 at 80 kg ha$^{-1}$ seed rate. This was followed by drilled NERICA 8 at 80 kg ha$^{-1}$ which gave a gross margin of ₦30, 832.50 and return on investment of ₦1.73. However, dibbled JAMILA at 120 kg ha$^{-1}$ brought a loss of ₦16, 716.50 and 0.62k was lost per every naira invested.

In 2012 and the combined, the result indicated that the highest gross margin was by drilled JAMILA at 80 kg ha$^{-1}$ seed rate. The gross margin of 2012 and the combined were ₦246, 166.50 with return on investment of₦ 6.72 and ₦182, 874.00 with return on investment of ₦5.30 respectively. This was followed by drilled JAMILA at 120 kg ha$^{-1}$ seed rate with gross margin of ₦198, 666.50 and return on investment of₦ 5.62, gross margin of ₦98, 547.50 and return on investment of ₦3.32 for 2012 and the combined data. However, the least gross margin was observed when NERICA 8 was dibbled at 40 kg ha$^{-1}$ seed rate in 2012 and the combined were ₦24, 375.15 with return on investment of ₦1.54 and ₦11, 187.50 with return on investment of ₦1.25. In paddy, a labour saving of 95-99 percent in broadcasting or drilling methods were observed as compared to bibbling method during both years of the study and total labour use mainly depends on the weed management. In present study we made two hand weeding to cope up with weeds as the herbicide used were not so effective, which ultimately resulted in more labour use and higher cost of production. In overall, broadcasting and drilling methods had more percent labour saving as compared to dibbling method in both years of the study. Sehrawat et al. (2010) also observed 13-16% labour saving in direct seeding as compared to manual puddled transplanted rice.

Economic analysis is the ultimate yard stick to recommend a production technology. The loss incurred at both locations in 2011 was due to low rainfall recorded in that year.

Generally, throughout the study, increase in total revenue resulted an increase in gross margin in all the three sowing methods used in the study. In 2011 at Samaru, the maximum gross margin obtained from broadcasting method was due to the fact that crops established with these methods matured early than drilling and dibbling method. This helped the crops to escape the stress as a result of late planting as well early cessation of rainfall in that year and thereby resulted in highest yield. Despite the higher cost of production in 2012 wet season and the combined at Samaru, the highest gross margin obtained from broadcasting and drilling method were due to higher grain yield as a result of better rain fall recorded in that year at the experimental sites. Generally, dibbling method consistently produced the least gross margin throughout the study at both locations. This was due to the low yield recorded when rice was sown using this method throughout the study period. In 2011 at both locations, the lowest gross margin was recorded as well as the loss incurred was due to low yield as a result of late planting. However, the highest gross margin in that year was obtained from the broadcasting method at 80 kg of NERICA 8. The differences in gross margin in the two years at both locations could be attributed to the differences in cost of production as well as the market price. In this country it was discovered that prices of materials at different locations increase as the years passed by without necessarily increase the price of the farm produce. In this study, the cost of production at Maigana was far cheaper than Samaru especially the cost of labour and as well as some inputs such as fertilizers.

In both locations, the cost of labour such as weed control in broadcasting and drilling methods were lower than that of dibbling method. The highest difference in labour cost was in crop establishment. This reduced need for labour not only saves time and money of farmers but also allows greater flexibility so that farmers can attend other activities either in farm or at home. Higher expenditures on seeds were expected in direct seeding methods because higher seeding rates are required for direct seeding relative to transplanting. Farmers who practiced direct seeding were more reliant on herbicides simply because they cannot rely on flooding to suppress weeds during the crucial initial period of crop establishment (Johnson 2006). On the other hand, expenditures on fertilizers, and rent in land preparation were lower for direct seeding than transplanting. This was because farmers were inclined to use more fertilizers as a treatment or preventive measure against transplanting shock.

Previous studies had found that direct seeding may obtain a lower yield due to the unstable establishment of rice seedlings and slow growth during the early growing stage (Kimio et al., 1999). Yield in direct seeding can be also reduced by weed problems. Yield losses (due to weeds) largely depend on season, weed species, weed density, rice cultivar, and growth rate and density of weeds and rice (Azmi et al., 2005). Another factor affecting yield in direct seeding is seed rate. A seed rate higher than the recommended rate can result in lower yield of direct seeding since it may lead to nitrogen deficiency, thus reducing tillering and increasing the proportion of ineffective tillers, to attacks of brown plant hoppers, and to crop lodging. However, yield of direct seeding is not always lower than that of transplanting. Rice yields of wet- or dry-seeded crops have been higher than those of transplanted crops, provided weeds are adequately controlled (Johnson et al., 2003) .Higher rice yield resulting from direct seeding is due to the shorter time it takes for direct seeding rice to reach maturity. This allows for on-time planting, thus saving farmers from a 1 percent (or more) reduction in yield per day (Hobbs 2001).

Rainfall at Samaru during the experimental period was 945.2 and 1333.3mm in 2011 and 2012 respectively (Table 2). Rain started in 2011 on 16th April and ended on 11th October with a total of 73 rainy days. In 2012, rain commenced on10th April and terminated on 24tth October with a total of 76 rainy days. Rainfall at Maigana during the experimental period was 636.6 and 762.9 in 2011 and 2012 respectively. Rain started in 2011 on 25th April and ended on 12th October with a total of 67 rainy days. In 2012, rain commenced on 17th April and ended on 22nd October with a total of 70 rainy days. Generally, rainfall was not evenly distributed during the conduct of this research. Rain was low at both locations in 2011 with Maigana recording lower rainfall in the two years of this study when compared with Samaru.

**Table – 1: Physical and Chemical Properties of soil (0-30cm) collected from the Experimental Sites**

| | Samaru | | Maigana | |
|---|---|---|---|---|
| **Soil Properties** | **2011** | **2012** | **2011** | **2012** |
| **Physical Properties** | | | | |
| Sand (g/kg) | 222.00 | 240.00 | 192.00 | 520.00 |
| Silt (g/kg) | 474.00 | 540.00 | 504.00 | 380.00 |
| Clay (g/kg) | 304.00 | 220.00 | 304.00 | 100.00 |
| Textural Class | Clay loam | Silt loam | Silty Clay loam | Sandy loam |
| **Chemical Properties** | | | | |
| pH in 0.01M $CaCL_2$ | 4.28 | 5.98 | 4.35 | 4.55 |
| Organic Carbon (g/kg) | 13.6 | 2.10 | 16.00 | 20.40 |
| Total Nitrogen (g/kg) | 1.70 | 9.00 | 2.20 | 10.10 |
| Available P mg/Kg | 3.05 | 1.42 | 4.48 | 2.84 |
| **Exchangeable Cation (cmol/kg)** | | | | |
| K | 0.15 | 0.22 | 0.17 | 0.22 |
| Mg | 0.59 | 2.00 | 0.63 | 1.76 |
| Ca | 0.59 | 3.00 | 0.61 | 5.00 |
| Na | 7.32 | 1.60 | 9.10 | 1.60 |
| CEC | 4.28 | 5.20 | 4.35 | 2.90 |

**Table – 2: Rainfall Distribution at Samaru and Maigana during 2011 and 2012 Cropping Seasons**

| | Samaru 2011 | | 2012 | | Maigana 2011 | | 2012 | |
|---|---|---|---|---|---|---|---|---|
| **MONTHS** | **Amount of Rainfall** | **Rainy days** | **Amount of Rainfall** | **Rainy days** | **Amount of Rainfall** | **Rainy days** | **Amount of Rainfall** | **Rainy days** |
| January | 0 | 0 | 0 | 0 | 0 | 0 | 0 | 0 |
| February | 0 | 0 | 0 | 0 | 0 | 0 | 0 | 0 |
| March | 0 | 0 | 0 | 0 | 0 | 0 | 0 | 0 |
| April | 27.8 | 4 | 7.3 | 3 | 36.4 | 1 | 21.1 | 4 |
| May | 123.3 | 11 | 263.4 | 15 | 63.4 | 8 | 109.6 | 14 |
| June | 162.5 | 12 | 120.7 | 8 | 102.5 | 15 | 145.7 | 13 |
| July | 223.9 | 14 | 165.3 | 12 | 113.2 | 12 | 150.4 | 15 |
| August | 239.9 | 16 | 426.7 | 16 | 165.2 | 16 | 130.5 | 19 |
| September | 113.8 | 12 | 270.3 | 17 | 89.1 | 19 | 191 | 18 |
| October | 54 | 4 | 79.6 | 5 | 66.8 | 5 | 14.6 | 3 |
| November | 0 | 0 | 0 | 0 | 0 | 0 | 0 | 0 |
| December | 0 | 0 | 0 | 0 | 0 | 0 | 0 | 0 |
| TOTAL | 945.2 | 73 | 1333.3 | 76 | 636.6 | 76 | 762.9 | 86 |

**Table - 3: Economic analysis of cost and return on investment of growing upland rice varieties using seed rate and sowing method at Samaru in 2011 wet season.**

| Variety | Sowing method | Seed rate (kg ha$^{-1}$) | Total yield (Kg ha$^{-1}$) | No. of bags (100kg) | Average Price/bag 100kgha[1] ₦ | Total revenue (TR) (₦ ha$^{-1}$) | Total variable cost (TVC) (₦ ha$^{-1}$) | Gross margin (TR- TVC) (₦ ha$^{-1}$) | Return on investment (ROI) ₦ |
|---|---|---|---|---|---|---|---|---|---|
| NERICA 4 | Broadcasting | 40 | 1088.67 | 10.89 | 4500 | 48990.15 | 43000 | 5990.15 | 0.14 |
| NERICA 4 | Broadcasting | 80 | 2413.08 | 24.13 | 4500 | 108588.60 | 43000 | 65588.60 | 1.53 |
| NERICA 4 | Broadcasting | 120 | 1395.77 | 13.96 | 4500 | 62809.65 | 43000 | 19809.65 | 0.46 |
| NERICA 4 | Drilling | 40 | 1517.92 | 15.18 | 4500 | 68306.40 | 48000 | 20306.40 | 0.42 |
| NERICA 4 | Drilling | 80 | 1710.75 | 17.11 | 4500 | 76983.75 | 48000 | 28983.75 | 0.60 |
| NERICA 4 | Drilling | 120 | 1895.43 | 18.95 | 4500 | 85294.35 | 48000 | 37294.35 | 0.78 |
| NERICA 4 | Dibbling | 40 | 1565.17 | 15.65 | 4500 | 70432.65 | 53000 | 17432.65 | 0.33 |
| NERICA 4 | Dibbling | 80 | 1699.48 | 16.99 | 4500 | 76476.60 | 53000 | 23476.60 | 0.44 |
| NERICA 4 | Dibbling | 120 | 1101.33 | 11.01 | 4500 | 49559.85 | 53000 | -3440.15 | -0.06 |
| NERICA 8 | Broadcasting | 40 | 1666.58 | 16.67 | 4500 | 74996.10 | 43000 | 31996.10 | 0.74 |
| NERICA 8 | Broadcasting | 80 | 3839.85 | 38.40 | 4500 | 172793.25 | 43000 | 129793.25 | 3.02 |
| NERICA 8 | Broadcasting | 120 | 3084.17 | 30.84 | 4500 | 138787.65 | 43000 | 95787.65 | 2.23 |
| NERICA 8 | Drilling | 40 | 2289.02 | 22.89 | 4500 | 103005.90 | 48000 | 55005.90 | 1.15 |
| NERICA 8 | Drilling | 80 | 2994.75 | 29.95 | 4500 | 134763.75 | 48000 | 86763.75 | 1.81 |
| NERICA 8 | Drilling | 120 | 2757.83 | 27.58 | 4500 | 124102.35 | 48000 | 76102.35 | 1.59 |
| NERICA 8 | Dibbling | 40 | 1876.5 | 18.77 | 4500 | 84442.50 | 53000 | 31442.50 | 0.59 |
| NERICA 8 | Dibbling | 80 | 2537.92 | 25.38 | 4500 | 114206.40 | 53000 | 61206.40 | 1.15 |
| NERICA 8 | Dibbling | 120 | 1279.67 | 12.80 | 4500 | 57585.15 | 53000 | 4585.15 | 0.09 |
| JAMILA | Broadcasting | 40 | 917.33 | 9.17 | 5000 | 45866.50 | 43000 | 2866.50 | 0.07 |
| JAMILA | Broadcasting | 80 | 1342.16 | 13.42 | 5000 | 67108.00 | 43000 | 24108.00 | 0.56 |
| JAMILA | Broadcasting | 120 | 969.58 | 9.70 | 5000 | 48479.00 | 43000 | 5479.00 | 0.13 |
| JAMILA | Drilling | 40 | 868.01 | 8.68 | 5000 | 43400.50 | 48000 | -4599.50 | -0.10 |
| JAMILA | Drilling | 80 | 1080.83 | 10.81 | 5000 | 54041.50 | 48000 | 6041.50 | 0.13 |
| JAMILA | Drilling | 120 | 1094.85 | 10.95 | 5000 | 54742.50 | 48000 | 6742.50 | 0.14 |
| JAMILA | Dibbling | 40 | 967.85 | 9.68 | 5000 | 48392.50 | 53000 | -4607.50 | -0.09 |
| JAMILA | Dibbling | 80 | 1014.55 | 10.15 | 5000 | 50727.50 | 53000 | -2272.50 | -0.04 |
| JAMILA | Dibbling | 120 | 701.35 | 7.01 | 5000 | 35067.50 | 53000 | -17932.50 | -0.34 |

Calculation of total revenue was based on market prevailing prices of ₦4,500.00 and ₦5000.00 per 100 kg bag of NERICAs and JAMILA respectively at Samaru and environs.

**Table – 4: Economic analysis of cost and return on investment of growing upland rice varieties using seed rate and sowing method at Samaru in 2012 wet season.**

| Variety | Sowing method | Seed rate kg ha$^{-1}$ | Total yield Kg ha$^{-1}$ | No. of bags (100kg) | Average Price/bag ₦ | Total revenue (TR) ₦ ha$^{-1}$ | Total variable cost (TVC) ₦ ha$^{-1}$ | Gross margin (TR- TVC) ₦ ha$^{-1}$ | Return on investment (ROI) ₦ |
|---|---|---|---|---|---|---|---|---|---|
| NERICA 4 | Broadcasting | 40 | 3125.00 | 31.25 | 4500 | 140625.00 | 45000 | 95625.00 | 2.13 |
| NERICA 4 | Broadcasting | 80 | 3500.00 | 35.00 | 4500 | 157500.00 | 45000 | 112500.00 | 2.50 |
| NERICA 4 | Broadcasting | 120 | 2833.33 | 28.33 | 4500 | 127499.85 | 45000 | 82499.85 | 1.83 |
| NERICA 4 | Drilling | 40 | 3541.67 | 35.42 | 4500 | 159375.15 | 50000 | 109375.15 | 2.19 |
| NERICA 4 | Drilling | 80 | 4208.33 | 42.08 | 4500 | 189374.85 | 50000 | 139374.85 | 2.79 |
| NERICA 4 | Drilling | 120 | 4208.33 | 42.08 | 4500 | 189374.85 | 50000 | 139374.85 | 2.79 |
| NERICA 4 | Dibbling | 40 | 2708.33 | 27.08 | 4500 | 121874.85 | 55000 | 66874.85 | 1.22 |
| NERICA 4 | Dibbling | 80 | 2583.33 | 25.83 | 4500 | 116249.85 | 55000 | 61249.85 | 1.11 |
| NERICA 4 | Dibbling | 120 | 2875.00 | 28.75 | 4500 | 129375.00 | 55000 | 74375.00 | 1.35 |
| NERICA 8 | Broadcasting | 40 | 3083.33 | 30.83 | 4500 | 138749.85 | 45000 | 93749.85 | 2.08 |
| NERICA 8 | Broadcasting | 80 | 4666.67 | 46.67 | 4500 | 210000.15 | 45000 | 165000.15 | 3.67 |
| NERICA 8 | Broadcasting | 120 | 3750.00 | 37.50 | 4500 | 168750.00 | 45000 | 123750.00 | 2.75 |
| NERICA 8 | Drilling | 40 | 3583.33 | 35.83 | 4500 | 161249.85 | 50000 | 111249.85 | 2.22 |
| NERICA 8 | Drilling | 80 | 5458.33 | 54.58 | 4500 | 245624.85 | 50000 | 195624.85 | 3.91 |
| NERICA 8 | Drilling | 120 | 3958.33 | 39.58 | 4500 | 178124.85 | 50000 | 128124.85 | 2.56 |
| NERICA 8 | Dibbling | 40 | 4125.00 | 41.25 | 4500 | 185625.00 | 55000 | 130625.00 | 2.38 |
| NERICA 8 | Dibbling | 80 | 4125.00 | 41.25 | 4500 | 185625.00 | 55000 | 130625.00 | 2.38 |
| NERICA 8 | Dibbling | 120 | 3000.00 | 30.00 | 4500 | 135000.00 | 55000 | 80000.00 | 1.45 |
| JAMILA | Broadcasting | 40 | 3208.00 | 32.08 | 5000 | 160400.00 | 45000 | 115400.00 | 2.56 |
| JAMILA | Broadcasting | 80 | 4875.00 | 48.75 | 5000 | 243750.00 | 45000 | 198750.00 | 4.42 |
| JAMILA | Broadcasting | 120 | 4791.67 | 47.92 | 5000 | 239583.50 | 45000 | 194583.50 | 4.32 |
| JAMILA | Drilling | 40 | 4875.00 | 48.75 | 5000 | 243750.00 | 50000 | 193750.00 | 3.88 |
| JAMILA | Drilling | 80 | 4208.33 | 42.08 | 5000 | 210416.50 | 50000 | 160416.50 | 3.21 |
| JAMILA | Drilling | 120 | 5150.00 | 51.50 | 5000 | 257500.00 | 50000 | 207500.00 | 4.15 |
| JAMILA | Dibbling | 40 | 3458.33 | 34.58 | 5000 | 172916.50 | 55000 | 117916.50 | 2.14 |
| JAMILA | Dibbling | 80 | 4833.33 | 48.33 | 5000 | 241666.50 | 55000 | 186666.50 | 3.39 |
| JAMILA | Dibbling | 120 | 3625.00 | 36.25 | 5000 | 181250.00 | 55000 | 126250.00 | 2.30 |

Calculation of total revenue was based on market prevailing prices of ₦4,500.00 and ₦5000.00 per 100 kg bag of NERICAs and JAMILA respectively at Samaru and environs.

**Table – 5: Combined economic analysis of cost and return on investment of growing upland rice varieties using seed rate and sowing method at Samaru in 2011 and 2012 wet seasons**

| Variety | Sowing method | Seed rate (kg ha$^{-1}$) | Total yield (Kg ha$^{-1}$) | No. of bags (100kg) | Average Price/ bag ₦ | Total revenue (TR) (₦ ha$^{-1}$) | Total variable cost (TVC) (₦ ha$^{-1}$) | Gross margin (TR- TVC) (₦ ha$^{-1}$) | Return on investment (ROI) ₦ |
|---|---|---|---|---|---|---|---|---|---|
| NERICA 4 | Broadcasting | 40 | 2106.83 | 21.07 | 4500 | 94807.35 | 44000 | 50807.35 | 2.15 |
| NERICA 4 | Broadcasting | 80 | 2956.54 | 29.57 | 4500 | 133044.30 | 44000 | 89044.30 | 3.02 |
| NERICA 4 | Broadcasting | 120 | 2114.55 | 21.15 | 4500 | 95154.75 | 44000 | 51154.75 | 2.16 |
| NERICA 4 | Drilling | 40 | 2529.79 | 25.30 | 4500 | 113840.55 | 49000 | 64840.55 | 2.32 |
| NERICA 4 | Drilling | 80 | 2959.54 | 29.60 | 4500 | 133179.30 | 49000 | 84179.30 | 2.72 |
| NERICA 4 | Drilling | 120 | 3051.88 | 30.52 | 4500 | 137334.60 | 49000 | 88334.60 | 2.80 |
| NERICA 4 | Dibbling | 40 | 2136.75 | 21.37 | 4500 | 96153.75 | 54000 | 42153.75 | 1.78 |
| NERICA 4 | Dibbling | 80 | 2141.41 | 21.41 | 4500 | 96363.45 | 54000 | 42363.45 | 1.78 |
| NERICA 4 | Dibbling | 120 | 1988.17 | 19.88 | 4500 | 89467.65 | 54000 | 35467.65 | 1.66 |
| NERICA 8 | Broadcasting | 40 | 2374.96 | 23.75 | 4500 | 106873.20 | 44000 | 62873.20 | 2.43 |
| NERICA 8 | Broadcasting | 80 | 4226.54 | 42.27 | 4500 | 190215.00 | 44000 | 146215.00 | 4.32 |
| NERICA 8 | Broadcasting | 120 | 3694.92 | 36.95 | 4500 | 166275.00 | 44000 | 122275.00 | 3.78 |
| NERICA 8 | Drilling | 40 | 2936.18 | 29.36 | 4500 | 132128.10 | 49000 | 83128.10 | 2.70 |
| NERICA 8 | Drilling | 80 | 4226.54 | 42.27 | 4500 | 190194.30 | 49000 | 141194.30 | 3.88 |
| NERICA 8 | Drilling | 120 | 3358.08 | 33.58 | 4500 | 151113.60 | 49000 | 102113.60 | 3.08 |
| NERICA 8 | Dibbling | 40 | 3000.75 | 30.01 | 4500 | 135033.75 | 54000 | 81033.75 | 2.50 |
| NERICA 8 | Dibbling | 80 | 3331.46 | 33.31 | 4500 | 149915.70 | 54000 | 95915.70 | 2.78 |
| NERICA 8 | Dibbling | 120 | 2139.83 | 21.40 | 4500 | 96292.35 | 54000 | 42292.35 | 1.78 |
| JAMILA | Broadcasting | 40 | 2062.83 | 20.63 | 5000 | 103141.50 | 44000 | 59141.50 | 2.34 |
| JAMILA | Broadcasting | 80 | 3066.92 | 30.67 | 5000 | 153346.00 | 44000 | 109346.00 | 3.49 |
| JAMILA | Broadcasting | 120 | 2922.29 | 29.22 | 5000 | 146114.50 | 44000 | 102114.50 | 3.32 |
| JAMILA | Drilling | 40 | 2871.51 | 28.72 | 5000 | 143575.50 | 49000 | 94575.50 | 2.93 |
| JAMILA | Drilling | 80 | 3804.58 | 38.05 | 5000 | 190229.00 | 49000 | 141229.00 | 3.88 |
| JAMILA | Drilling | 120 | 3122.43 | 31.22 | 5000 | 156121.50 | 49000 | 107121.50 | 3.19 |
| JAMILA | Dibbling | 40 | 2213.09 | 22.13 | 5000 | 110654.50 | 54000 | 56654.50 | 2.05 |
| JAMILA | Dibbling | 80 | 2923.94 | 29.24 | 5000 | 146197.00 | 54000 | 92197.00 | 2.71 |
| JAMILA | Dibbling | 120 | 2163.18 | 21.63 | 5000 | 108159.00 | 54000 | 54159.00 | 2.002 |

Calculation of total revenue was based on market prevailing prices of ₦4,500.00 and ₦5000.00 per 100 kg bag of NERICAs and JAMILA respectively at Samaru and environs.

**Table – 6: Economic analysis of cost and return on investment of growing upland rice varieties using seed rate and sowing methods at Maigana in 2011 wet season.**

| Variety | Sowing method | Seed rate (kg ha$^{-1}$) | Total yield (Kg ha$^{-1}$) | No. of bags (100kg) | Average Price/ bag ₦ | Total revenue (TR) (₦ ha$^{-1}$) | Total variable cost (TVC) (₦ ha$^{-1}$) | Gross margin (TR- TVC) (₦ ha$^{-1}$) | Return on investment (ROI) ₦ |
|---|---|---|---|---|---|---|---|---|---|
| NERICA 4 | Broadcasting | 40 | 718.67 | 7.19 | 4500 | 32340.15 | 40000 | -7659.85 | -0.81 |
| NERICA 4 | Broadcasting | 80 | 1451.20 | 14.51 | 4500 | 65304.00 | 40000 | 25304.00 | 1.63 |
| NERICA 4 | Broadcasting | 120 | 946.10 | 9.46 | 4500 | 42574.50 | 40000 | 2574.50 | 1.06 |
| NERICA 4 | Drilling | 40 | 1199.73 | 11.10 | 4500 | 53987.85 | 42000 | 11987.85 | 1.29 |
| NERICA 4 | Drilling | 80 | 1372.73 | 13.73 | 4500 | 61772.85 | 42000 | 19772.85 | 1.47 |
| NERICA 4 | Drilling | 120 | 732.80 | 7.33 | 4500 | 32976.00 | 42000 | -9024.00 | -0.79 |
| NERICA 4 | Dibbling | 40 | 818.72 | 8.19 | 4500 | 36842.40 | 44000 | -7157.60 | -0.84 |
| NERICA 4 | Dibbling | 80 | 1621.83 | 16.22 | 4500 | 72982.35 | 44000 | 28982.35 | 1.66 |
| NERICA 4 | Dibbling | 120 | 1119.62 | 11.20 | 4500 | 50382.90 | 44000 | 6382.90 | 1.15 |
| NERICA 8 | Broadcasting | 40 | 1143.77 | 11.44 | 4500 | 51469.65 | 40000 | 11469.65 | 1.29 |
| NERICA 8 | Broadcasting | 80 | 3231.63 | 32.32 | 5000 | 161581.50 | 42000 | 119581.50 | 3.85 |
| NERICA 8 | Broadcasting | 120 | 637.02 | 6.37 | 4500 | 28665.90 | 40000 | -11334.10 | -0.72 |
| NERICA 8 | Drilling | 40 | 1246.91 | 12.47 | 4500 | 56110.95 | 42000 | 14110.95 | 1.34 |
| NERICA 8 | Drilling | 80 | 1618.5 | 16.19 | 4500 | 72832.50 | 42000 | 30832.50 | 1.73 |
| NERICA 8 | Drilling | 120 | 634.83 | 6.35 | 4500 | 28567.35 | 42000 | -13432.65 | -0.68 |
| NERICA 8 | Dibbling | 40 | 933.33 | 9.33 | 4500 | 41999.85 | 44000 | -2000.15 | -0.95 |
| NERICA 8 | Dibbling | 80 | 1253.13 | 12.53 | 4500 | 56390.85 | 44000 | 12390.85 | 1.28 |
| NERICA 8 | Dibbling | 120 | 810.13 | 8.10 | 4500 | 36455.85 | 44000 | -7544.15 | -0.83 |
| JAMILA | Broadcasting | 40 | 691.83 | 6.92 | 5000 | 34591.50 | 40000 | -5408.50 | -0.86 |
| JAMILA | Broadcasting | 80 | 873.33 | 8.73 | 5000 | 43666.50 | 40000 | 3666.50 | 1.09 |
| JAMILA | Broadcasting | 120 | 1000.00 | 10.00 | 5000 | 50000.00 | 40000 | 10000.00 | 1.25 |
| JAMILA | Drilling | 40 | 674.17 | 6.74 | 5000 | 33708.50 | 42000 | -8291.50 | -0.80 |
| JAMILA | Drilling | 80 | 1380.97 | 13.81 | 5000 | 69050.00 | 42000 | 27050.00 | 1.55 |
| JAMILA | Drilling | 120 | 808.57 | 8.09 | 5000 | 40428.50 | 42000 | -1571.50 | -0.96 |
| JAMILA | Dibbling | 40 | 873.00 | 8.73 | 5000 | 43650.00 | 44000 | -350.00 | -0.99 |
| JAMILA | Dibbling | 80 | 978.80 | 9.79 | 5000 | 48940.00 | 44000 | 4940.00 | 1.11 |
| JAMILA | Dibbling | 120 | 545.67 | 5.46 | 5000 | 27283.50 | 44000 | -16716.50 | -0.62 |

Calculation of total revenue was based on market prevailing prices of ₦4,500.00 and ₦5000.00 per 100 kg bag of NERICAs and JAMILA respectively at Samaru and environs.

**Table – 7: Economic analysis of cost and return on investment of growing upland rice varieties using seed rate and sowing method at Maigana in 2012 wet season.**

| Variety | Sowing method | Seed rate (kg ha$^{-1}$) | Total yield (Kg ha$^{-1}$) | No .of bags (100kg) | Average Price/ bag ₦ | Total revenue (TR) (₦ ha$^{-1}$) | Total variable cost (TVC) (₦ ha$^{-1}$) | Gross margin (TR- TVC) (₦ ha$^{-1}$) | Return on investment (ROI) ₦ |
|---|---|---|---|---|---|---|---|---|---|
| NERICA 4 | Broadcasting | 40 | 2500.00 | 25.00 | 4500 | 112500.00 | 41000 | 71500 00 | 2.74 |
| NERICA 4 | Broadcasting | 80 | 2916.67 | 29.17 | 4500 | 131250.15 | 41000 | 90250.15 | 3.20 |
| NERICA 4 | Broadcasting | 120 | 3916.67 | 39.17 | 4500 | 176250.15 | 41000 | 135250.15 | 4.30 |
| NERICA 4 | Drilling | 40 | 1666.67 | 16.67 | 4500 | 75000.15 | 43000 | 32000.15 | 1.74 |
| NERICA 4 | Drilling | 80 | 2625.00 | 26.25 | 4500 | 118125.00 | 43000 | 75125.00 | 2.75 |
| NERICA 4 | Drilling | 120 | 3425.00 | 34.25 | 4500 | 154125.00 | 43000 | 111125.00 | 3.58 |
| NERICA 4 | Dibbling | 40 | 2075.00 | 20.75 | 4500 | 93375.00 | 45000 | 48375.00 | 2.08 |
| NERICA 4 | Dibbling | 80 | 1750.00 | 17.50 | 4500 | 78750.00 | 45000 | 33750.00 | 1.75 |
| NERICA 4 | Dibbling | 120 | 1875.00 | 18.75 | 4500 | 84375.00 | 45000 | 39375.00 | 1.88 |
| NERICA 8 | Broadcasting | 40 | 2000.00 | 20.00 | 4500 | 90000.00 | 41000 | 49000.00 | 2.20 |
| NERICA 8 | Broadcasting | 80 | 4166.67 | 41.67 | 4500 | 187500.15 | 41000 | 146500.15 | 4.57 |
| NERICA 8 | Broadcasting | 120 | 3375.00 | 33.75 | 4500 | 151875.00 | 41000 | 110875.00 | 3.70 |
| NERICA 8 | Drilling | 40 | 1625.00 | 16.25 | 4500 | 73125.00 | 43000 | 30125.00 | 1.71 |
| NERICA 8 | Drilling | 80 | 3000.00 | 30.00 | 4500 | 135000.00 | 43000 | 92000.00 | 3.14 |
| NERICA 8 | Drilling | 120 | 3458.33 | 34.58 | 4500 | 155624.85 | 43000 | 112624.85 | 3.62 |
| NERICA 8 | Dibbling | 40 | 1541.67 | 15.42 | 4500 | 69375.15 | 45000 | 24375.15 | 1.54 |
| NERICA 8 | Dibbling | 80 | 2416.67 | 24.17 | 4500 | 108750.15 | 45000 | 63750.15 | 2.42 |
| NERICA 8 | Dibbling | 120 | 2375.00 | 23.75 | 4500 | 106875.00 | 45000 | 61875.00 | 2.38 |
| JAMILA | Broadcasting | 40 | 2583.33 | 25.83 | 5000 | 129166.50 | 41000 | 88166.50 | 3.15 |
| JAMILA | Broadcasting | 80 | 4333.33 | 43.33 | 5000 | 216666.50 | 41000 | 175666.50 | 5.28 |
| JAMILA | Broadcasting | 120 | 3541.67 | 35.42 | 5000 | 177083.50 | 41000 | 136083.50 | 4.32 |
| JAMILA | Drilling | 40 | 2291.67 | 22.92 | 5000 | 114583.50 | 43000 | 71583.50 | 2.66 |
| JAMILA | Drilling | 80 | 5783.33 | 57.83 | 5000 | 289166.50 | 43000 | 246166.50 | 6.72 |
| JAMILA | Drilling | 120 | 4833.33 | 48.33 | 5000 | 241666.50 | 43000 | 198666.50 | 5.62 |
| JAMILA | Dibbling | 40 | 2916.67 | 29.17 | 5000 | 145833.50 | 45000 | 100833.50 | 3.24 |
| JAMILA | Dibbling | 80 | 3041.67 | 30.42 | 5000 | 152083.50 | 45000 | 107083.50 | 3.38 |
| JAMILA | Dibbling | 120 | 4375.00 | 43.75 | 5000 | 218750.00 | 45000 | 173750.00 | 4.86 |

Calculation of total revenue was based on market prevailing prices of ₦4,500.00 and ₦5000.00 per 100 kg bag of NERICAs and JAMILA respectively at Samaru and environs.

**Table – 8: Combined economic analysis of cost and return on investment of growing upland rice varieties using seed rate and sowing method at Maigana in 2011 and 2012 wet seasons**

| Variety | Sowing method | Seed rate (kg ha$^{-1}$) | Total yield (Kg ha$^{-1}$) | No. of bags (100kg) | Average Price/bag ₦ | Total revenue (TR) (₦ ha$^{-1}$) | Total variable cost (TVC) (₦ ha$^{-1}$) | Gross margin (TR- TVC) (₦ ha$^{-1}$) | Return on investment (ROI) ₦ |
|---|---|---|---|---|---|---|---|---|---|
| NERICA 4 | Broadcasting | 40 | 1609.33 | 16.09 | 4500 | 72419.85 | 40500 | 31919.85 | 1.79 |
| NERICA 4 | Broadcasting | 80 | 2233.93 | 22.34 | 4500 | 100526.85 | 40500 | 60026.85 | 2.48 |
| NERICA 4 | Broadcasting | 120 | 2431.38 | 24.31 | 4500 | 109412.10 | 40500 | 68912.10 | 2.70 |
| NERICA 4 | Drilling | 40 | 1433.20 | 14.33 | 4500 | 64494.10 | 42500 | 21994.00 | 1.52 |
| NERICA 4 | Drilling | 80 | 1998.87 | 19.99 | 4500 | 89949.15 | 42500 | 47449.15 | 2.12 |
| NERICA 4 | Drilling | 120 | 2078.90 | 20.79 | 4500 | 93550.50 | 42500 | 51050.50 | 2.20 |
| NERICA 4 | Dibbling | 40 | 1446.86 | 14.47 | 4500 | 65108.70 | 44500 | 20608.70 | 1.46 |
| NERICA 4 | Dibbling | 80 | 1685.92 | 16.86 | 4500 | 75866.40 | 44500 | 31366.40 | 1.70 |
| NERICA 4 | Dibbling | 120 | 1497.31 | 14.97 | 4500 | 67378.95 | 44500 | 22878.95 | 1.51 |
| NERICA 8 | Broadcasting | 40 | 1571.88 | 15.72 | 4500 | 70734.60 | 40500 | 30234.60 | 1.75 |
| NERICA 8 | Broadcasting | 80 | 2773.82 | 27.74 | 4500 | 124821.90 | 40500 | 84321.90 | 3.08 |
| NERICA 8 | Broadcasting | 120 | 2006.01 | 20.06 | 4500 | 90270.45 | 40500 | 49770.45 | 2.23 |
| NERICA 8 | Drilling | 40 | 1435.95 | 14.36 | 4500 | 64617.75 | 42500 | 22117.75 | 1.52 |
| NERICA 8 | Drilling | 80 | 2309.25 | 23.09 | 4500 | 103916.25 | 42500 | 61416.25 | 2.45 |
| NERICA 8 | Drilling | 120 | 2046.58 | 20.47 | 4500 | 92096.10 | 42500 | 49596.10 | 2.17 |
| NERICA 8 | Dibbling | 40 | 1237.50 | 12.38 | 4500 | 55687.50 | 44500 | 11187.50 | 1.25 |
| NERICA 8 | Dibbling | 80 | 1834.90 | 18.35 | 4500 | 82570.50 | 44500 | 38070.50 | 1.86 |
| NERICA 8 | Dibbling | 120 | 1592.57 | 15.93 | 4500 | 71665.65 | 44500 | 27165.65 | 1.61 |
| JAMILA | Broadcasting | 40 | 1637.58 | 16.38 | 5000 | 81879.00 | 40500 | 41379.00 | 2.02 |
| JAMILA | Broadcasting | 80 | 2207.50 | 22.08 | 5000 | 110375.00 | 40500 | 69875.00 | 2.73 |
| JAMILA | Broadcasting | 120 | 2666.67 | 26.67 | 5000 | 133333.50 | 40500 | 92833.50 | 3.29 |
| JAMILA | Drilling | 40 | 1482.92 | 14.83 | 5000 | 74146.00 | 42500 | 31646.00 | 1.74 |
| JAMILA | Drilling | 80 | 4507.48 | 45.07 | 5000 | 225374.00 | 42500 | 182874.00 | 5.30 |
| JAMILA | Drilling | 120 | 2820.95 | 28.21 | 5000 | 141047.50 | 42500 | 98547.50 | 3.32 |
| JAMILA | Dibbling | 40 | 1894.83 | 18.95 | 5000 | 94741.50 | 44500 | 50241.50 | 2.13 |
| JAMILA | Dibbling | 80 | 2010.23 | 20.10 | 5000 | 100511.50 | 44500 | 56011.50 | 2.26 |
| JAMILA | Dibbling | 120 | 2460.33 | 24.60 | 5000 | 123016.50 | 44500 | 78516.50 | 2.76 |

Calculation of total revenue was based on market prevailing prices of ₦4,500.00 and ₦5000.00 per 100 kg bag of NERICAs and JAMILA respectively at Samaru and environs.

## Conclusion

In this study, direct seeding of rice practice was evaluated and it is clear that direct seeded rice practices may not perform similarly in all agro ecological conditions because of rainfall distribution. At both location, the result indicated that the highest gross margin was by drilled JAMILA at 80 kg ha$^{-1}$ seed rate. The economic analysis of upland rice production at both locations indicated that production of NERICA 8 and JAMILA by either broadcasting or drilling method at the seed rate of 80 kg ha$^{-1}$ gave the highest gross margin as well as return on investment. The result revealed that NERICA 8 and JAMILA sown by broadcasting method at 80 kg ha$^{-1}$ was the most profitable with Gross margain of ₦246, 166.50

with return on investment of ₦6.72. This was followed by broadcasting JAMILA at 120 kg $ha^{-1}$ seed rate which gave a gross margin of ₦194, 583.50 and return on investment of ₦4.32. However, the least gross margin of ₦61, 249.85 was observed when NERICA 4 was dibbled at 120 kg $ha^{-1}$ which brought a loss of ₦16, 716.50 and ₦ 0.62k was lost per every naira invested. The data presented in the study shows that broadcasting and drilling method can also be a viable solution under scarcity of labour and water. but, there is need to develop proper weed management practices and requires further study to access the long term effects of herbicides on soil, water and development of weed flora.

## References


Anoop D, Khurana R, Jastarn S and Gususahib SD, 2007. Farm Power and Machinery, Punjab Agricultural University, Ludhiana 141001. Comparative performance of Different Paddy Transplanter Development in India, A review. Agric. Rev. 28(4): 262-269.

Azmi M, Chin DV, Vongsaroj P and Johnson DE, 2005. Emerging Issues in Weed Management of Direct-Seeded Rice in Malaysia, Vietnam, and Thailand. In Rice is Life: Scientific Perspectives for the 21st Century. K. Toriyama, K.L. Heong (eds), Tokyo and Tsukuba, Japan. Proceedings of the World Rice Research Conference pp. 196-198.

Bhusan L, Ladha JK, Gupta RK, Singh S Tirole-Padre A, Sehrawat YS, Gathala M and Pathak H, 2007. Saving of water and labour in rice-wheat system with no tillage and direct seeding technologies. Agron. J. 99: 1288-1296.

Chuhan BS, 2012. Weed ecology and weed management strategy for dry seeded rice in Asia. Weed Technol. 26: 1-13.

Estorninos Jr. LE and Moody K, 1988. Evaluation of herbicides for weed control in dry-seeded wetland rice (*Oryza sativa* L.). Philipp. J. Weed Sci. 15: 50–58.

FAO, 2009. FAOSTAT Database FAO, Rome www.faostat.fao.org (accessed in June 2010).

Hira GS, 2009. Water Management in Northern and food Security of Indian. J. Crop Improve. 23: 36-47.

Inyat Ali R, Nadeem I, Usman MS and Akhtar M, 2012. Effect of different planting methods on economic yield and grain quality of rice. Int. J. Agri. Sci. 4(1): 28-34.

Kumar V, Ladha JK and Gathala MK, 2009. Direct drill-seeded rice: A need of the day. In Annual Meeting of Agronomy Society of America, Pittsburgh, November 1–5, 2009, http://a-c-s.confex.com/crops/2009am/webprogram/Paper53386.html

Kumar V and Ladha JK, 2011 Direct seeding of rice: Recent developments and future research needs. Advances in Agron. 111: 297-413.

Kumar R. 2011. Comparative performance evaluation of mechanical transplanting and direct seeding of rice under puddled and unpuddled condition. MTech Thesis, Deptt. of Farm Machinery and Power Engineering, CCS HAU, Hisar, Haryana, India.

Mahajan G, Bharaj TS and Timsina J, 2009. Yield and water productivity of rice as affected by time of transplanting in Punjab, India. Agri. Water Manage. 96: 525-532.

Kimio I, Ko N and Hiromitsu K, 1999. An Automatic Irrigation System for Direct Seeding. Rice Cultivation. J. Agri. Meteorol. 55(2): 127–135.

Johnson D, 2006. The Direct Approach. Rice Today 5(2). International Rice Research Institute Rogers, E.M. 1995. Diffusion of Innovations. 4th ed. New York: The Free Press.

Rao AN, Johnson DE, Shivaprasad B, Ladha JK and Mortimer AM, 2007. Weed management in direct-seeded rice. Adv. Agron. 93: 153-255.

Reeves T, 2009. The impacts of climate change on wheat production in India – adaptation, mitigation and future directions. Food and Agriculture Organization, Rome. Hobbs, P Present Practices and Future Options. J. Crop Prod. 4(1): 1–22.

Saharawat YS, Singh B, Malik RK, Ladha JK, Gathala M, Jat ML and Kumar V, 2010. Evaluation of alternative tillage and crop establishment methods in a rice–wheat rotation in North Western IGP. Field Crops Res. 116: 260–267.

Khush GS, 1997. Origin, dispersal, cultivation and variation of rice. Plant Mol. Biol. 35: 25-34.

Santhi PK, Pannuswamy and Kempuchetty N, 1998. A labour saving techniques in direct sown and transplanted rice. Int. Rice Res. Notes. 23: 35-36.

Singh Y, Singh B, Nayyar VK and Singh J, 2003. Nutrient management for sustainable rice-wheat cropping system. NATP, ICAR, New Delhi and PAU, Ludhiana, Punjab, India.

Olukosi JO and Erhabor PO, 1988. Introduction to Farm Management Economics. Ajitab Publishers, Zaria, pp. 35-36 and 71-80.

Von Braun J and Bos MS, 2004. The changing economics and politics of rice: Implications for food security, globalization, and environmental sustainability. In "Rice Is Life: Scientific Perspectives for the 21st Century" (K. Toriyama, K. L. Heong, and B. Hardy, Eds.), pp. 7–20. International Rice Research Institute, Los Ban ˜os, Philippines and Japan International Research Center for Agricultural Sciences, Tsukuba, Japan.

Yamano T, Aroma A, Labarta RA, Huelgas ZM and Mohanty S, 2016. Adoption and impacts of International rice research technologies. Global Food Sec. 8: 1-8.

# Permissions

All chapters in this book were first published in AJAB, by Life Sciences Society; hereby published with permission under the Creative Commons Attribution License or equivalent. Every chapter published in this book has been scrutinized by our experts. Their significance has been extensively debated. The topics covered herein carry significant findings which will fuel the growth of the discipline. They may even be implemented as practical applications or may be referred to as a beginning point for another development.

The contributors of this book come from diverse backgrounds, making this book a truly international effort. This book will bring forth new frontiers with its revolutionizing research information and detailed analysis of the nascent developments around the world.

We would like to thank all the contributing authors for lending their expertise to make the book truly unique. They have played a crucial role in the development of this book. Without their invaluable contributions this book wouldn't have been possible. They have made vital efforts to compile up to date information on the varied aspects of this subject to make this book a valuable addition to the collection of many professionals and students.

This book was conceptualized with the vision of imparting up-to-date information and advanced data in this field. To ensure the same, a matchless editorial board was set up. Every individual on the board went through rigorous rounds of assessment to prove their worth. After which they invested a large part of their time researching and compiling the most relevant data for our readers.

The editorial board has been involved in producing this book since its inception. They have spent rigorous hours researching and exploring the diverse topics which have resulted in the successful publishing of this book. They have passed on their knowledge of decades through this book. To expedite this challenging task, the publisher supported the team at every step. A small team of assistant editors was also appointed to further simplify the editing procedure and attain best results for the readers.

Apart from the editorial board, the designing team has also invested a significant amount of their time in understanding the subject and creating the most relevant covers. They scrutinized every image to scout for the most suitable representation of the subject and create an appropriate cover for the book.

The publishing team has been an ardent support to the editorial, designing and production team. Their endless efforts to recruit the best for this project, has resulted in the accomplishment of this book. They are a veteran in the field of academics and their pool of knowledge is as vast as their experience in printing. Their expertise and guidance has proved useful at every step. Their uncompromising quality standards have made this book an exceptional effort. Their encouragement from time to time has been an inspiration for everyone.

The publisher and the editorial board hope that this book will prove to be a valuable piece of knowledge for researchers, students, practitioners and scholars across the globe.

# List of Contributors

**Ayesha Aihetasham, Aysha Anayat and Muhammad Xaaceph**
Department of Zoology, University of the Punjab, Lahore, Pakistan

**Norsyazwani Mohammad, Emilia Zainal Abidin, Nor'Aqilah Amani Zainal Mubarik, Vivien How, Sarva Mangala Praveena and Zailina Hashim**
Department of Environmental and Occupational Health, Faculty of Medicine and Health Sciences, Universiti Putra Malaysia, 43400 UPM Serdang, Selangor, Malaysia

**Waqas Ashraf**
Department of Plant Pathology, University College of Agriculture and Environmental Sciences, The Islamia University of Bahawalpur, Bahawalpur, Pakistan

**Shahbaz Talib Sahi and Amer Habib**
Department of Plant Pathology, University of Agriculture, Faisalabad, Pakistan

**Atta Ur Rehman Khan**
Department of Plant Pathology, PMAS-Arid Agriculture University Rawalpindi, Pakistan

**Muhammad Ahmad Zeshan and Anum Intisar**
Department of Plant Pathology, University College of Agriculture, University of Sargodha, Sargodha, Pakistan

**Absar Ahmad**
University of Veterinary and Animal Sciences, Lahore, Pakistan

**Salman Ghuffar, Muhammad Zeshan Ahmed and Luqman Amrao**
Department of Plant Pathology, University of Agriculture Faisalabad, Pakistan

**Muhammad Farooq Aslam and Sajjad Hyder**
Department of Plant Pathology, PMAS-Arid Agriculture University Rawalpindi, Pakistan

**Thura Alyasiri and Ithar AlMayaly**
Department of Biology, College of Science, Baghdad University, Iraq

**Salah Alchalabi**
Biotechnology Research Center, Al-Nahrain University, Iraq

**Muhammad Abbas, Muhammad Irfan, Javaid Ahmed Shah and Muhammad Yousuf Memon**
Soil and Environmental Sciences Division, Nuclear Institute of Agriculture (NIA), Tando Jam, Sindh, Pakistan

**Rukhsana Kausar, Muhammad Amjad Bashir, Muhammad Shahid Nisar and Sagheer Atta**
Department of Plant Protection, Faculty of Agricultural Sciences, Ghazi University DG Khan, Punjab, Pakistan

**Muhammad Ishaq Asif Rehmani**
Department of Agronomy, Faculty of Agricultural Sciences, Ghazi University DG Khan, Punjab, Pakistan

**Muhammad Sadiq Hashami**
Department of Economics, Institute of Southern Punjab, Multan, Pakistan

**Saboor Ahmad and Muhammad Asif Aziz**
Department of Entomology, PMAS Arid Agriculture University, Rawalpindi, Pakistan

**Ayesha Aihetasham, Muhammad Saeed Akhtar and Tayyaba Bibi**
Department of Zoology, University of the Punjab, Lahore-54590, Pakistan

**Imran Bodlah**
Department of Entomology, PMAS-Arid Agriculture University, Rawalpindi, Pakistan

**Noor Asyraf Noor Azman, Nik Ahmad Nizam Nik Malek, Norfarah Syuhaidah Mohd Noor and Muhammad Arshad Javed**
Department of Biotechnology and Medical Engineering, Faculty of Biosciences and Medical Engineering, Universiti Teknologi Malaysia, 81310 UTM, Skudai, Johor, Malaysia

**Asma Tariq, Sumera Afsheen and Mobushir Hussain**
Department of Zoology, University of Gujrat, Gujrat, Punjab, Pakistan

**Syed Shakeel Shah and Sabila Afzal**
Department of Zoology, University of Gujrat, Sub-Campus Narowal, Pakistan

**Ahmed Zia**
National Insect Museum, National Agriculture Research Centre, Islamabad, Pakistan

**Inamullah Khan**
Department of Plant Protection, Agricultural University Peshawar, KPK, Pakistan

**Yousaf Hayat**
Department of Statistics, Mathematics and Computer Science, University of Agriculture, Peshawar, KPK, Pakistan

**I Ketut Widnyana, I Wayan Seputra Kuspianta and Putu Lasmi Yulianthi Sapanca**
Faculty of Agriculture, Mahasaraswati University of Denpasar, Bali, Indonesia

**Dalia Sukmawati, Nur Saidah, Tri Handayani K and Sri Rahayu**
Department of Biology, Faculty of Mathematics and Natural Sciences, Universitas Negeri Jakarta, Campus B, Jl. Pemuda No. 10 Rawamangun, East Java, Indonesia

**Mubashar Hussain, Iqra Mustafa, Muhammad Faheem Malik and Sharon Zulifqar**
Department of Zoology

**Zaheer Abbas**
Department of Statistics, University of Gujrat, Punjab, Pakistan

**Eloiza May Galon**
AgriVet Sciences Institute, De La Salle Araneta University, Upper Ciudad Real, City of San Jose Del Monte, Bulacan, 3023 Philippines
National Research Center for Protozoan Diseases, Obihiro University of Agriculture and Veterinary Medicine, Obihiro, Hokkaido, 080-8555 Japan

**Dante Lorenzo**
AgriVet Sciences Institute, De La Salle Araneta University, Upper Ciudad Real, City of San Jose Del Monte, Bulacan, 3023 Philippines
Department of Animal Science, College of Agriculture, Central Luzon State University, Science City of Muñoz, Nueva Ecija, 3119 Philippines

**Florencia Claveria**
AgriVet Sciences Institute, De La Salle Araneta University, Upper Ciudad Real, City of San Jose Del Monte, Bulacan, 3023 Philippines
Biology Department, College of Science, De La Salle University, 2401 Taft Avenue, 0922 Manila, Philippines

**Rabiya Ikram and Basharat Ali**
Department of Microbiology and Molecular Genetics, University of the Punjab, Lahore, Pakistan

**Aqsa Sattar, Iqra Azam, Muhammad Kaleem Sarwar and Afifa Amjad**
Department of Zoology, University of Gujrat, Sub-Campus Narowal, Narowal, Pakistan

**Muhammad Faheem Malik**
Department of Zoology, University of Gujrat, Hafiz Hayat Campus, Gujrat, Pakistan

**Sehrish Akram and Fakhar- un- Nisa**
Govt. College Women University Lahore, Pakistan

**Shakil Ahmad Khan**
Sericulture Research Laboratory, Lahore, Pakistan

**Mubashar Hussain, Maimoona Kanwal and Fatima Zafar**
University of Gujrat, Pakistan

**Fahad J. Alatawi, Syed Zain ul Abidin, Jawwad H. Mirza and Muhammad Kamran**
Department of Plant Protection, College of Food and Agriculture Sciences, King Saud University, KSA

**Raheel Atif Hameed and Shazia Ajum**
Cholistan Institute of Desert Studies, The Islamia University of Bahawalpur

**Muhammad Naveed Afzal**
Central Cotton research Institute Multan. Punjab, Pakistan

**Dalia Sukmawati and Mieke Miarsyah**
Department of Biology, Faculty of Mathematics and Natural Sciences, Universitas Negeri Jakarta, Campus A, Jl. Rawamangun Muka East Java. Hasyim Ashari Building, 9th floor, Indonesia

**Adnan A. Lahuf, Ola H. Jaafar and Muhassen Al-mosoy**
Department of Plant protection, College of Agriculture, University of Kerbala, Kerbala, Iraq

**Zainab L. Hameed**
Department of Field crops, College of Agriculture, University of Kerbala, Kerbala, Iraq

**Junman Li**
Institute of Virology and Biotechnology, Zhejiang Academy of Agricultural Sciences, Hangzhou, Zhejiang, China

**Maqshoof Ahamd, Azhar Hussain, Muhammad Fakhar-U-Zaman Akhtar and Zafar Iqbal**
Department of Soil Science, University College of Agriculture and Environmental Sciences, The Islamia University of Bahawalpur, Bahawalpur, Pakistan

**Muhammad Zafar-Ul-Hye**
Department of Soil Science, College of Agriculture, Bahauddin Zakariya University Multan

**Tayyaba Naz**
Institute of Soil and Environmental Sciences, University of Agriculture, Faisalabad, Pakistan

**Muhammad Mazhar Iqbal**
Soil and Water Testing Laboratory for Research, Chiniot, Department of Agriculture, Government of Punjab-Pakistan

**Priscilla Hope Swann and Florencia G. Claveria**
Biology Department, College of Sciences, De La Salle University, 2401 Taft Avenue, Manila Philippines

**Muhammad Imran and Umer Ayyaz Aslam Sheikh**
Department of Entomology, Faculty of Agriculture, The University of Poonch, Rawalakot, AJ&K, Pakistan

**Kanwal Hanif, Munir Ahmad and Muhammad Nasir**
Department of Entomology, Pir Mehr Ali Shah, Arid Agriculture University, Rawalpindi, Pakistan

**Farheen Nazli and Muhammad Ramzan Kashif**
Pesticide Quality Control Laboratory Bahawalpur-63100, Pakistan

**Bushra**
Institute of Soil and Environmental Sciences, University of Agriculture Faisalabad, Pakistan

**Muhammad Mazhar Iqbal**
Soil and Water Testing Laboratory, Chiniot, Pakistan

**Fatima Bibi**
Mango Research Station, Multan, Pakistan

**Zafar-ul-Hye**
Department of Soil Science, College of Agriculture, Bahauddin Zakariya University, Multan, Pakistan

**Maqshoof Ahmad**
Department of Soil and Environmental Sciences, the Islamia University of Bahawalpur, Bahawalpur-63100, Pakistan

**Shahbaz Khan and Faisal Mehmood**
Department of Agronomy, University of Agriculture, Faisalabad, Pakistan

**Sohail Irshad**
In-Service Agricultural Training Institute, Rahim Yar Khan, Pakistan

**Muhammad Nawaz**
Department of Agronomy, College of Agriculture, BZU, Bahadur Campus Layyah, Pakistan

# Index

www.ingramcontent.com/pod-product-compliance
Lightning Source LLC
LaVergne TN
LVHW060544210926
848485LV00005B/1254
* 9 7 8 1 6 3 9 8 9 4 6 0 4 *